Lecture Notes in Applied and Computational Mechanics

Volume 76

Series editors

Friedrich Pfeiffer, Technische Universität München, Garching, Germany
e-mail: pfeiffer@amm.mw.tum.de

Peter Wriggers, Universität Hannover, Hannover, Germany
e-mail: wriggers@ikm.uni-hannover.de

About this Series

This series aims to report new developments in applied and computational mechanics—quickly, informally and at a high level. This includes the fields of fluid, solid and structural mechanics, dynamics and control, and related disciplines. The applied methods can be of analytical, numerical and computational nature.

More information about this series at http://www.springer.com/series/4623

Kazumi Watanabe

Integral Transform Techniques for Green's Function

Second Edition

Springer

Kazumi Watanabe
Yamagata University
Yonezawa
Japan

ISSN 1613-7736 ISSN 1860-0816 (electronic)
Lecture Notes in Applied and Computational Mechanics
ISBN 978-3-319-34587-1 ISBN 978-3-319-17455-6 (eBook)
DOI 10.1007/978-3-319-17455-6

Springer Cham Heidelberg New York Dordrecht London

Softcover reprint of the hardcover 2nd edition 2015

Printed on acid-free paper

Springer International Publishing AG Switzerland is part of Springer Science+Business Media (www.springer.com)

Dedicated to my teachers,
Dr. Akira Atsumi (Late Professor, Tohoku University)

and

Dr. Kyujiro Kino (Late Professor, Osaka Institute of Technology)

Preface to the Second Edition

After publishing the original version, the author noticed that more detailed mathematical techniques should be included so that the young reader could learn the traditional analytical techniques without any mathematical skip. That is, the discussion on square root functions. In many dynamic/wave problems, we frequently encounter the square root function which is the typical multi-valued function and have to introduce branch cuts in the complex plane for the inversion integral. A simple elementary technique for the introduction of the branch cut and the discussion on the argument of the square root function along the cut are included in Chap. 1 as Sect. 1.3. This branch cut is employed throughout the book and applied to the inversion integrals in Sect. 2.5 and other sections. Due to the introduction of the unified branch cut, Sect. 2.5 for the time-harmonic Green's function is wholly rewritten.

In the revising process the author also noticed five exact closed-form solutions: three are Green's functions for torsion problems, the fourth one is for the reflection problem and the last is for the scattering problem. Green's functions for the torsion problem are inserted in Chaps. 3 and 7. SH-wave reflection at a moving boundary in Sect. 7.4 is slightly rewritten in order to include the closed-form solution. Section 7.5 is newly inserted and shows the exact closed-form solution for a wave scattering problem in an inhomogeneous elastic solid. Further, employing the branch cut described in Chap. 1, an excellent application technique of the complex integral is explained in the last Sect. 7.6. It is the transformation of a semi-infinite integral to a finite one that is suitable for numerical computations. Needless to say, many errors and mistakes in the original version have also been corrected.

The author hopes the young reader can learn one of the traditional analytical techniques, especially the application of the complex integral for the integral transform. Thus, the present revised version is more instructive than the original one, and every question and inquiry via email "kazy@yz.yamagata-u.ac.jp" is welcome.

Hikoshima Island, Japan, January 2015 Kazumi Watanabe

Preface

When I was a senior student, I found a book on the desk of my advisor professor and asked him how to get it. His answer was negative, saying its content was too hard, even for a senior student. Some weeks later, I found it again in a book store, the biggest one in Osaka. This was my first encounter with "Fourier Transforms" written by the late Prof. I. N. Sneddon. Since then, I have learned the power of integral transform, i.e. the principle of superposition.

All phenomena, regardless of their fields of event, can be described by differential equations. The solution of the differential equation contains the crucial information to understand the essential feature of the phenomena. Unfortunately, we cannot solve every differential equation, and almost all phenomena are governed by nonlinear differential equations, of which most are not tractable. The differential equations that can be solved analytically are limited to a very small number. But their solutions give us the essence of the event. The typical partial differential equations that can be solved exactly are the Laplace, the diffusion and the wave equations. These three partial differential equations, which are linearized for simplicity, govern many basic phenomena in physical, chemical and social events. In addition to single differential equations, some coupled linear partial differential equations, which govern somewhat complicated phenomena, are also solvable and their solutions give much information about, for example, the deformation of solid media, propagation of seismic and acoustic waves, and fluid flows.

In a case where phenomena are described by linear differential equations, the solutions can be expressed by superposition of basic/fundamental solutions. The integral transform technique is a typical superposition technique. The integral transform technique does not require any previous knowledge for solving differential equations. It simply transforms partial or ordinary differential equations to reduced ordinary differential equations or to simple algebraic equations. However, a substantial difficulty is present regarding the inversion process. Many inversion integrals are tabulated in various formula books, but typically, this is not enough. If a suitable integration formula cannot be found, the complex integral must be considered and Cauchy's integral theorem is applied to the inversion integral. Thus, integral transform techniques are intrinsically connected with the theory of complex integrals.

The present book intends to show how to apply integral transforms to partial differential equations and how to invert the transformed solution into the actual space-time domain. Not only the use of integration formula tabulated in books, but also the application of Cauchy's integral theorem for the inversion integrals are described concisely and in detail. A particular solution for a differential equation with a nonhomogeneous term of a point source is called the "Green's function." The Green's functions for coupled differential equations are called "Green's dyadic." The Green's function and Green's dyadic are the basic and fundamental solution of the differential equation and give the principal features of the event. Furthermore, these Green's functions and dyadics have many applications for numerical computation techniques such as the Boundary Element Method. However, the Green's function and Green's dyadic have been scattered in many branches of applied mechanics and thus, their solution methods are not unified. This book intends to present and illustrate a unified solution method, namely the method of integral transform for the Green's function and Green's dyadic. Thus, the fundamental Green's function for the Laplace and wave equations and the Green's dyadic for elasticity equations are gathered in this single book so that the reader can have access to a proper Green's function and understand the mathematical process for its derivation.

Chapter 1 describes roughly the definition of the integral transforms and the distributions to be used throughout the book. Chapter 2 shows how to apply an integral transform for solving a single partial differential equation such as the Laplace equation and the wave equation. The basic technique of the integral transform method is demonstrated. Especially, in the case of the time-harmonic response for the wave equation, the integration path for the inversion integral is discussed in detail. At the end of the chapter, the obtained Green's functions are listed in a table so that the reader can easily find the difference of the functional form among the Green's functions. An evaluation technique for a singular inversion integral which arises in a 2D static problem of Laplace equation is also developed.

The Green's dyadic for 2D and 3D elastodynamic problems are discussed in Chap. 3. Three basic responses, impulsive, time-harmonic and static responses, are obtained by the integral transform method. The time-harmonic response is derived by the convolution integral of the impulsive response without solving the differential equations for the time-harmonic source.

Chapter 4 presents the governing equations for acoustic waves in a viscous fluid. Introducing a small parameter, the nonlinear field equations are linearized and reduced to a single partial differential equation for velocity potential or pressure deviation. The Green's function which gives the acoustic field in a uniform flow is derived by the method of integral transform. A conversion technique for the inversion integral is demonstrated. That is, to transform an inversion integral along the complex line to that along the real axis in the complex plane. It enabled us to apply the tabulated integration formula.

Chapter 5 presents Green's functions for beams and plates. The dynamic response produced by a point load on the surface of a beam and a plate is discussed. The impulsive and time-harmonic responses are derived by the integral transform method. In addition to the tabulated integration formulas, the inversion integrals are evaluated by application of complex integral theory.

Chapter 6 presents a powerful inversion technique for transient problems of elastodynamics, namely the Cagniard-de Hoop method. Transient response of an elastic half space to a point impulsive load is discussed by the integral transform method. Applying Cauchy's complex integral theorem, the Fourier inversion integral is converted to an integral of the Laplace transform and then its Laplace inversion is carried out by inspection without using any integration formula. The Green's function for an SH-wave and Green's dyadics for P, SV and SH-waves are obtained.

The last Chap. 7 presents three special Green's functions/dyadics. The 2D static Green's dyadic for an orthotropic elastic solid and that for an inhomogeneous solid are derived. In the last section, a moving boundary problems is discussed. Two different Laplace transforms are applied for a single problem, and a conversion formula between two Laplace transforms is developed with use of Cauchy's theorem. This conversion enables us to apply the integral transform technique to a moving boundary problem.

The integral transform technique has been used for many years. The inversion process inevitably requires a working knowledge of the theory of complex functions. The author finds the challenge of a complex integral amusing, especially the challenge of choosing the right contour for the inversion integral. He hopes that young researchers will join the fun and carry on with the inversion techniques. In this respect it must be mentioned that he feels a lack of mathematical skill in the recent research activities, since some researchers tend to use numerical techniques without considering the possibility of an analytical solution. The more mathematical techniques expand the horizon of the differential equations wider and one can extract more firm knowledge from the nature which is described by the differential equations. The author hopes that the present book gives one more technique to the younger researchers.

Finally, the author wishes to express his sincere thanks to Dr. Mikael A. Langthjem, Associate Professor of Yamagata University, for his advice and nice comments.

Yonezawa, Japan, January 2013 Kazumi Watanabe

Contents

Chapter 1
Definition of Integral Transforms and Distributions

This first chapter describes a brief definition of integral transforms, such as Laplace and Fourier transforms, a rough definition of delta and step functions which are frequently used as the source function, and a concise introduction of the branch cut for a multi-valued square root function. The multiple integral transforms and their notations are also explained. The introduction of the branch cut and the discussion on the argument of the root function along the branch cut will be helpful for applying the complex integral to the inverse transform. The last short comment lists some important formula books which are crucial for the inverse transform, i.e. the evaluation of the inversion integral.

1.1 Integral Transforms

For a well-defined function $f(x),\, x \in (a, b)$, when the integral with the kernel function $K(\xi, x)$,

$$F(\xi) = \int_a^b K(\xi, x) f(x) dx \tag{1.1.1}$$

has its inverse integral with another kernel function $K^*(\xi, x)$,

$$f(x) = \int_L K^*(\xi, x) F(\xi) d\xi \tag{1.1.2}$$

we call this integration pair an "integral transform." The function $f(x)$ is an original function and the function $F(\xi)$ is the "image or transformed function" in the transformed domain. If the reciprocal $f(x) \Leftrightarrow F(\xi)$ holds, we call $F(\xi)$ the "integral

K. Watanabe, *Integral Transform Techniques for Green's Function*,
Lecture Notes in Applied and Computational Mechanics 76,
DOI 10.1007/978-3-319-17455-6_1

transform" of $f(x)$ and the two-variable-functions $K(\xi, x)$ and $K^*(\xi, x)$ the kernels of the integral transform.

We have already learned many integral transforms which are classified and named depending on the kernel function and the integration range. A well-known integral transform is the Laplace transform. The (one-sided) Laplace transform is, in the present book, defined for the time-variable function $f(t)$, $t \in [0, \infty)$ as

$$f^*(s) = \int_0^\infty f(t) \exp(-st) dt \tag{1.1.3}$$

where "s" is the transform parameter and the transform kernel is $\exp(-st)$. The inverse transform is also defined by the integral along the complex line,

$$f(t) = \frac{1}{2\pi i} \int_{c-i\infty}^{c+i\infty} f^*(s) \exp(st) ds \tag{1.1.4}$$

where the integration path from $c - i\infty$ to $c + i\infty$ is called the "Bromwich line." The real constant "c" must be larger than the real part of any singular point of the transformed function $f^*(s)$. Thus this line is placed at far right from all singular points in the complex s-plane.

In these definitions, the variable s is called the transform parameter and the two kernels for the transform and the inverse transform are exponential functions:

$$K(s, t) = \exp(-st), \quad K^*(s, t) = \frac{1}{2\pi i} \exp(st) \tag{1.1.5}$$

Further, the integration ranges are also different from each other. The transform integral is carried out along the semi-infinite real line $[0, \infty)$ for the time and the inverse transform is carried out along an infinite line $(c - i\infty, c + i\infty)$ in the complex s-plane.

Since any notation for the transform parameter is available, one should be aware of the notation of the parameter since some authors use "p" instead of "s." When the Laplace inversion is carried out by using some inversion formulas in a reference book, not performing the inversion integral in the complex plane, the symbolic form of the Laplace inversion

$$f(t) = L^{-1}[f^*(s)] \tag{1.1.6}$$

is used for the sake of simplicity. The present book also employs frequently this simple expression for the Laplace inversion.

So far, many integral transform pairs have been found and defined. We choose and use one suitable integral transform depending on the geometry (integration range) and the simplicity of its application. The followings are typical integral transforms which are much used in applications.

(1) Finite Fourier transform (Fourier series): $f(x), x \in [-\pi, \pi]$

When an original function is defined within a finite region, the Fourier finite transform, i.e. Fourier series is used and its transform pair can be defined as follows:

(1.1) Complex Fourier series

$$f_n = \int_{-\pi}^{\pi} f(x)\exp(\pm inx)dx, \quad f(x) = \frac{1}{2\pi}\sum_{n=-\infty}^{\infty} f_n \exp(\mp inx) \tag{1.1.7}$$

where the index "n" is an integer.

(1.2) Fourier cosine series

$$a_0 = \frac{1}{2}\int_{-\pi}^{\pi} f(x)dx, \quad a_n = \int_{-\pi}^{\pi} f(x)\cos(nx)dx, \quad f(x) = \frac{1}{\pi}\sum_{n=0}^{\infty} a_n \cos(nx) \tag{1.1.8}$$

(1.3) Fourier sine series

$$b_n = \int_{-\pi}^{\pi} f(x)\sin(nx)dx, \quad f(x) = \frac{1}{\pi}\sum_{n=1}^{\infty} b_n \sin(nx) \tag{1.1.9}$$

(2) Complex Fourier transform: $f(x), x \in (-\infty, \infty)$

When the function is defined in an infinite region, the complex Fourier transform pair is defined by

$$\bar{f}(\xi) = \int_{-\infty}^{\infty} f(x)\exp(\pm i\xi x)dx, \quad f(x) = \frac{1}{2\pi}\int_{-\infty}^{\infty} \bar{f}(\xi)\exp(\mp i\xi x)d\xi \tag{1.1.10}$$

(2.1) Fourier transform with a non-uniform parameter.

The Fourier transform with a non-uniform parameter is defined as

$$\bar{f}(\xi) = \int_{a}^{b} f(x)\exp\{+ixh(\xi)\}dx \tag{1.1.11a}$$

$$f(x) = \frac{1}{2\pi} \int_{-\infty}^{+\infty} \bar{f}(\xi) \exp\{-ixh(\xi)\} h'(\xi) d\xi \tag{1.1.11b}$$

where $h(\xi)$ is a monotonically increasing/decreasing function, and the lower and upper integration limits, a and b, are arbitrary constants.

Let us verify the pair of Eq. (1.1.11), briefly. Since $h(\xi)$ is the monotonically increasing or decreasing function, we can define its inverse function uniquely as

$$\eta = h(\xi) \quad \Leftrightarrow \quad \xi = h^{-1}(\eta) \tag{1.1.12}$$

Using the above definition, Eq. (1.1.11a) is rewritten in the form of the regular Fourier transform,

$$\bar{f}(h^{-1}(\eta)) = \int_a^b f(x) \exp(+ix\eta) dx \tag{1.1.13}$$

Applying the inversion formula of Fourier transform in Eq. (1.1.10) to the above equation, the Fourier inversion of Eq. (1.1.13) is given by

$$f(x) = \frac{1}{2\pi} \int_{-\infty}^{+\infty} \bar{f}(h^{-1}(\eta)) \exp(-ix\eta) d\eta; \quad a < x < b \tag{1.1.14}$$

As we have defined the inverse function in Eq. (1.1.12), the integration variable η is returned to the original variable ξ,

$$\xi = h^{-1}(\eta) \quad \Rightarrow \quad \eta = h(\xi), \quad d\eta = h'(\xi) d\xi \tag{1.1.15}$$

Then, Eq. (1.1.14) is rewritten as

$$f(x) = \frac{1}{2\pi} \int_{-\infty}^{+\infty} \bar{f}(\xi) \exp\{-ixh(\xi)\} h'(\xi) d\xi \tag{1.1.16}$$

This is the second of the transform pair Eq. (1.1.11b). Therefore, the pair of the Fourier transform with the non-uniform parameter $h(\xi)$ is verified.

(3) Fourier cosine/sine transform: $f(x),\ x \in [0, \infty)$

When an original function is even or odd in an infinite region, or the function is defined in a semi-infinite region, we employ Fourier cosine or sine transform. They are defined as follows:

(3.1) Fourier cosine transform

$$\bar{f}^{(C)}(\xi) = \int_0^\infty f(x)\cos(\xi x)dx, \quad f(x) = \frac{2}{\pi}\int_0^\infty \bar{f}^{(C)}(\xi)\cos(\xi x)d\xi \qquad (1.1.17)$$

(3.2) Fourier sine transform

$$\bar{f}^{(S)}(\xi) = \int_0^\infty f(x)\sin(\xi x)dx, \quad f(x) = \frac{2}{\pi}\int_0^\infty \bar{f}^{(S)}(\xi)\sin(\xi x)d\xi \qquad (1.1.18)$$

(4) Hankel transform: $f(x),\ x \in [0, \infty)$

Another semi-infinite integral transform is Hankel transform (Sneddon 1951, p. 48) defined by

$$\tilde{f}_n(\xi) = \int_0^\infty x f(x) J_n(\xi x)dx, \quad f(x) = \int_0^\infty \xi \tilde{f}_n(\xi) J_n(\xi x)d\xi \qquad (1.1.19)$$

where $J_n(z)$ is the n-th order Bessel function of the first kind.

In order to guarantee the application of the integral transform, each integral must converge. For example, the Fourier transform requires a convergence condition at infinity,

$$f(x) \underset{x\to\pm\infty}{\Rightarrow} O(|x|^{-\nu}),\ \nu > 1 \qquad (1.1.20)$$

for the original function. However, if we employ the Hankel transform, it is allowable for the original function to be finite since the Bessel function decays with the order of the inverse square root at the infinity.

In some applications, where we encounter multi-variable functions, or where a multiple integral transform is employed, a different mark for each transform is defined such as $\bar{f}, \tilde{f}, f^*$, and the multiple transform is denoted by piling up the transform marks, like $\bar{\tilde{f}}^*$. In the present book, we employ the Laplace transform with respect to the time variable t as defined in Eqs. (1.1.3) and (1.1.4). For the space variables (x, y, z), we apply three Fourier transforms with the transform parameters (ξ, η, ζ) respectively. They are defined as

$$\bar{f}(\xi) = \int_{-\infty}^\infty f(x)\exp(+i\xi x)dx, \quad f(x) = \frac{1}{2\pi}\int_{-\infty}^\infty \bar{f}(\xi)\exp(-i\xi x)d\xi \qquad (1.1.21a)$$

$$\tilde{f}(\eta) = \int_{-\infty}^{\infty} f(y)\exp(+i\eta y)dy, \quad f(y) = \frac{1}{2\pi}\int_{-\infty}^{\infty} \tilde{f}(\eta)\exp(-i\eta y)d\eta \tag{1.1.21b}$$

$$\hat{f}(\zeta) = \int_{-\infty}^{\infty} f(z)\exp(+i\zeta z)dz, \quad f(z) = \frac{1}{2\pi}\int_{-\infty}^{\infty} \hat{f}(\zeta)\exp(-i\zeta z)d\zeta \tag{1.1.21c}$$

Please remember that the pair of the space variable and transform parameter is fixed throughout the present book, such as the pair, (x, ξ), (y, η) and (z, ζ).

1.2 Distributions and Their Integration Formulas

For modeling engineering phenomena, many mathematical functions are used. Elementary and some special functions are used for continuous phenomena. But for discontinuous phenomena, distributions such as delta and step functions are frequently used. This subsection explains briefly the definition of three distributions: Dirac's delta function, Heaviside's unit step function and Heisenberg's delta function.

(1) Heaviside's unit step function: $H(x-a)$

Heaviside's unit step function is defined by the graphical form in Fig. 1.1. This function takes the value 0 when its argument is negative and +1 when the argument is positive,

$$H(x-a) = \begin{cases} +1; & x > a \\ 0; & x < a \end{cases} \tag{1.2.1}$$

Due to this definition, the zero argument at $x = a$ is a discontinuous point. So, it takes two limiting values from the positive and negative sides of the discontinuous point,

$$H(0^+) = +1, \quad H(0^-) = 0 \tag{1.2.2}$$

Then, we have to understand that the step function is not defined at $x = a$.

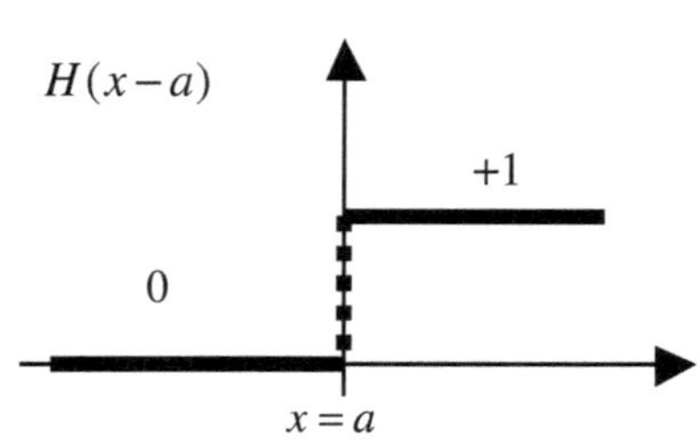

Fig. 1.1 Heaviside's unit step function

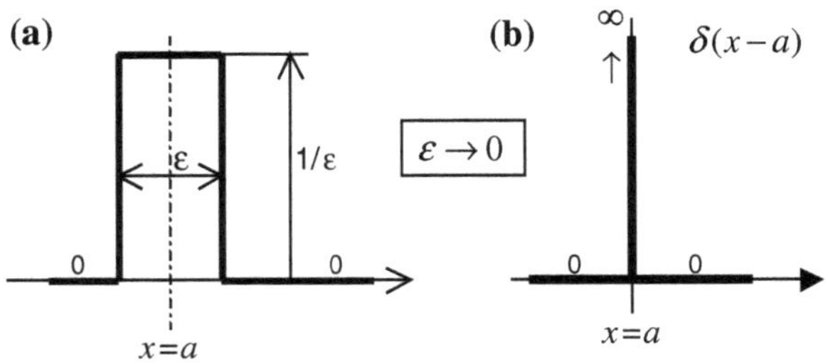

Fig. 1.2 Schematic definition of Dirac's delta function **a** rectangular pulse **b** Dirac's delta function

(2) Dirac's delta function: $\delta(x-a)$

Dirac's delta function is defined as the limit $\varepsilon \to 0$ of a rectangular pulse with width ε and height $1/\varepsilon$ as shown in Fig. 1.2.
The center of the rectangular distribution is fixed at $x = a$ in the limiting process and we understand the point as an application point of the delta function. It is denoted by $\delta(x-a)$. In the limit, the width of the delta function vanishes and the functional value becomes infinite. But, its internal invisible area is one due to the definition $(1/\varepsilon) \cdot \varepsilon = 1$. This nature makes the evaluation of the integral very simple, and we have the formula where the integrand includes the delta function as

$$\int_a^b f(x)\delta(x-c)dx = \begin{cases} f(c); & a < c < b \\ 0; & c < a \text{ or } b < c \end{cases} \tag{1.2.3}$$

Using this simple integration formula, we can obtain an integral representation for the delta function. Applying Fourier transform defined by Eq. (1.1.10) to the delta function, $f(x) = \delta(x-c)$, we use the integration formula (1.2.3). The transform integral is evaluated as

$$\bar{f}(\xi) = \int_{-\infty}^{\infty} \delta(x-c)\exp(+i\xi x)dx = \exp(+i\xi c) \tag{1.2.4}$$

The inverse Fourier transform is also applied to the above $\bar{f}(\xi)$ and its integration range is reduced to the semi-infinite,

$$f(x) = \delta(x-c) = \frac{1}{2\pi}\int_{-\infty}^{\infty} \exp\{-i\xi(x-c)\}d\xi = \frac{1}{\pi}\int_0^{\infty} \cos\{\xi(x-c)\}d\xi \tag{1.2.5}$$

Then, we have the integral representation for Dirac's delta function, i.e.

$$\delta(x-c) = \frac{1}{\pi}\int_{0}^{\infty} \cos\{\xi(x-c)\}d\xi \tag{1.2.6}$$

Finally, we would like to add one useful formula between delta and step functions. That is

$$\delta(x-a) = \frac{d}{dx}H(x-a) \tag{1.2.7}$$

This relation will be understood from the graphical discussion on delta and step functions.

(3) Heisenberg's delta function: $\delta_{\pm}(x)$

In many applications, we have to express a semi-infinite distribution of a physical quantity, such as the uniform load over the surface. In order to treat the semi-infinite distribution, another delta function is defined. That is Heisenberg's delta function which is the Fourier transform of the semi-infinite distribution.

Let us consider Fourier transform of Heaviside's unit step function,

$$H(x) = \begin{cases} 1; & 0 < x < +\infty \\ 0; & -\infty < x < 0 \end{cases} \tag{1.2.8}$$

Its formal Fourier transform is

$$\bar{h}(\xi) = \int_{-\infty}^{\infty} H(x)\exp(+i\xi x)dx = \int_{0}^{\infty} \exp(+i\xi x)dx \tag{1.2.9}$$

This integral cannot be evaluated in the regular sense of calculus. Instead, we shall look for the transformed image function $\bar{h}(\xi)$ so that its Fourier inversion integral results in the step function.

Remember the delta function whose Fourier transform is a constant, and assume that the image function is the sum of delta function and a newly introduced unknown function $\bar{h}_1(\xi)$,

$$\bar{h}(\xi) = \pi\delta(\xi) + \bar{h}_1(\xi) \tag{1.2.10}$$

Let us apply the Fourier inversion integral to the above function and look for the suitable form of the unknown function $\bar{h}_1(\xi)$. The inversion integral is processed as

$$H(x) = \frac{1}{2\pi}\int_{-\infty}^{\infty} \bar{h}(\xi)\exp(-i\xi x)d\xi = \frac{1}{2\pi}\int_{-\infty}^{\infty} \{\pi\delta(\xi) + \bar{h}_1(\xi)\}\exp(-i\xi x)d\xi$$
$$= \frac{1}{2} + \frac{1}{2\pi}\int_{-\infty}^{\infty} \bar{h}_1(\xi)\exp(-i\xi x)d\xi \overset{\text{must be}}{\Rightarrow} \begin{cases} 1; & x > 0 \\ 0; & x < 0 \end{cases} \tag{1.2.11}$$

From the last line in the above equation, we learn that the Fourier transform (inversion) integral of the unknown function $\bar{h}_1(\xi)$ must be $+1/2$ in $x > 0$ and $-1/2$ in $x < 0$, i.e.

$$\frac{1}{2\pi}\int_{-\infty}^{\infty} \bar{h}_1(\xi)\exp(-i\xi x)d\xi = \begin{cases} +1/2; & x > 0 \\ -1/2; & x < 0 \end{cases} \tag{1.2.12}$$

We remember the integration formula that gives $\pm 1/2$ in each semi-infinite x-region (Erdélyi 1954, vol I, pp. 64, 3), that is

$$\frac{1}{\pi}\int_{0}^{\infty} \frac{\sin(x\xi)}{\xi}d\xi = \frac{1}{2\pi}\int_{-\infty}^{\infty} \frac{i}{\xi}\exp(-ix\xi)d\xi = \begin{cases} +1/2; & x > 0 \\ -1/2; & x < 0 \end{cases} \tag{1.2.13}$$

Comparing Eq. (1.2.12) with Eq. (1.2.13), we learn that the whole integrand in Eq. (1.2.12) must be the function

$$\frac{\sin(x\xi)}{\xi} \tag{1.2.14}$$

and thus we can guess that the suitable functional form for $\bar{h}_1(\xi)$ is

$$\bar{h}_1(\xi) = \frac{i}{\xi} \tag{1.2.15}$$

We shall now examine whether this form is really suitable or not. Substitute the above Eq. (1.2.15) into the last line in Eq. (1.2.11) and split the integral into real and imaginary parts, we get

$$\frac{1}{2} + \frac{1}{2\pi}\int_{-\infty}^{\infty} \frac{i}{\xi}\exp(-i\xi x)dx = \frac{1}{2} + \frac{1}{2\pi}\int_{-\infty}^{\infty} \frac{1}{\xi}\sin(\xi x)dx + \frac{i}{2\pi}\int_{-\infty}^{\infty} \frac{1}{\xi}\cos(\xi x)dx \tag{1.2.16}$$

The third term on the right hand side vanishes due to the anti-symmetric nature of the integrand. The second term is evaluated with the aid of formula (1.2.13) as

$$\frac{1}{2}+\frac{1}{2\pi}\int_{-\infty}^{\infty}\frac{\sin(\xi x)}{\xi}d\xi=\frac{1}{2}+\frac{1}{\pi}\int_{0}^{\infty}\frac{\sin(\xi x)}{\xi}d\xi=\frac{1}{2}+\frac{1}{2}\begin{cases}+1; & x>0\\ -1; & x<0\end{cases}$$
$$=\begin{cases}1; & x>0\\ 0; & x<0\end{cases} \tag{1.2.17}$$

Then, we learn that the supposed function $\bar{h}_1(\xi)$ in Eq. (1.2.15) is correct and that the Fourier transform of the step function is given by

$$\bar{h}(\xi)=\pi\delta(\xi)+\frac{i}{\xi} \tag{1.2.18}$$

This function is called "Heisenberg's delta function" and its transform paír is

$$\delta_+(\xi)=\pi\delta(\xi)+\frac{i}{\xi}=\int_0^{\infty}\exp(+i\xi x)dx,\quad H(x)=\frac{1}{2\pi}\int_{-\infty}^{\infty}\delta_+(\xi)\exp(-i\xi x)d\xi \tag{1.2.19}$$

Heisenberg's delta function $\delta_+(\xi)$ is the Fourier transform of the step function $H(x)$. If we have the another step function $H(-x)$, its transform pair is also given by

$$\delta_-(\xi)=\pi\delta(\xi)-\frac{i}{\xi}=\int_{-\infty}^{0}\exp(+i\xi x)dx,\quad H(-x)=\frac{1}{2\pi}\int_{-\infty}^{\infty}\delta_-(\xi)\exp(-i\xi x)d\xi \tag{1.2.20}$$

Then, Heisenberg's delta function has two definitions as

$$\delta_\pm(\xi)=\pi\delta(\xi)\pm\frac{i}{\xi},\quad H(\pm x)=\frac{1}{2\pi}\int_{-\infty}^{\infty}\delta_\pm(\xi)\exp(-i\xi x)d\xi \tag{1.2.21}$$

As a byproduct, we can obtain a new formula. Two operations, addition and subtraction of Eqs. (1.2.19) and (1.2.20) give us the formulas as

$$\int_0^{\infty}\cos(\xi x)dx=\pi\delta(\xi),\quad \int_0^{\infty}\sin(\xi x)dx=\frac{1}{\xi} \tag{1.2.22}$$

The formula on the left is the same as the integral in Eq. (1.2.6), but with $c=0$.

1.3 Branch Cut and Argument of Square Root Functions

When we apply the multiple integral transform to the dynamic problem, some root (radical) functions of the complex variable appear. The square root function is one of multi-valued functions and we have to introduce branch cuts in order to make the root function single-valued. Many textbooks for the complex analysis describe the branch cut/line for very simple multi-valued functions. However, in the practical applications, we have to consider the branch cut for a little bit complicated function. In this subsection, we show how to introduce the branch cut for the fundamental square root function following the method of Ewing et al. (1957, Sect. 2.5, pp. 44–49). This method is very simple and is easily understood even for college students. After introducing the branch cut, we also show the argument of the root function along the branch cut and that our introduction of the branch cut satisfies the regular convergence condition in the complex plane. This condition is crucial for the inversion integral.

We shall discuss the branch cut and the argument of two square root (radical) functions: $g(z) = \sqrt{z - z_0}, \sqrt{z^2 - z_0^2}$ where z_0 is a complex constant. The branch cut is introduced under the conditional: the real part of the root function is positive in the complex z-plane, i.e. $\mathrm{Re}\{g(z)\} \geq 0$. Discussions on these two root functions are carried out, separately.

1.3.1 Square Root Function 1: $g(z) = \sqrt{z - z_0}$

Let us consider the introduction of the branch cut/line for a simple square root function which is defined by

$$g(z) = \sqrt{z - z_0} \tag{1.3.1}$$

where $z(= x + iy)$ is the complex variable and $z_0(= \alpha + i\beta)$ is an arbitrary complex constant. We impose the conditional

$$\mathrm{Re}(\sqrt{z - z_0}) \geq 0 \tag{1.3.2}$$

on the root function.

It is easily found that the branch point is $z = z_0$ which is derived from $g(z) = 0$. In order to discuss the branch cut, we introduce the capital letters, X and Y, which denote the real and imaginary parts of the root function, respectively. The root function is rewritten in the detailed form,

$$X + iY = \sqrt{x + iy - (\alpha + i\beta)} \tag{1.3.3}$$

Squaring both sides, and equating the real and imaginary parts, we have the simultaneous equations for two unknowns, X and Y, as

$$\begin{cases} X^2 - Y^2 = x - \alpha \\ XY = (y - \beta)/2 \end{cases} \tag{1.3.4}$$

Since the conditional is equivalent to

$$X = \mathrm{Re}(\sqrt{z - z_0}) \geq 0 \tag{1.3.5}$$

the border for this conditional is $X = 0$ and then, from the second equation in (1.3.4), we have

$$y - \beta = 0 \tag{1.3.6}$$

On the other hand, substituting $X = 0$ into the first equation in (1.3.4), we have

$$-Y^2 = x - \alpha \tag{1.3.7}$$

Since $-Y^2 < 0$, x must be smaller than α,

$$x < \alpha \tag{1.3.8}$$

The two conditions, Eqs. (1.3.6) and (1.3.8), give one semi-infinite line $(x < \alpha, \quad y = \beta)$ in the complex z-plane. This semi-infinite horizontal line with the edge $(x = \alpha, \ y = \beta)$ is called the branch cut or line. The conditional $\mathrm{Re}\left(\sqrt{z - z_0}\right) = 0$ is satisfied on this branch cut which is shown by the thick line PQ in Fig. 1.3.

Now, we shall examine whether the conditional $\mathrm{Re}\left(\sqrt{z - z_0}\right) \geq 0$ is satisfied or not in the whole complex z-plane. We introduce the polar coordinates (r, θ) for the complex variable as

$$z - z_0 = r\exp(+i\theta) \tag{1.3.9}$$

and measure its argument θ from the horizontal line PC as shown in Fig. 1.3. Then the root function is expressed in the polar form as

$$\sqrt{z - z_0} = \sqrt{r}\exp(+i\theta/2) \tag{1.3.10}$$

If the argument θ is defined within $-\pi \leq \theta \leq +\pi$, the argument of the root function is in the range, $-\pi/2 \leq \theta/2 \leq +\pi/2$, and thus the real part of the root function is positive in the whole complex z-plane. Based on this argument definition, we can specify the argument of the root function on the line along the branch cut. The argument of the root function on the upper horizontal line AB in Fig. 1.3 is measured from the horizontal line PC and is $\arg\left(\sqrt{z - z_0}\right) = +\pi/2$. On the lower

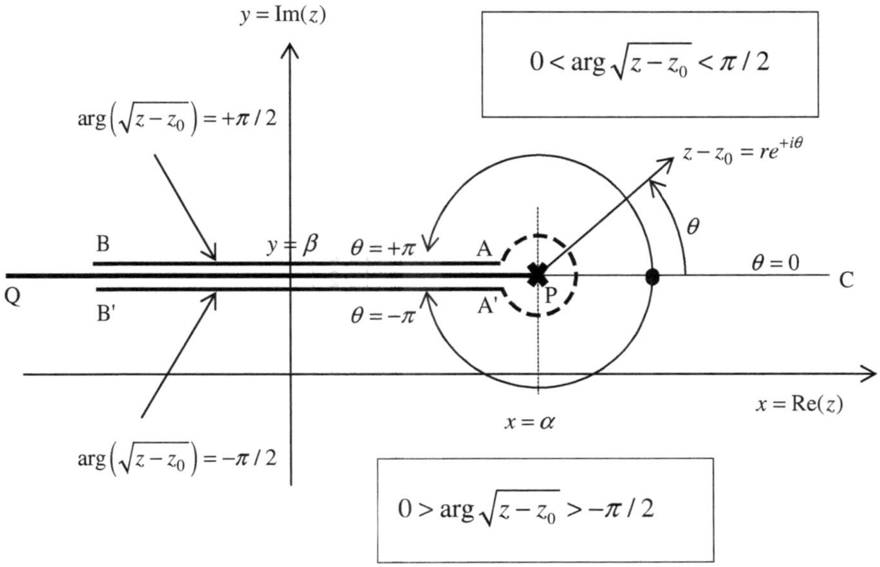

Fig. 1.3 Definition of the arguments of the variable $z - z_0$ and its square root function $\sqrt{z - z_0}$

line A′B′, the argument is $\arg\left(\sqrt{z - z_0}\right) = -\pi/2$. Thus, the argument of the root function is in the range

$$-\pi/2 \leq \arg(\sqrt{z - z_0}) \leq +\pi/2 \tag{1.3.11}$$

Therefore, the real part of the root function is positive in the whole complex z-plane.

In the above discussion, the argument of the variable $z - z_0$ is restricted in the range $-\pi \leq \theta \leq +\pi$. However, if we take the positive argument only, i.e. $0 \leq \theta$, we have to introduce two Riemann sheets. The two sheets are connected along the branch cut as shown in Fig. 1.4 and the argument of the variable $z - z_0$ is measured from the horizontal line as that in Fig. 1.3, but, when the argument exceeds $+\pi$ (cross the cut), we measure the argument in the lower sheet. After rounding one anti-clockwise rotation in the lower sheet, the argument is then measured in the upper sheet. The variation of the argument is shown by the solid and dotted thin curves in Fig. 1.4, schematically. Due to this definition, the argument of the variable is $\arg(z - z_0) = +\pi$ on the upper line AB, and $\arg(z - z_0) = +3\pi$ on the lower line A′B′. Thus, the argument of the square root function is positive in the upper Riemann sheet, but, unfortunately, it is not continuous across the horizontal line PC in Fig. 1.3, i.e. $\theta = 0$ and $\theta = +4\pi$ in Fig. 1.4. Due to this discontinuous nature of the argument, the former definition is preferable for the argument of the root function as defined in Fig. 1.3.

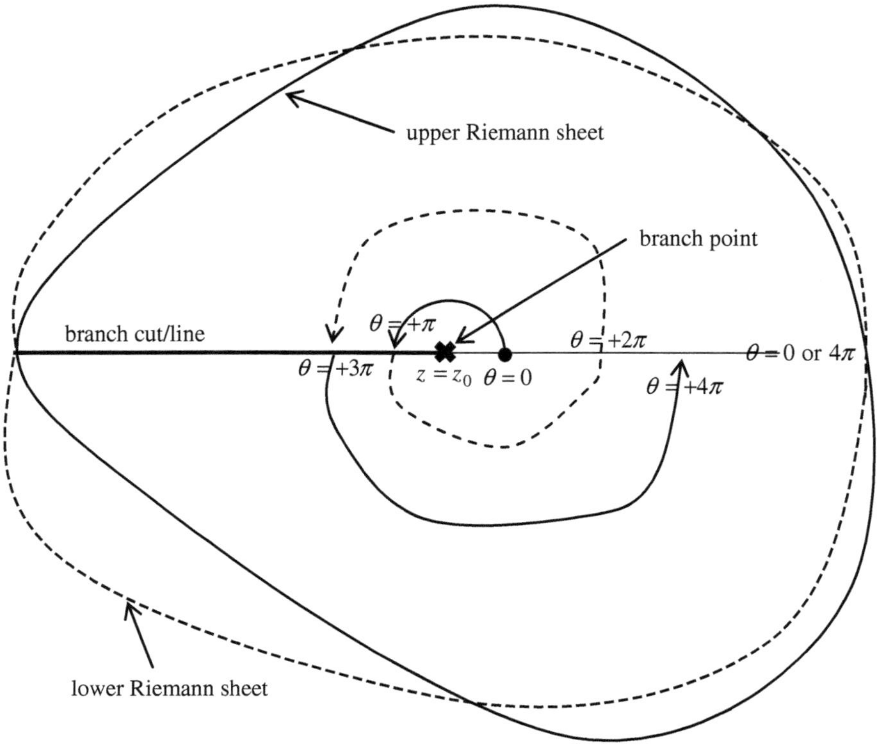

Fig. 1.4 Two Riemann sheets and argument variation of the variable $z - z_0$ (upper sheet: *solid curve*, lower sheet: *dotted curve*)

1.3.2 Square Root Function 2: $g(z) = \sqrt{z^2 - z_0^2}$

This square root function appears frequently in many dynamic problems. The conditional, $\mathrm{Re}\left(\sqrt{z^2 - z_0^2}\right) \geq 0$ is also employed in order to guarantee the convergence at infinity. We shall discuss the introduction of the branch cut under this conditional.

(1) Branch cut

The branch cut and the argument of a little bit complicated square root function, $g(z) = \sqrt{z^2 - z_0^2}$, is discussed. The conditional imposed on the function is similar to that in the previous subsection, i.e. $\mathrm{Re}\left(\sqrt{z^2 - z_0^2}\right) \geq 0$. It should be understood that the variable is $z = x + iy$ and the complex constant is $z_0 = \alpha + i\beta$ where α and β are positive. It is easily found that two branch points of this root function are at $z = \pm(\alpha + i\beta)$. In order to discuss the branch cut, we introduce the capital X and

Y for the real and imaginary parts of the function and rewrite the root function in the explicit form,

$$X + iY = \sqrt{(x + iy)^2 - (\alpha + i\beta)^2} \tag{1.3.12}$$

Squaring the both sides, and equating the real and imaginary parts, we have the simultaneous equations for the unknowns, X and Y,

$$\begin{cases} X^2 - Y^2 = x^2 - y^2 - (\alpha^2 - \beta^2) \\ XY = xy - \alpha\beta \end{cases} \tag{1.3.13}$$

Since the conditional is $X = \mathrm{Re}\left(\sqrt{z^2 - z_0^2}\right) \geq 0$, its border is $X = 0$. Substituting $X = 0$ into the second of Eq. (1.3.13), we have the simple equation for the hyperbola in the complex z-plane,

$$y = \frac{\alpha\beta}{x} \tag{1.3.14}$$

This is the border curve for the conditional, but its range must be specified. So, we substitute $X = 0$ into the first of Eq. (1.3.13),

$$-Y^2 = x^2 - y^2 - (\alpha^2 - \beta^2) \tag{1.3.15}$$

As the left side of the above equation is negative, the right side also must be negative,

$$x^2 - y^2 - (\alpha^2 - \beta^2) \leq 0 \tag{1.3.16}$$

Substituting the Eq. (1.3.14) into the above equation, we have the quadratic inequality for x^2,

$$x^4 - (\alpha^2 - \beta^2)x^2 - \alpha^2\beta^2 = (x^2 - \alpha^2)(x^2 + \beta^2) \leq 0 \tag{1.3.17}$$

Solving for x^2, we have

$$-\beta^2 \leq x^2 \leq +\alpha^2 \tag{1.3.18}$$

Then, the admissible range for the variable x in the complex z-plane is

$$0 \leq |x| \leq \alpha \tag{1.3.19}$$

The combination of two Eqs. (1.3.14) and (1.3.19) determines the branch cut which satisfies the conditional $\mathrm{Re}\left(\sqrt{z^2 - z_0^2}\right) = 0$. This shows two semi-infinite hyperbolic curves and their edges are corresponding to the branch points as shown in Fig. 1.5.

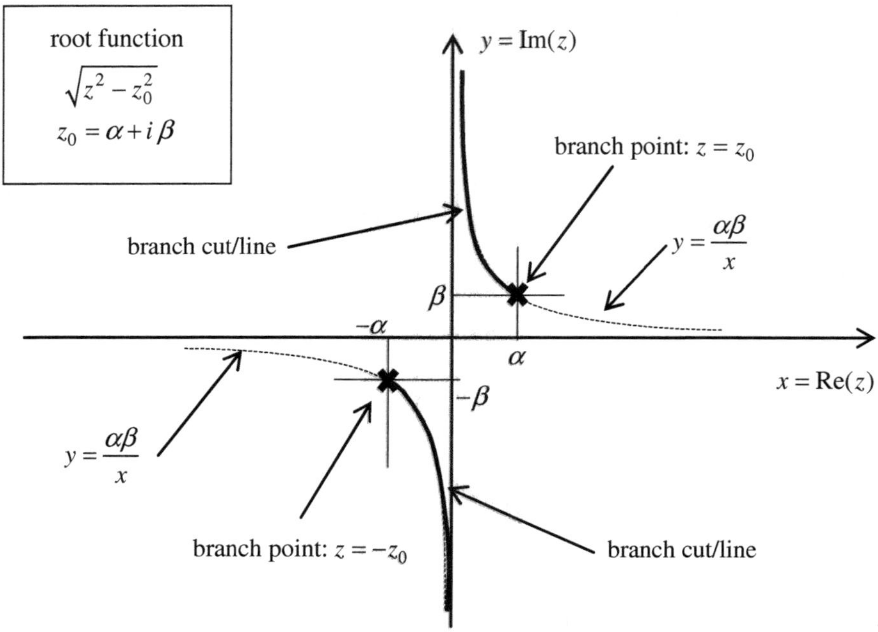

Fig. 1.5 Branch cut/line for the square root function $\sqrt{z^2 - z_0^2}$ with $z_0 = \alpha + i\beta$

When the real or imaginary part of the constant z_0 vanishes, the branch cut becomes a simple line. In the case of the pure real constant, $z_0 = \alpha$, the branch cut for the root function $\sqrt{z^2 - \alpha^2}$ is the line along the real and imaginary axes as shown in Fig. 1.6a, where the cut along the real axis is slightly shifted from the axis for the visual explanation. On the other hand, in the case of the pure imaginary constant $z_0 = +i\beta$, the branch cut for the root function $\sqrt{z^2 + \beta^2}$ is the semi-infinite straight line along the imaginary axis as shown in Fig. 1.6b.

It is easily anticipated that the branch cut/curve depends on the sign of the imaginary part of the complex constant z_0. The former discussion is carried out for the positive imaginary part of the constant. So, we shall reconsider the branch cut in the case of the negative imaginary part of the constant, i.e. $z_0 = \alpha - i\beta$. Employing the same conditional and procedure as those in the former, the branch points for the root function $\sqrt{z^2 - z_0^2}$ are $z = \pm(\alpha - i\beta)$ and the branch cuts are given by

$$y = -\frac{\alpha\beta}{x}; \quad 0 \leq |x| \leq \alpha \tag{1.3.20}$$

This is the semi-hyperbolic curve shown in Fig. 1.7. Two cuts are introduced in the second and the fourth quadrants in the complex z-plane and the edge of the cut is corresponding to each branch point, $z = \pm(\alpha - i\beta)$. When the imaginary part of the constant vanishes, $z_0 = \alpha$; $\beta = 0$, the branch cut for the root function $\sqrt{z^2 - \alpha^2}$

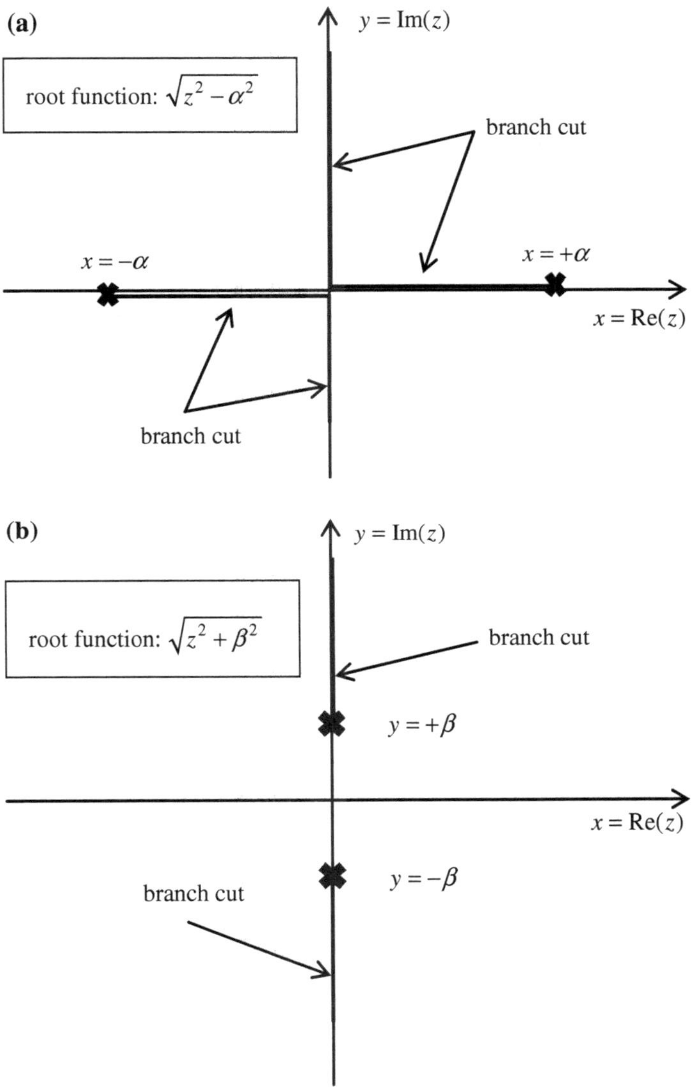

Fig. 1.6 **a** Branch cut for the root function $\sqrt{z^2 - \alpha^2}$. **b** Branch cut for the root function $\sqrt{z^2 + \beta^2}$

becomes the line on the real and imaginary axes as shown in Fig. 1.8a. On the other hand, when the real part vanishes, $z_0 = -i\beta$; $\alpha = 0$, the branch cut for the root function $\sqrt{z^2 + \beta^2}$ becomes the two semi-infinite lines on the imaginary axis as shown in Fig. 1.8b.

Comparing Fig. 1.8a with Fig. 1.6a, we learn that the branch cut on the real axis is different in spite of the same form of the root function $\sqrt{z^2 - \alpha^2}$. This difference

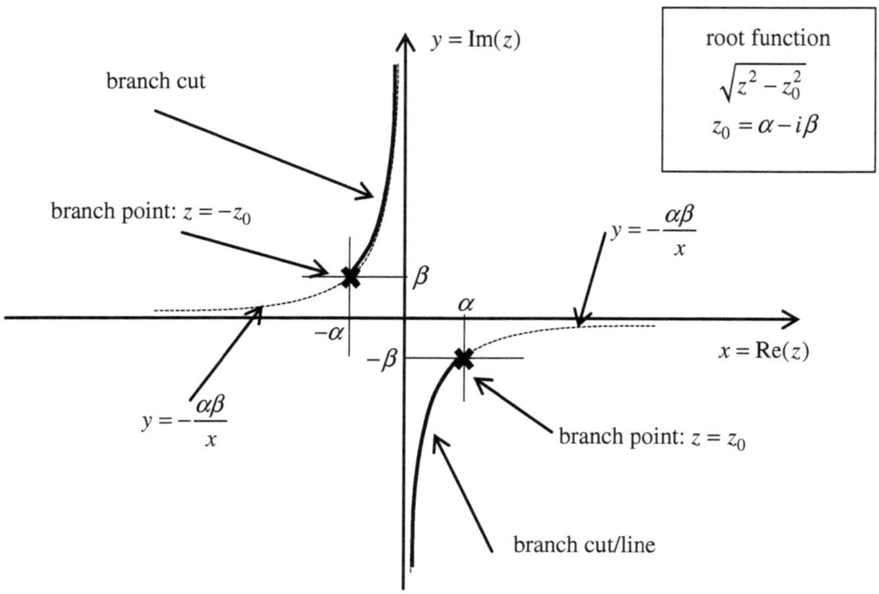

Fig. 1.7 Branch cut/line for the square root function $\sqrt{z^2 - z_0^2}$ with $z_0 = \alpha - i\beta$

comes from the approaching direction of the zero imaginary part of the constant z_0. When the imaginary part approaches to zero from above, the branch cut for the root function $\sqrt{z^2 - \alpha^2}$ is given by the lines in Fig. 1.6a. But, when the imaginary part approaches to zero from below, the branch cut is given by the lines in Fig. 1.8a. In the practical applications, the real constant α expresses the frequency ω of the time-harmonic vibration and the root function is given as $\sqrt{z^2 - \omega^2}$. Therefore, we are puzzled which is the suitable branch cut. This puzzle is solved by the sign of the exponential time factor $\exp(\pm i\omega t)$ and physical conditions. However the answer and verification are given in the later chapters, a brief conclusion is stated here. The branch cut in Fig. 1.6a is employed when the time factor is negative, i.e. $\exp(-i\omega t)$, and the cut in Fig. 1.8a is employed when the time factor is positive, $\exp(+i\omega t)$.

When the real part of the complex constant vanishes, $\alpha \to 0$, the branch cut lays on the imaginary axis as shown in Figs. 1.6b and 1.8b. These two figures are same. Thus, the branch cut for the square root function $\sqrt{z^2 + \beta^2}$ is unchanged regardless of the sign of the imaginary part β.

(2) Argument

We have just determined the branch cut for the square root function under the conditional $\mathrm{Re}\left(\sqrt{z^2 - z_0^2}\right) = 0$. In this subsection, we examine whether the real part of the root function is positive or not in the whole complex plane. In order to

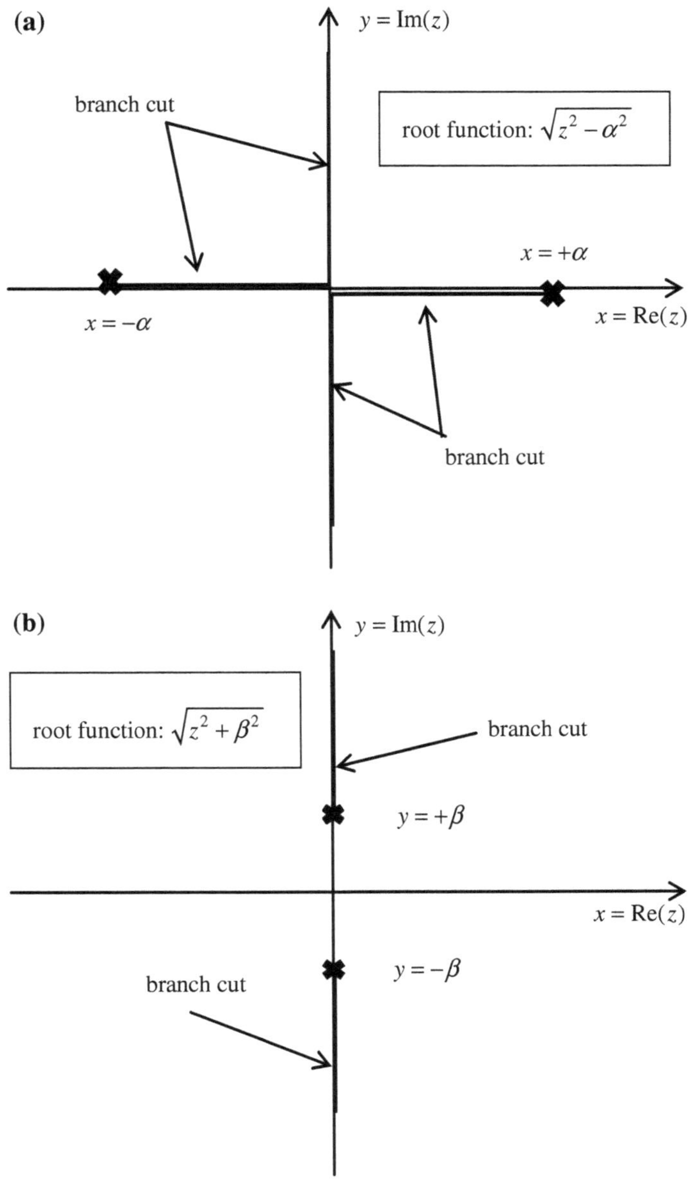

Fig. 1.8 **a** Branch cut for the root function $\sqrt{z^2-\alpha^2}$. **b** Branch cut for the root function $\sqrt{z^2+\beta^2}$

discuss the argument of the root function, we factorize the function as the product of two simple root functions as

$$\sqrt{z^2 - z_0^2} = \sqrt{z - z_0}\sqrt{z + z_0} \tag{1.3.21}$$

and introduce the polar coordinate (r_j, θ_j); $j = 1, 2$ for each variable as

$$z - z_0 = r_1 \exp(+i\theta_1), \quad z + z_0 = r_2 \exp(+i\theta_2) \tag{1.3.22}$$

where r_j is the radial arm/distance from each branch point, and the arguments θ_j is measured from the horizontal line parallel to the positive real axis. Then, the root function is rewritten in the polar form,

$$\sqrt{z^2 - z_0^2} = \sqrt{r_1 r_2} \exp\{+i(\theta_1 + \theta_2)/2\} \tag{1.3.23}$$

and the discussion for the argument of the square root function is carried out on its polar angle $(\theta_1 + \theta_2)/2$. We assume that the real and imaginary parts of the constant α and β are positive. Therefore, the branch points and cuts are in the first and third quadrants in the complex z-plane as shown in Fig. 1.5. The argument of the root function is discussed on three paths: (1) path along the branch cut, (2) path on the non-branch cut hyperbola, and (3) the line which connects two branch points. We shall discuss their arguments, separately.

(2.1) Right side of the upper branch cut

Firstly, we consider the argument on the right side of the upper branch cut in Fig. 1.9a, whose branch cuts are the same as those in Fig. 1.5. The mark "P_1" denotes an arbitrary point on the upper branch cut and its coordinate is (x_P, y_P) in the complex plane. The arguments θ_1 of the radial arm $r_1 = O_1P_1$ is measured from the horizontal line O_1A_1. Since the branch point is at (α, β), the complementary angle ϕ_1 of the radial arm r_1 is introduced as

$$\tan \phi_1 = \frac{y_P - \beta}{\alpha - x_P} \tag{1.3.24}$$

and thus the argument θ_1 is expressed by

$$\theta_1 = \pi - \phi_1 \tag{1.3.25}$$

On the other hand, the argument θ_2 of the radial arm $r_2 = O_2P_1$ is measured from the horizontal line O_2A_2. Since the lower branch point is at $(-\alpha, -\beta)$, the argument of the radial arm, i.e. θ_2, is given by

$$\tan \theta_2 = \frac{\beta + y_P}{\alpha + x_P} \tag{1.3.26}$$

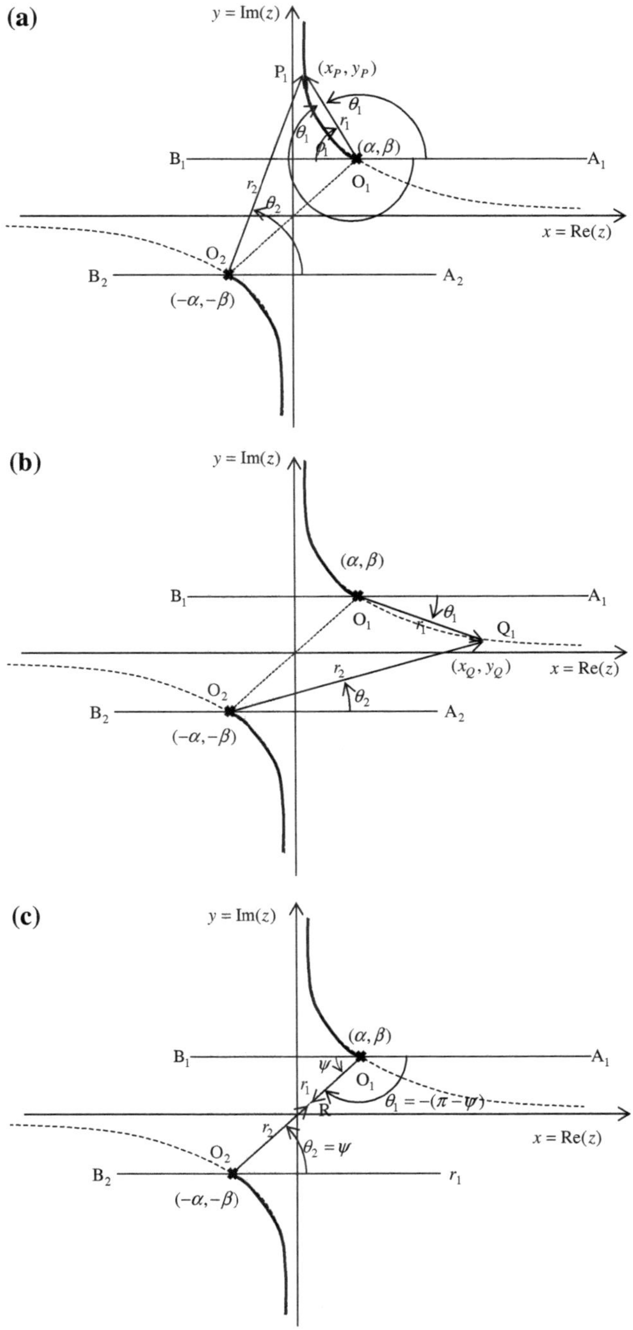

Fig. 1.9 Arguments, θ_1 and θ_2, **a** along the upper branch cut, **b** on the upper non-branch cut hyperbola, **c** on the line connecting two branch points, **d** along the lower branch cut and **e** on the lower non-branch cut hyperbola

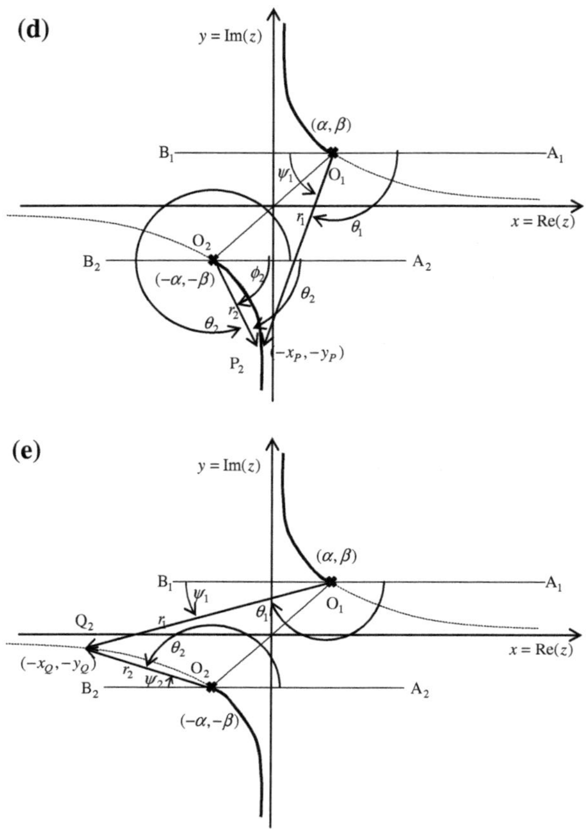

Fig. 1.9 (continued)

Two Eqs. (1.3.24) and (1.3.26) are looks different, but, the point P_1 is on the hyperbolic curve at the limit. The hyperbolic curve gives the relation between x_P and y_P,

$$y_P = \frac{\alpha\beta}{x_P} \tag{1.3.27}$$

Substituting the above relation into the two Eqs. (1.3.24) and (1.3.26),

$$\tan\theta_2 = \tan\phi_1 = \frac{\beta}{x_P} \tag{1.3.28}$$

we find that $\theta_2 = \phi_1$. The argument of the root function at the right side of the upper branch cut is thus given by

$$(\theta_1 + \theta_2)/2 = (\pi - \phi_1 + \phi_1)/2 = +\pi/2 \tag{1.3.29}$$

This shows that the value of the square root function $\sqrt{z^2 - z_0^2}$ at the right side of the upper branch cut is positive imaginary.

(2.2) Left side of the upper branch cut

We now consider the argument of the root function at the left side of the upper branch cut at P_1. The argument θ_1 must be measured in the clockwise direction from the horizontal line O_1A_1, since the arm r_1 cannot cross its own branch cut. However, the argument θ_2 can cross the cut, since the cut is not for the lower branch point. Thus, from the geometry in Fig. 1.9a, we have

$$\theta_1 = -(\pi + \phi_1), \quad \theta_2 = \phi_1 \tag{1.3.30}$$

and the argument of the root function is

$$(\theta_1 + \theta_2)/2 = -\pi/2 \tag{1.3.31}$$

Then, the value of the root function at the left side of the upper branch cut is negative imaginary.

(2.3) On the upper non-branch cut hyperbola

One arbitrary point Q_1 is taken on the non-branch cut hyperbola as shown in Fig. 1.9b and we assume that its coordinate is (x_Q, y_Q) in the complex plane. The arm r_1 from the upper branch point is denoted by the line O_1Q_1 and the arm r_2 from the lower branch point is O_2Q_1. They are shown in Fig. 1.9b. The argument of the arm r_1 is measured from the horizontal line O_1A_1 in the clockwise direction and that of the arm r_2 is measured from the horizontal line O_2A_2 in the anti-clockwise direction. A simple analytical geometry shows that the arguments θ_1 and θ_2 are

$$\tan\theta_1 = -\frac{\beta - y_Q}{x_Q - \alpha}, \quad \tan\theta_2 = \frac{\beta + y_Q}{\alpha + x_Q} \tag{1.3.32}$$

Substituting the equation for the hyperbola $y_Q = \alpha\beta/x_Q$ into the above equations, we have

$$\tan\theta_1 = -\frac{\beta}{x_Q}, \quad \tan\theta_2 = \frac{\beta}{x_Q} \tag{1.3.33}$$

Thus, $\theta_2 = -\theta_1$, and the argument of the root function on the non-branch cut hyperbola is

$$(\theta_1 + \theta_2)/2 = 0 \tag{1.3.34}$$

This shows the value of the root function $\sqrt{z^2 - z_0^2}$ is positive real.

(2.4) On the line connecting two branch points

We take an arbitrary point R on the line O_1O_2, which connects two branch points, as shown in Fig. 1.9c. Since the argument of each arm from the branch point is unchanged on this line, it is no need to specify the location of the point R. Introducing the complementary angle ψ for the acute angle $\angle B_1O_1O_2$, the argument θ_2 for the lower arm r_2 is expressed by

$$\theta_2 = \psi \tag{1.3.35}$$

and that for the upper arm r_1 is

$$\theta_1 = -(\pi - \psi) \tag{1.3.36}$$

Then, the argument of the root function is given by

$$(\theta_1 + \theta_2)/2 = -(\pi/2 - \psi) \tag{1.3.37}$$

This shows that the argument of the root function is negative on the connecting line, since $0 < \psi < \pi/2$. Thus, the real part of the square root function is positive.

(2.5) Right side of the lower branch cut

We take an arbitrary point P_2 on the lower branch cut as shown in Fig. 1.9d, and consider the argument of the root function on the right side of the cut. Assuming the location of the point P_2 as $(-x_P, -y_P)$, the complementary angle $\psi_1 (= \angle B_1O_1P_2)$ is given by

$$\tan \psi_1 = \frac{\beta + y_P}{\alpha + x_P} \tag{1.3.38}$$

Then, the argument of the radial arm r_1 is measured in the clockwise direction,

$$\theta_1 = -(\pi - \psi_1) \tag{1.3.39}$$

The argument θ_2 of the arm $r_2 (= O_2P_2)$ is also measured in the clockwise direction. It is given by

$$\tan \theta_2 = -\frac{y_P - \beta}{\alpha - x_P} \tag{1.3.40}$$

Substituting the equation for the hyperbola, $y_P = \alpha\beta/x_P$, into Eqs. (1.3.38) and (1.3.40), we have

$$\tan\psi_1 = \frac{\beta}{x_P}, \quad \tan\theta_2 = -\frac{\beta}{x_P} \tag{1.3.41}$$

Then,

$$\theta_2 = -\psi_1 \tag{1.3.42}$$

Two Eqs. (1.3.39) and (1.3.42) give the argument of the root function as

$$(\theta_1 + \theta_2)/2 = -\pi/2 \tag{1.3.43}$$

This shows that the value of the root function on the right side of the lower branch cut is negative imaginary.

(2.6) Left side of the lower branch cut

When we consider the argument on the left side of the lower branch cut, an attention must be paid to the argument θ_2 since the arm r_2 from the lower branch point cannot cross its own branch cut. As shown in Fig. 1.9d, the argument θ_2 is measured from the horizontal line O_2A_2 in the counterclockwise direction. Introducing the complementary angle ϕ_2 as shown in Fig. 1.9d, the argument θ_2 is given by

$$\theta_2 = 2\pi - \phi_2 \tag{1.3.44}$$

On the other hand, the arm r_1 can cross the lower branch cut and its argument is

$$\theta_1 = -(\pi - \psi_1) \tag{1.3.45}$$

The two complementary angles are given by

$$\tan\psi_1 = \frac{\beta + y_P}{\alpha + x_P}, \quad \tan\phi_2 = \frac{y_P - \beta}{\alpha - x_P} \tag{1.3.46}$$

Substituting the equation for the hyperbola, $y_P = \alpha\beta/x_P$, into the above, we find that $\psi_1 = \phi_2$. Thus, the argument of the root function is

$$(\theta_1 + \theta_2)/2 = +\pi/2 \tag{1.3.47}$$

This shows that the value of the root function on the left side of the lower branch cut is positive imaginary.

(2.7) On the lower non-branch cut hyperbola

We consider the argument on the lower non-branch cut hyperbola and take an arbitrary point Q_2 on it, as shown in Fig. 1.9e. Assuming the point Q_2 is at $(-x_Q, -y_Q)$, we introduce two complementary angles $\psi_1 (= \angle B_1O_1Q_2)$ and ψ_2 $(= \angle B_2O_2Q_2)$. They are

$$\tan \psi_1 = \frac{\beta + y_Q}{\alpha + x_Q}, \quad \tan \psi_2 = \frac{\beta - y_Q}{x_Q - \alpha} \tag{1.3.48}$$

Since the point Q_2 is on the hyperbola, we substitute $y_Q = \alpha\beta / x_Q$ and then find that $\psi_1 = \psi_2$. The two arguments are

$$\theta_1 = -(\pi - \psi_1), \quad \theta_2 = \pi - \psi_2 \tag{1.3.49}$$

Thus, the argument of the root function is

$$(\theta_1 + \theta_2)/2 = 0 \tag{1.3.50}$$

This shows the value of the root function on the non-branch cut hyperbola is positive real.

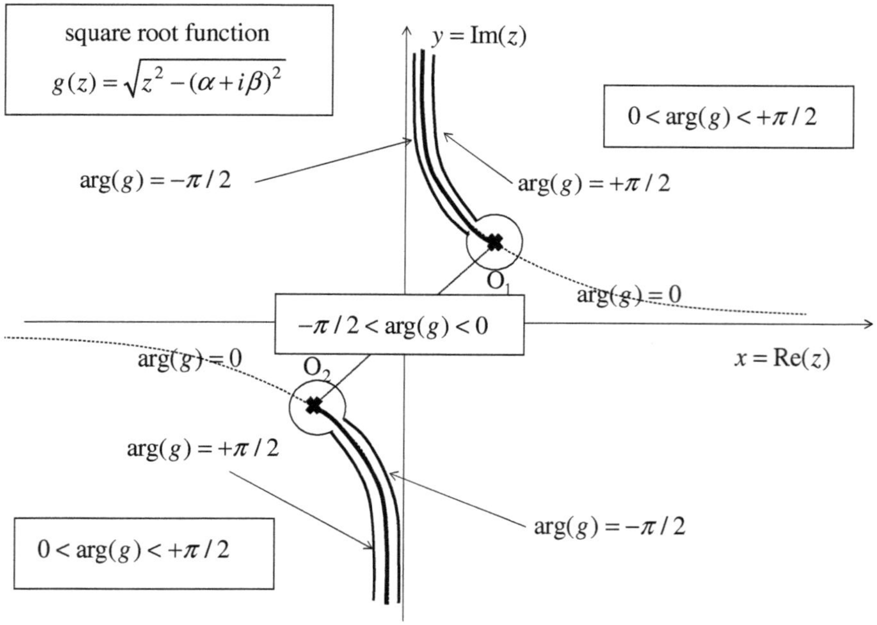

Fig. 1.10 Argument of the square root function $\sqrt{z^2 - (\alpha + i\beta)^2}$ (The imaginary part of the constant is positive)

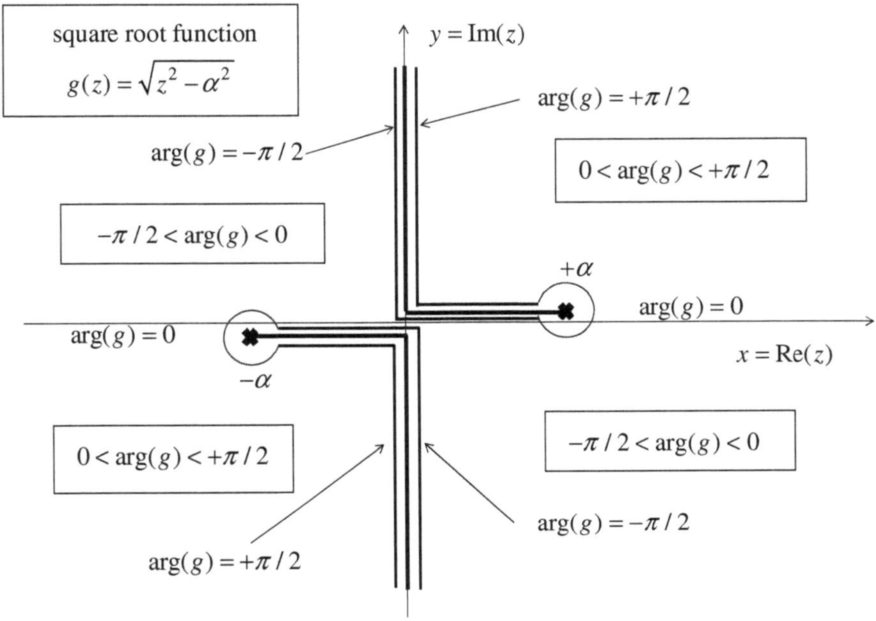

Fig. 1.11 Argument of the square root function $\sqrt{z^2-\alpha^2}$ (The positive imaginary part of the complex constant vanishes)

Summarizing the above discussions in (2.1)–(2.7), the argument of the square root function in the whole complex plane is shown schematically in Fig. 1.10. However the argument in the non-specific point/region was not discussed, we can easily guess the argument from our result. It is shown in the square box in the figure. When the imaginary part of the complex constant vanishes, $\beta \to 0$, Fig. 1.10 is reduced to Fig. 1.11. On the other hand, when the real part of the complex constant vanishes, $\alpha \to 0$, Fig. 1.10 is reduced to Fig. 1.12. In any case, the argument of the square root function in the whole complex z-plane is in the range

$$-\pi/2 \leq (\theta_1+\theta_2)/2 \leq +\pi/2 \tag{1.3.51}$$

This shows that the conditional $\mathrm{Re}\left(\sqrt{z^2-z_0^2}\right)\geq 0$ is satisfied in the whole complex plane and thus we have introduced the branch cuts and defined the argument of the square root function, correctly.

If we employ the complex constant with the negative imaginary part, $z_0=\alpha-i\beta$, the branch cuts are in the second and the fourth quadrants in the complex z-plane as shown in Fig. 1.7. The discussion in the above subsections can be applied to the argument of the root function. Although the detailed discussions for the argument are not repeated here, the final results are schematically shown in Figs. 1.13, 1.14 and 1.15.

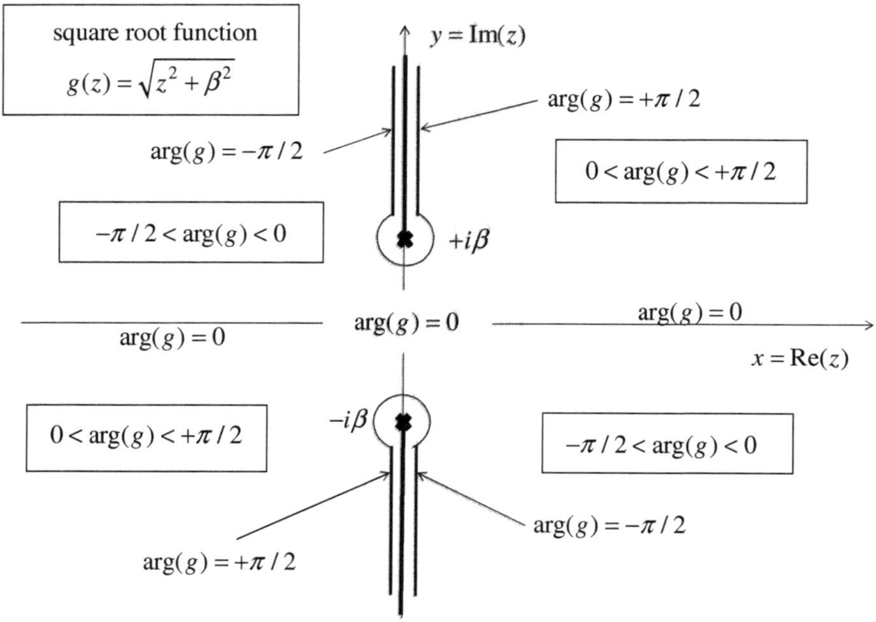

Fig. 1.12 Argument of the square root function $\sqrt{z^2+\beta^2}$ in the complex plane (The real part of the complex constant vanishes)

One comment should be repeated here. Comparing Fig. 1.11 with Fig. 1.14 in the case of vanishing imaginary part of the complex constant, the square root function $\sqrt{z^2-\alpha^2}$ on the real axis $x=\mathrm{Re}(z)$ is the positive or negative purely imaginary as

$$\sqrt{x^2-\alpha^2}=\begin{cases}+i\sqrt{\alpha^2-x^2}\\ -i\sqrt{\alpha^2-x^2}\end{cases};\quad 0\leq|x|<\alpha \tag{1.3.52}$$

The sign depends on the vanishing direction of the imaginary part of the complex constant z_0. Thus, as was mentioned before, we have to pay special attention to the square root function with the real constant $\sqrt{z^2-\alpha^2}$.

1.4 Comments on Inversion Techniques and Integration Formulas

The integral transform technique is a powerful tool for solving linear differential equations. However, its success is up to the evaluation of the inversion integral. So far, many integration formulas have been found and published. The most comprehensive formula books are Erdélyi (1954), Gradshteyn and Ryzhik (1980),

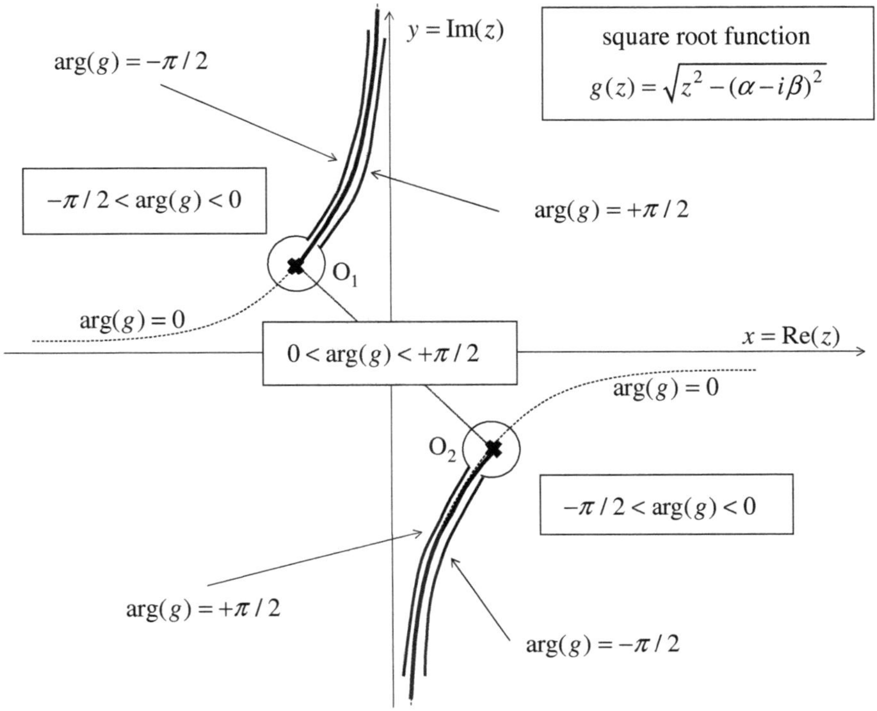

Fig. 1.13 Argument of the square root function $\sqrt{z^2-(\alpha-i\beta)^2}$ (The imaginary part of the complex constant is negative)

Magnus et al. (1966), Watson (1966), Moriguchi et al. (1972) and Titchmarsh (1948).

The books from Erdélyi (1954) to Watson (1966) are well-known and it is not difficult to obtain access to them. The book Watson (1966) deals solely with Bessel functions and not with integral transforms, but it gives many integration formulas for the Hankel transform. The book Moriguchi et al. (1972), written in Japanese, is very compact and is separated into three small handy books. In spite of its compactness, the principal formulas which are included in Erdélyi (1954), Gradshteyn and Ryzhik (1980), Magnus et al. (1966) and Watson (1966) are cited. The author believes that the three handy books are most convenient as a “first aid.” The last book Titchmarsh (1948) describes the mathematics of the theory of Fourier transform. When someone needs more detailed mathematics for the integral transform, this book will give proper answers.

If a desired formula cannot be found in these books, the complex integral is employed to evaluate the inversion integral. The complex integral based on Cauchy’s integral theorem is the most useful evaluation technique and the discussion on

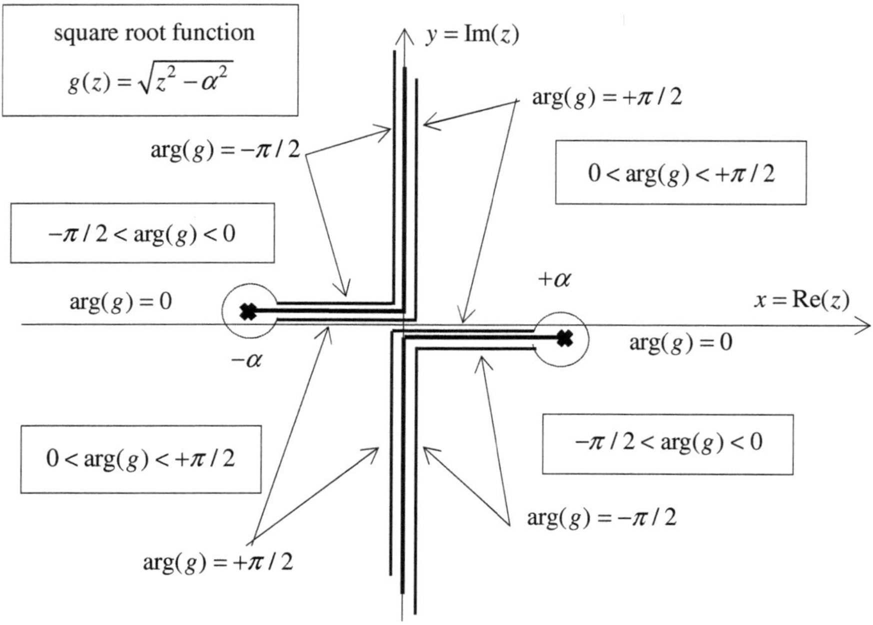

Fig. 1.14 Argument of the square root function $\sqrt{z^2-\alpha^2}$ in the complex plane (The negative imaginary part of the complex constant vanishes)

the branch cut in the previous subsections will be helpful. If the complex integral does not give any compact result, the inversion integral is left in its definition form or is converted to the numerically tractable form by the complex integral.

Exercises

(1.1) Apply the finite Fourier transform, complex Fourier series defined by Eq. (1.1.6), to Dirac's delta function which is defined in a finite region $(-\pi,+\pi)$ as

$$\delta(x-a)\ ;\quad -\pi<a<+\pi \tag{a}$$

and then show the series form of the delta function,

$$\delta(x-a)=\frac{1}{2\pi}\left[1+2\sum_{n=1}^{\infty}\cos\{n(x-a)\}\right] \tag{b}$$

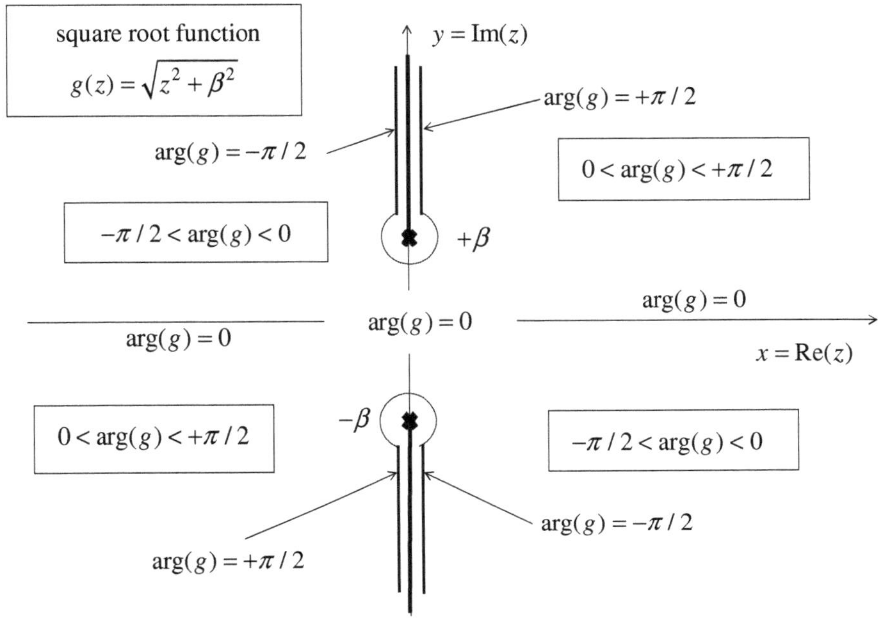

Fig. 1.15 Argument of the square root function $\sqrt{z^2+\beta^2}$ in the complex plane (The real part of the complex constant vanishes)

Fig. 1.16 Two branch cuts for the root function $\sqrt{z^2+a^2}$

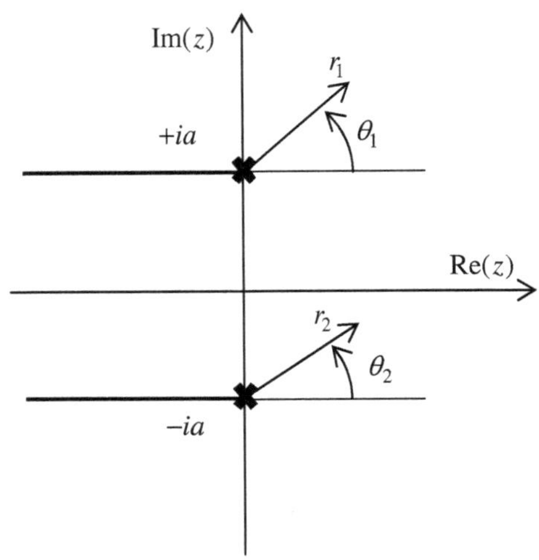

(1.2) If we expand the x-range to $(-\infty, +\infty)$, the Fourier series in the above equation (b) gives an infinite sequence of delta function, i.e.

$$\frac{1}{2\pi}\left[1 + 2\sum_{n=1}^{\infty} \cos\{n(x-a)\}\right] = \sum_{m=-\infty}^{+\infty} \delta(x - a - 2m\pi) \qquad \text{(c)}$$

Explain why the infinite sequence?

(1.3) For the root function $\sqrt{z^2 + a^2}$, where a is a positive real constant, if we introduce two branch cuts as shown in Fig. 1.16, how is the argument of the root function along the branch cut?

References

Erdélyi A (ed) (1954) Tables of integral transforms, vol I and II. McGraw-Hill, New-York

Ewing WM, Jardetzky WS, Press F (1957) Elastic waves in layered media. McGraw-Hill, New York

Gradshteyn IS, Ryzhik IM (Jefferey A, ed) (1980) Table of integrals, series, and products, 5th edn. Academic Press, San Diego

Magnus W, Oberhettinger F, Soni RP (1966) Formulas and theorems for the special functions of mathematical physics. Springer, New York

Moriguchi S, Udagawa K, Ichimatsu S (1972) Mathematical formulas, vols I–III (in Japanese). Iwanami, Tokyo

Sneddon IN (1951) Fourier transforms. McGraw-Hill, New York

Titchmarsh EC (1948) Introduction to the theory of Fourier integrals, 2nd edn. Oxford, London

Watson GN (1966) A treatise on the theory of Bessel functions. Cambridge University Press, Cambridge

Chapter 2
Green's Functions for Laplace and Wave Equations

This chapter shows the solution method for Green's functions of 1, 2 and 3D Laplace and wave equations. Lengthy and detailed explanations are given in order to instruct the basic technique of the integral transform. Especially, Fourier inversion integral for the time-harmonic Green's function is discussed in detail, and three evaluation techniques are introduced in Sect. 2.5.

2.1 1D Impulsive Source

We start from the simplest wave equation that has two variables: a single space variable x and the time t,

$$\frac{\partial^2 \phi}{\partial x^2} = \frac{1}{c^2}\frac{\partial^2 \phi}{\partial t^2} - P\delta(x)\delta(t) \tag{2.1.1}$$

The nonhomogeneous term represents a wave source with magnitude P. The two Dirac's delta functions, $\delta(x)$ and $\delta(t)$, show the location and the impulsive nature of the source. The Green's function is a particular solution of the differential equation corresponding to the impulsive source. The Green's function is sought under the quiescent condition at an initial time,

$$\phi|_{t=0} = \left.\frac{\partial \phi}{\partial t}\right|_{t=0} = 0 \tag{2.1.2}$$

and the convergence condition at infinity,

$$\phi|_{x\to\pm\infty} = \left.\frac{\partial \phi}{\partial x}\right|_{x\to\pm\infty} = 0 \tag{2.1.3}$$

To obtain the particular solution of the wave equation (2.1.1), we apply the integral transforms and reduce the differential equation to an algebraic equation in the transformed domain. Since the unknown function $\phi(x,t)$ has two variables, we

K. Watanabe, *Integral Transform Techniques for Green's Function*,
Lecture Notes in Applied and Computational Mechanics 76,
DOI 10.1007/978-3-319-17455-6_2

apply the double transform: Laplace transform with respect to the time variable t: $0 \leq t < +\infty$,

$$f^*(s) = L[f(t)] = \int_0^{+\infty} f(t)\exp(-st)dt \tag{2.1.4}$$

and Fourier transform with respect to the space variable x: $-\infty < x < +\infty$,

$$\bar{f}(\xi) = \int_{-\infty}^{+\infty} f(x)\exp(+i\xi x)dx, \qquad f(x) = \frac{1}{2\pi}\int_{-\infty}^{+\infty} \bar{f}(\xi)\exp(-i\xi x)d\xi \tag{2.1.5}$$

Firstly, we multiply the kernel of the Laplace transform $\exp(-st)$ to both sides of the differential equation (2.1.1),

$$\int_0^{\infty}\left(\frac{\partial^2\phi}{\partial x^2} = \frac{1}{c^2}\frac{\partial^2\phi}{\partial t^2} - P\delta(x)\delta(t)\right)\exp(-st)dt \tag{2.1.6}$$

and perform the Laplace transform integral term by term. The order of integration and differentiation are interchanged for the first term in the left hand side of the equation. The first term in the right hand side is left in its order and then integrated by parts. The last nonhomogeneous term, which has the delta function, is evaluated by using the formula (1.2.3). Then, Eq. (2.1.6) is rewritten as

$$\frac{d^2}{dx^2}\int_0^{\infty}\phi\exp(-st)dt = \frac{1}{c^2}\int_0^{\infty}\frac{\partial^2\phi}{\partial t^2}\exp(-st)dt - P\delta(x)\int_0^{\infty}\delta(t)\exp(-st)dt \tag{2.1.7}$$

We define the Laplace transform of the unknown function as

$$\phi^* = \int_0^{\infty}\phi\exp(-st)dt \tag{2.1.8}$$

The time-derivative term is integrated by parts,

$$\int_0^{\infty}\frac{\partial^2\phi}{\partial t^2}\exp(-st)dt = \left[\frac{\partial\phi}{\partial t}\exp(-st)\right]_{t=0}^{t\to\infty} + [s\phi\exp(-st)]_{t=0}^{t\to\infty} + s^2\int_0^{\infty}\phi\exp(-st)dt = s^2\phi^* \tag{2.1.9}$$

where the quiescent condition at the initial time is incorporated. To the last non-homogeneous term, the simple integration formula for the delta function,

$$\int_0^{\infty} \delta(t)\exp(-st)dt = 1 \tag{2.1.10}$$

is applied. Then, we have the ordinary differential equation for the function ϕ^* in the transformed domain,

$$\frac{d^2\phi^*}{dx^2} = (s/c)^2\phi^* - P\delta(x) \tag{2.1.11}$$

It is possible to obtain the exact solution for the above ordinary differential equation by the elementary method. However, in order to demonstrate the integral transform technique, we further apply Fourier transform to the ordinary differential equation (2.1.11). The Fourier transform defined by Eq. (2.1.5) is applied to Eq. (2.1.11),

$$\int_{-\infty}^{+\infty} \left(\frac{d^2\phi^*}{dx^2} = (s/c)^2\phi^* - P\delta(x)\right)\exp(+i\xi x)dx \tag{2.1.12}$$

The Fourier transform integral is applied to each term as

$$\int_{-\infty}^{+\infty} \frac{d^2\phi^*}{dx^2}\exp(+i\xi x)dx = (s/c)^2\int_{-\infty}^{+\infty} \phi^*\exp(+i\xi x)dx - P\int_{-\infty}^{+\infty} \delta(x)\exp(+i\xi x)dx \tag{2.1.13}$$

The convergence condition is also transformed, as

$$\int_0^{\infty} \left(\phi|_{x\to\pm\infty} = \left.\frac{\partial\phi}{\partial x}\right|_{x\to\pm\infty} = 0\right)\exp(-st)dt \quad \Rightarrow \quad \phi^*|_{x\to\pm\infty} = \left.\frac{d\phi^*}{dx}\right|_{x\to\pm\infty} = 0 \tag{2.1.14}$$

Defining the Fourier transform of the Laplace transformed unknown function,

$$\bar{\phi}^* = \int_{-\infty}^{+\infty} \phi^*\exp(+i\xi x)dx \tag{2.1.15}$$

the transform of the space-derivative in Eq. (2.1.13) is carried out with the aid of the convergence condition,

$$\int_{-\infty}^{+\infty} \frac{d^2\phi^*}{dx^2}\exp(+i\xi x)dx = \left[\frac{d\phi^*}{dx}\exp(+i\xi x)\right]_{x\to-\infty}^{x\to+\infty} - i\xi[\phi^*\exp(+i\xi x)]_{x\to-\infty}^{x\to+\infty} - \xi^2\int_{-\infty}^{+\infty}\phi^*\exp(+i\xi x)dx$$
$$= -\xi^2\tilde{\phi}^* \tag{2.1.16}$$

The integration formula for the delta function

$$\int_{-\infty}^{+\infty} \delta(x)\exp(+i\xi x)dx = 1 \tag{2.1.17}$$

is also used for evaluating the last term. The Fourier transform of the ordinary differential equation (2.1.12) becomes a simple algebraic equation for the double transformed unknown function $\bar{\phi}^*$,

$$-\xi^2\bar{\phi}^* = \frac{s^2}{c^2}\bar{\phi}^* - P \tag{2.1.18}$$

Then we have the exact expression for the double transformed function $\bar{\phi}^*$,

$$\bar{\phi}^* = \frac{P}{\xi^2 + (s/c)^2} \tag{2.1.19}$$

The unknown function has just been determined explicitly in the transformed domain. We shall carry out two inverse transforms successively. As the first inversion, we apply the Fourier inversion integral which is defined by the second of Eq. (2.1.5). The formal Fourier inversion is given by the integral

$$\phi^* = \frac{1}{2\pi}\int_{-\infty}^{+\infty} \frac{P}{\xi^2 + (s/c)^2}\exp(-i\xi x)d\xi \tag{2.1.20}$$

Due to the symmetric nature of the integrand, the integral is reduced to the simpler semi-infinite integral,

$$\phi^* = \frac{P}{\pi}\int_{0}^{\infty} \frac{1}{\xi^2 + (s/c)^2}\cos(\xi x)d\xi \tag{2.1.21}$$

This is a simple integral and the integration formula (Erdélyi 1954, vol. I, pp. 8, 11),

$$\int_0^{+\infty} \frac{1}{x^2 + a^2} \cos(xy) dx = \frac{\pi}{2a} \exp(-a|y|) \tag{2.1.22}$$

can be applied. Then, Eq. (2.1.21) gives

$$\phi^* = \frac{P}{2(s/c)} \exp(-s|x|/c) \tag{2.1.23}$$

The next step is to carry out the Laplace inversion. The symbolical form of the Laplace inversion is given by

$$\phi = L^{-1}[\phi^*] = \frac{cP}{2} L^{-1}\left[\frac{1}{s}\exp(-s|x|/c)\right] \tag{2.1.24}$$

Fortunately, we have the Laplace inversion formula (Erdélyi 1954, vol. I, pp. 241, 1),

$$L^{-1}\left[\frac{1}{s}\exp(-as)\right] = \begin{cases} 0; & t < a \\ 1; & t > a \end{cases} = H(t-a) \tag{2.1.25}$$

where $H(.)$ is Heaviside's unit step function. Applying this formula to Eq. (2.1.24), the solution for the non-homogeneous 1D wave equation is obtained as

$$\phi = \frac{cP}{2} H(t - |x|/c) = \frac{cP}{2} H(ct - |x|) \tag{2.1.26}$$

This solution shows an expanding (or out-going) 1D wave with the uniform amplitude $cP/2$ and with the velocity c as shown in Fig. 2.1. Consequently, we get the Green's function for the 1D wave equation,

$$\frac{\partial^2 \phi}{\partial x^2} = \frac{1}{c^2}\frac{\partial^2 \phi}{\partial t^2} - P\delta(x)\delta(t) \quad \Rightarrow \quad \phi(x,t) = \frac{cP}{2} H(ct - |x|) \tag{2.1.27}$$

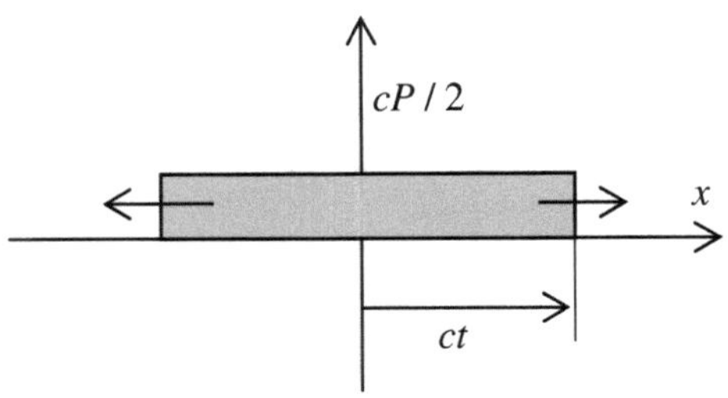

Fig. 2.1 1D expanding wave from a source point

2.2 1D Time-Harmonic Source

When the source is time-harmonic, the nonhomogeneous term in Eq. (2.1.1) is replaced with a harmonic function, but the source location is unchanged. Thus, the wave equation with a time-harmonic source is given by

$$\frac{\partial^2 \phi}{\partial x^2} = \frac{1}{c^2}\frac{\partial^2 \phi}{\partial t^2} - Q\delta(x)\exp(+i\omega t) \tag{2.2.1}$$

where Q is the source magnitude and ω the frequency of the time-harmonic vibration. We assume that its solution satisfies the convergence condition at infinity, i.e.

$$\phi|_{x\to\pm\infty} = \left.\frac{\partial \phi}{\partial x}\right|_{x\to\pm\infty} = 0 \tag{2.2.2}$$

As the first step of the solution method, we assume that the solution is also time-harmonic,

$$\phi(x,t) = \phi^{\#}(x)\exp(+i\omega t) \tag{2.2.3}$$

where $\phi^{\#}$ is called the "amplitude function." Due to this assumption, the convergence condition (2.2.2) is rewritten for the amplitude function,

$$\phi^{\#}|_{x\to\pm\infty} = \left.\frac{d\phi^{\#}}{dx}\right|_{x\to\pm\infty} = 0 \tag{2.2.4}$$

Substituting the time-harmonic assumption of Eq. (2.2.3) into the wave equation (2.2.1), we have the ordinary differential equation for the amplitude function,

$$\frac{d^2\phi^{\#}}{dx^2} + (\omega/c)^2\phi^{\#} = -Q\delta(x) \tag{2.2.5}$$

The exact solution of this ordinary differential equation can also be obtained by the elementary method. However, in order to demonstrate the integral transform technique, we apply the Fourier transform, which is defined by Eq. (2.1.5), to the ordinary differential equation (2.2.5),

$$\int_{-\infty}^{+\infty}\left(\frac{d^2\phi^{\#}}{dx^2} + (\omega/c)^2\phi^{\#} = -Q\delta(x)\right)\exp(+i\xi x)dx \tag{2.2.6}$$

Defining the Fourier transform of the amplitude function as

$$\bar{\phi}^{\#} = \int_{-\infty}^{+\infty} \phi^{\#} \exp(+i\xi x)dx \tag{2.2.7}$$

the space derivative term in Eq. (2.2.6) is integrated by parts as

$$\begin{aligned} \int_{-\infty}^{+\infty} \frac{d^2\phi^{\#}}{dx^2} \exp(+i\xi x)dx = & \left[\frac{d\phi^{\#}}{dx} \exp(+i\xi x)\right]_{x\to-\infty}^{x\to+\infty} - i\xi\left[\phi^{\#} \exp(+i\xi x)\right]_{x\to-\infty}^{x\to+\infty} \\ & - \xi^2 \int_{-\infty}^{+\infty} \phi^{\#} \exp(+i\xi x)dx \end{aligned} \tag{2.2.8}$$

We apply the convergence condition of Eq. (2.2.4) and the definition of the Fourier transform (2.2.7) to the above equation. The Fourier transform of the double derivative is then reduced to

$$\int_{-\infty}^{+\infty} \frac{d^2\phi^{\#}}{dx^2} \exp(+i\xi x)dx = -\xi^2 \bar{\phi}^{\#} \tag{2.2.9}$$

The nonhomogeneous term is evaluated as

$$-Q \int_{-\infty}^{+\infty} \delta(x) \exp(+i\xi x)dx = -Q \tag{2.2.10}$$

Then, Eq. (2.2.6) gives a simple algebraic equation for the Fourier transformed amplitude function,

$$-\xi^2 \bar{\phi}^{\#} + (\omega/c)^2 \bar{\phi}^{\#} = -Q \tag{2.2.11}$$

The Fourier transformed amplitude is determined completely,

$$\bar{\phi}^{\#} = \frac{Q}{\xi^2 - (\omega/c)^2} \tag{2.2.12}$$

Our next task is to invert the transformed amplitude. Applying the formal Fourier inversion integral to Eq. (2.2.12), we get

$$\phi^{\#} = \frac{1}{2\pi} \int_{-\infty}^{+\infty} \frac{P}{\xi^2 - (\omega/c)^2} \exp(-i\xi x)d\xi \tag{2.2.13}$$

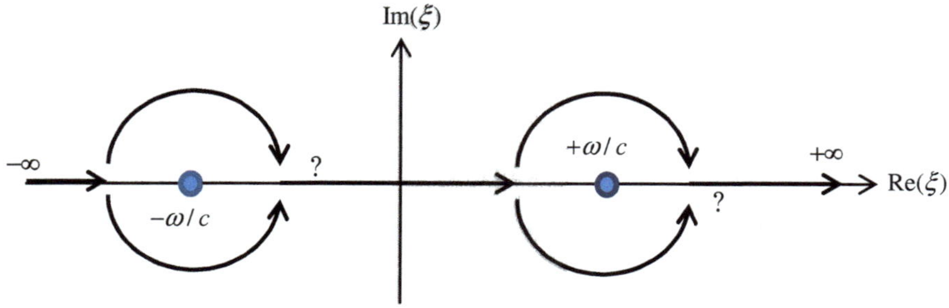

Fig. 2.2 Possible deformations of the integration path around the pole

Inspecting the integrand, we see that it has two simple poles at $\xi = \pm(\omega/c)$, that is to say, the poles are located on the integration path (real axis in the complex ξ-plane). Since the integration cannot be performed through these singular points, we have to distort the integration path around the poles. There are two ways of deforming the path. One is through an upper semi-circle, another is through a lower semi-circle as shown in Fig. 2.2. We have to determine which semi-circle is suitable. Discussing the nature of the initial wave equation (2.2.1), we learn that the wave will expand to the outer region from the source point, i.e. wave radiation from the source. Therefore, we have to choose the path so that the inversion integral results in a radiating (out-going) wave from the source.

Still, it is somewhat complicated to explain the path selection. To aid the understanding, two integrals with complex frequency are considered. After discussing the wave nature derived from each integral, we will determine and understand the path distortion.

Let us introduce and add a small imaginary number ε to the frequency in Eq. (2.2.13) so that the poles are shifted from the real axis and are not on the integration path. The complex frequency is considered in two ways, positive and negative imaginary parts, $\omega \to \omega \pm i\varepsilon$. Employing the complex frequency ϖ, we consider the complex integral,

$$\Phi = \frac{1}{2\pi}\int_C \frac{P}{\xi^2 - (\varpi/c)^2}\exp(-i\xi x)d\xi \qquad (2.2.14)$$

where the integrand is the same as that in the Fourier inversion integral (2.2.13), but the frequency is complex, $\varpi = \omega \pm i\varepsilon$. The integration loop C for the two cases of complex frequency, with positive and negative imaginary parts, is discussed separately.

(1) Small positive imaginary part: $\varpi = \omega + i\varepsilon$

When the frequency has a small positive imaginary part, the poles are shifted from the real axis. The integration path C is chosen so that the integrand vanishes on the large semi-circle with infinite radius. We employ the lower closed loop $C^{(-)}$ in the case of positive x and the upper loop $C^{(+)}$ in that of negative x as shown in Fig. 2.3a.

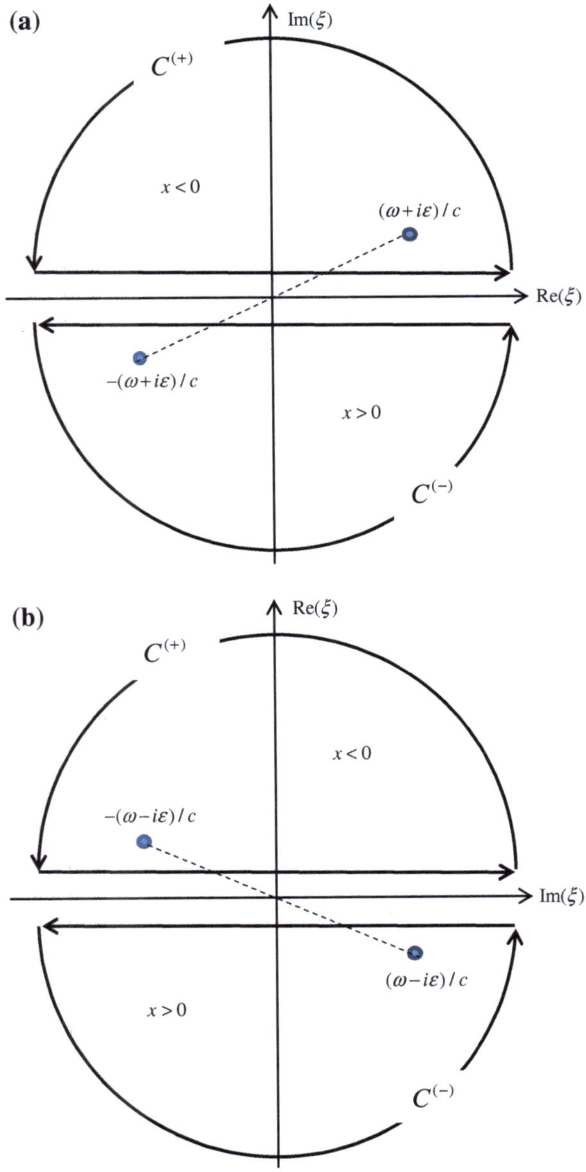

Fig. 2.3 Integration loop for the complex integral Φ in the case of **a** the positive imaginary part of the frequency, **b** the negative imaginary part of the frequency

Applying Cauchy's integral theorem (Jordan's lemma) to the complex integral Φ with the loop $C^{(-)}$ in Fig. 2.3a, the integral along the real axis is evaluated as the residue at the lower pole $\xi = -\varpi/c = -(\omega + i\varepsilon)/c$,

$$-\frac{1}{2\pi}\int_{-\infty}^{+\infty} \frac{P}{\xi^2 - (\varpi/c)^2}\exp(-i\xi x)d\xi = \frac{2\pi i}{2\pi}\left[P\frac{\xi + \varpi/c}{\xi^2 - (\varpi/c)^2}\exp(-i\xi x)\right]_{\xi=-\varpi/c} \tag{2.2.15}$$

Rewriting the above equation, and taking the limit $\varepsilon \to 0$, we have in the case of positive x,

$$\frac{1}{2\pi}\int_{-\infty}^{+\infty} \frac{P}{\xi^2 - (\omega/c)^2}\exp(-i\xi x)d\xi = \frac{iP}{2(\omega/c)}\exp(+i\omega x/c); \quad x > 0 \tag{2.2.16}$$

On the other hand, when $x < 0$, we employ the upper loop $C^{(+)}$ for the complex integral Φ and have

$$\frac{1}{2\pi}\int_{-\infty}^{+\infty} \frac{P}{\xi^2 - (\varpi/c)^2}\exp(-i\xi x)d\xi = \frac{2\pi i}{2\pi}\left[P\frac{\xi - \varpi/c}{\xi^2 - (\varpi/c)^2}\exp(-i\xi x)\right]_{\xi=\varpi/c} \tag{2.2.17}$$

Then, we take the limit $\varepsilon \to 0$,

$$\frac{1}{2\pi}\int_{-\infty}^{+\infty} \frac{P}{\xi^2 - (\omega/c)^2}\exp(-i\xi x)d\xi = \frac{iP}{2(\omega/c)}\exp(-i\omega x/c); \quad x<0 \tag{2.2.18}$$

Unifying the two Eqs. (2.2.16) and (2.2.18), we have for the Fourier inversion integral, where the positive imaginary part of the complex frequency tends to zero, i.e.

$$\frac{1}{2\pi}\int_{-\infty}^{+\infty} \frac{P}{\xi^2 - (\omega/c)^2}\exp(-i\xi x)d\xi = \frac{iP}{2(\omega/c)}\exp(+i\omega|x|/c); \quad \omega + i\varepsilon|_{\varepsilon\to 0} \tag{2.2.19}$$

(2) Small negative imaginary part: $\varpi = \omega - i\varepsilon$

When the imaginary part of the complex frequency is negative, the poles are also shifted from the real axis as shown in Fig. 2.3b. In order to guarantee the convergence on the large semi-circle, the lower loop $C^{(-)}$ is employed in the case of positive x, and the upper $C^{(+)}$ in that of negative x.

When $x > 0$, we employ the loop $C^{(-)}$ and apply Jordan's lemma to the complex integral Φ in Eq. (2.2.14). The integral along the real axis is converted to the residue at the lower pole, $\xi = \varpi(= \omega - i\varepsilon)/c$,

$$-\frac{1}{2\pi}\int_{-\infty}^{+\infty}\frac{P}{\xi^2-(\varpi/c)^2}\exp(-i\xi x)d\xi = \frac{2\pi i}{2\pi}\left[P\frac{\xi-\varpi/c}{\xi^2-(\varpi/c)^2}\exp(-i\xi x)\right]_{\xi=\varpi/c} \quad (2.2.20)$$

Rewriting the above and taking the limit $\varepsilon \to 0$, we have in the case of positive x,

$$\frac{1}{2\pi}\int_{-\infty}^{+\infty}\frac{P}{\xi^2-(\omega/c)^2}\exp(-i\xi x)d\xi = -\frac{iP}{2(\omega/c)}\exp(-i\omega x/c); \quad x > 0 \quad (2.2.21)$$

Similarly, when we employ the upper loop $C^{(+)}$ in the case of $x < 0$,

$$\frac{1}{2\pi}\int_{-\infty}^{+\infty}\frac{P}{\xi^2-(\varpi/c)^2}\exp(-i\xi x)d\xi = \frac{2\pi i}{2\pi}\left[P\frac{\xi+\varpi/c}{\xi^2-(\varpi/c)^2}\exp(-i\xi x)\right]_{\xi=-\varpi/c} \quad (2.2.22)$$

Taking the limit $\varepsilon \to 0$, the Fourier inversion integral is evaluated as

$$\frac{1}{2\pi}\int_{-\infty}^{+\infty}\frac{P}{\xi^2-(\omega/c)^2}\exp(-i\xi x)d\xi = -\frac{iP}{2(\omega/c)}\exp(+i\omega x/c); \quad x < 0 \quad (2.2.23)$$

Unifying two Eqs. (2.2.21) and (2.2.23), we have the unified expression when the negative imaginary part of the frequency approached to zero,

$$\frac{1}{2\pi}\int_{-\infty}^{+\infty}\frac{P}{\xi^2-(\omega/c)^2}\exp(-i\xi x)d\xi = -\frac{iP}{2(\omega/c)}\exp(-i\omega|x|/c); \quad \omega - i\varepsilon|_{\varepsilon\to 0} \quad (2.2.24)$$

(3) Selection of the imaginary part

Two expressions are obtained for the single Fourier inversion integral. They are Eqs. (2.2.19) and (2.2.24) and are summarized as

$$\phi^{\#} = \begin{cases} -\frac{iP}{2(\omega/c)}\exp(-i\omega|x|/c); & \omega - i\varepsilon|_{\varepsilon\to 0} \\ +\frac{iP}{2(\omega/c)}\exp(+i\omega|x|/c); & \omega + i\varepsilon|_{\varepsilon\to 0} \end{cases} \quad (2.2.25)$$

The wave nature of the two expressions is discussed by multiplying the time factor,

$$\phi = \begin{cases} -\frac{iP}{2(\omega/c)}\exp\{+i\omega(t-|x|/c)\}; & \omega - i\varepsilon|_{\varepsilon\to 0} \\ +\frac{iP}{2(\omega/c)}\exp\{+i\omega(t+|x|/c)\}; & \omega + i\varepsilon|_{\varepsilon\to 0} \end{cases} \tag{2.2.26}$$

Inspecting the argument of the exponential function in the above equation, the upper solution shows an out-going (radiation) wave from the source, since the argument of the exponential function is $t-|x|/c$. On the other hand, the lower includes $t+|x|/c$ and shows an incoming wave, coming from infinity. Since the suitable solution must be the out-going wave, we employ the upper formula in Eqs. (2.2.25) and (2.2.26). Thus, the complex frequency with the negative imaginary part is the right assumption. Then, the suitable integration contour for the Fourier inversion integral is that of Fig. 2.3b and the final result for the time-harmonic response is given by

$$\frac{\partial^2\phi}{\partial x^2} = \frac{1}{c^2}\frac{\partial^2\phi}{\partial t^2} - Q\delta(x)\exp(+i\omega t) \quad \Rightarrow \quad \phi = -\frac{iQ}{2(\omega/c)}\exp\{+i\omega(t-|x|/c)\} \tag{2.2.27}$$

We have just learned that the right selection for the complex frequency is $\varpi = \omega - i\varepsilon$, i.e. negative imaginary part, and the suitable integration loop is $C^{(\pm)}$ in Fig. 2.3b. If we do not introduce the small imaginary part and keep the integration path on the real axis, the poles is on the real axis and the integration path around the pole must be deformed by a small semi-circle shown in Fig. 2.4a. This deformation is valid only in the case of the positive time factor, $\exp(+i\omega t)$. If we assume the negative time factor $\exp(-i\omega t)$, the integration path on the real axis is deformed as that shown in Fig. 2.4b. Thus, the selection of the deformed path around the pole depends on the sign of the imaginary part of the complex frequency. Therefore, we could answer to the initial question about the deformation of the integration path for the Fourier inversion integral.

2.3 2D Static Source

Let us consider Green's function for a typical partial differential equation, the so-called Laplace equation. The Laplace equation with a source S is the nonhomogeneous differential equation,

$$\frac{\partial^2\phi}{\partial x^2} + \frac{\partial^2\phi}{\partial y^2} = -S\delta(x)\delta(y) \tag{2.3.1}$$

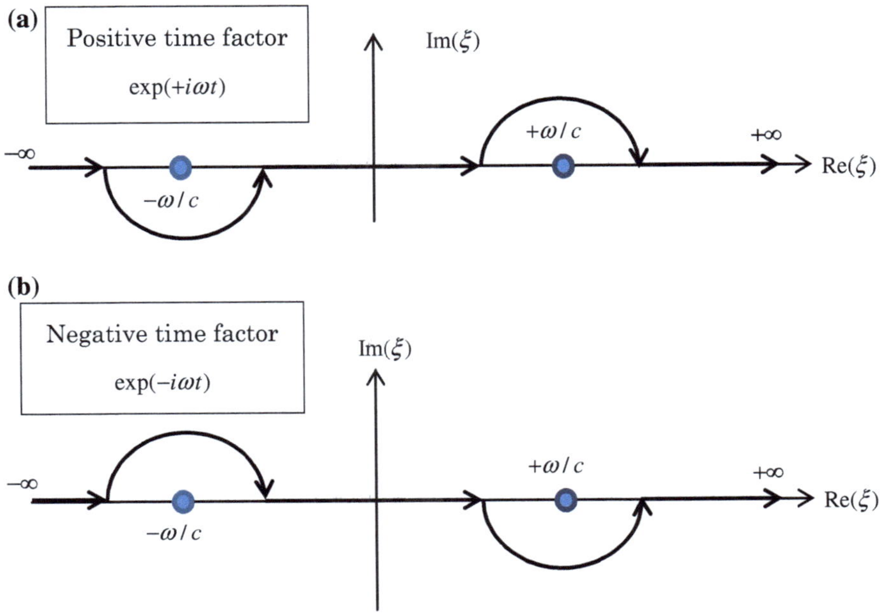

Fig. 2.4 Path deformation for the inversion integral. **a** Positive time factor, **b** Negative time factor

The product of two delta functions in the nonhomogeneous term shows the location of the source, i.e. the source S is placed at the coordinate origin (0, 0) in (x, y)-plane. The convergence condition at infinity,

$$\phi|_{\sqrt{x^2+y^2}\to\infty} = \left.\frac{\partial\phi}{\partial x}\right|_{\sqrt{x^2+y^2}\to\infty} = \left.\frac{\partial\phi}{\partial y}\right|_{\sqrt{x^2+y^2}\to\infty} = 0 \tag{2.3.2}$$

is also imposed.

Now, we apply the integral transforms to Eq. (2.3.1). Since the unknown function ϕ has two space variables, we apply the double Fourier transform defined by

$$\bar{\phi}(\xi) = \int_{-\infty}^{+\infty} \phi(x)\exp(+i\xi x)dx, \qquad \phi(x) = \frac{1}{2\pi}\int_{-\infty}^{+\infty} \bar{\phi}(\xi)\exp(-i\xi x)d\xi \tag{2.3.3}$$

$$\tilde{\phi}(\eta) = \int_{-\infty}^{+\infty} \phi(y)\exp(+i\eta y)dy, \qquad \phi(y) = \frac{1}{2\pi}\int_{-\infty}^{+\infty} \tilde{\phi}(\eta)\exp(-i\eta y)d\eta \tag{2.3.4}$$

to the differential equation (2.3.1) successively, as

$$\int_{-\infty}^{+\infty}\left(\int_{-\infty}^{+\infty}\left(\frac{\partial^2\phi}{\partial x^2}+\frac{\partial^2\phi}{\partial y^2}=-S\delta(x)\delta(y)\right)\exp(+i\xi x)dx\right)\exp(+i\eta y)dy \quad (2.3.5)$$

With the aid of the convergence condition (2.3.2), each term is transformed as follows:

$$\begin{aligned}
&\int_{-\infty}^{+\infty}\left(\int_{-\infty}^{+\infty}\frac{\partial^2\phi}{\partial x^2}\exp(+i\xi x)dx\right)\exp(+i\eta y)dy=-\xi^2\bar{\bar{\phi}}\\
&\int_{-\infty}^{+\infty}\left(\int_{-\infty}^{+\infty}\frac{\partial^2\phi}{\partial y^2}\exp(+i\xi x)dx\right)\exp(+i\eta y)dy=-\eta^2\bar{\bar{\phi}}\\
&\int_{-\infty}^{+\infty}\left(\int_{-\infty}^{+\infty}S\delta(x)\delta(y)\exp(+i\xi x)dx\right)\exp(+i\eta y)dy=S
\end{aligned} \quad (2.3.6)$$

Then, we have the simple algebraic equation $-(\xi^2+\eta^2)\bar{\bar{\phi}}=-S$ for the double transformed function and its solution is given by

$$\bar{\bar{\phi}}=\frac{S}{\xi^2+\eta^2} \quad (2.3.7)$$

The reader will find that the partial differential equation (2.3.1) is transformed to the simple algebraic equation. There is thus no need of solving the differential equation directly. The subsequent inversion process is however crucial for the solution in the actual space. The formal Fourier inversion integral with respect to the parameter η,

$$\bar{\phi}=\frac{1}{2\pi}\int_{-\infty}^{+\infty}\frac{S}{\xi^2+\eta^2}\exp(-i\eta y)d\eta=\frac{S}{\pi}\int_{0}^{\infty}\frac{1}{\xi^2+\eta^2}\cos(\eta y)d\eta \quad (2.3.8)$$

is evaluated with the aid of the formula (2.1.22) and yields

$$\bar{\phi}=\frac{S}{2|\xi|}\exp(-|\xi||y|) \quad (2.3.9)$$

The next inversion is to evaluate the Fourier inversion integral with respect to the parameter ξ,

$$\phi = \frac{1}{2\pi}\int_{-\infty}^{+\infty} \frac{S}{2|\xi|}\exp(-|\xi||y|)\exp(-i\xi x)d\xi \tag{2.3.10}$$

Inspecting the integrand, the singular point at $\xi = 0$ lies on the real axis, i.e. on the integration path. It is impossible to evaluate the integral in the regular sense. So, we have to deform the integration path around the pole as that in the previous section. But it was somewhat complicated to determine the path deformation. In order to avoid this troublesome work, we employ a simpler way for evaluating the integral.

Since the trouble stems from the singular point at $\xi = 0$, in order to avoid the trouble, we differentiate equation (2.3.10) with respect to the space variables, x and y, respectively,

$$\frac{\partial\phi}{\partial x} = -\frac{S}{4\pi}\int_{-\infty}^{+\infty} \frac{i\xi}{|\xi|}\exp(-|\xi||y|)\exp(-i\xi x)d\xi \tag{2.3.11}$$

$$\frac{\partial\phi}{\partial y} = -\frac{S}{4\pi}\mathrm{sgn}(y)\int_{-\infty}^{+\infty} \exp(-|\xi||y|)\exp(-i\xi x)d\xi \tag{2.3.12}$$

where sgn(.) is the sign function defined by

$$\mathrm{sgn}(y) = \begin{cases} +1; & y > 0 \\ -1; & y < 0 \end{cases} \tag{2.3.13}$$

Using the symmetry of the integrand in Eqs. (2.3.11) and (2.3.12), the integrals are reduced to the real valued semi-infinite integrals,

$$\frac{\partial\phi}{\partial x} = -\frac{S}{2\pi}\int_{0}^{\infty} \exp(-\xi|y|)\sin(\xi x)d\xi \tag{2.3.14}$$

$$\frac{\partial\phi}{\partial y} = -\frac{S}{2\pi}\mathrm{sgn}(y)\int_{0}^{\infty} \exp(-\xi|y|)\cos(\xi x)d\xi \tag{2.3.15}$$

The two integrals in the above equations are well-known from Calculus and we have the formulas,

$$\int_0^\infty \exp(-\xi y)\sin(\xi x)d\xi = \frac{x}{x^2+y^2}, \quad \int_0^\infty \exp(-\xi y)\cos(\xi x)d\xi = \frac{y}{x^2+y^2} \tag{2.3.16}$$

Then, Eqs. (2.3.14) and (2.3.15) are evaluated as

$$\frac{\partial\phi}{\partial x} = -\frac{S}{2\pi}\frac{x}{x^2+y^2}, \qquad \frac{\partial\phi}{\partial y} = -\frac{S}{2\pi}\frac{y}{x^2+y^2} \tag{2.3.17}$$

The derivative of ϕ is completely determined. We integrate the above two equations with respect to x and y, respectively. This integration leads to two expressions for the single function ϕ as

$$\begin{aligned} \frac{\partial\phi}{\partial x} \quad &\Rightarrow \quad \phi = -\frac{S}{4\pi}\log\left(x^2+y^2\right) + c'(y) \\ \frac{\partial\phi}{\partial y} \quad &\Rightarrow \quad \phi = -\frac{S}{4\pi}\log\left(x^2+y^2\right) + c''(x) \end{aligned} \tag{2.3.18}$$

Since the above two equations must be equal, two constant terms should be identical,

$$c'(y) = c''(x) \tag{2.3.19}$$

This condition is satisfied only when the two terms are pure constant and do not include any space variable:

$$c'(y) = c''(x) = \text{constant} \tag{2.3.20}$$

Then, we have Green's function for the Laplace equation,

$$\frac{\partial^2\phi}{\partial x^2} + \frac{\partial^2\phi}{\partial y^2} = -S\delta(x)\delta(y) \quad \Rightarrow \quad \phi = -\frac{S}{4\pi}\log\left(x^2+y^2\right) + \text{arbitrary constant} \tag{2.3.21}$$

This Green's function does not satisfy the convergence condition at infinity, since it has the constant. This is because we could not carry out the Fourier inversion integral of Eq. (2.3.10) directly.

2.4 2D Impulsive Source

The 2D wave equation with an impulsive source is given by

$$\frac{\partial^2 \phi}{\partial x^2} + \frac{\partial^2 \phi}{\partial y^2} = \frac{1}{c^2} \frac{\partial^2 \phi}{\partial t^2} - P\delta(x)\delta(y)\delta(t) \tag{2.4.1}$$

The nonhomogeneous term states that the source with magnitude P is placed at the coordinate origin and is impulsive in time. The quiescent condition at an initial time,

$$\phi|_{t=0} = \left.\frac{\partial \phi}{\partial t}\right|_{t=0} = 0 \tag{2.4.2}$$

and the convergence condition at infinity,

$$\phi|_{\sqrt{x^2+y^2}\to\infty} = \left.\frac{\partial \phi}{\partial x}\right|_{\sqrt{x^2+y^2}\to\infty} = \left.\frac{\partial \phi}{\partial y}\right|_{\sqrt{x^2+y^2}\to\infty} = 0 \tag{2.4.3}$$

are also employed. As the wave equation (2.4.1) has two space variables x and y, and one time variable t, the triple integral transform is applied to the differential equation (2.4.1): the Laplace transform with respect to the time variable,

$$\phi^*(s) = \int_0^\infty \phi(t) \exp(-st) dt \tag{2.4.4}$$

and the double Fourier transform with respect to the space variables,

$$\bar{\phi}(\xi) = \int_{-\infty}^{+\infty} \phi(x) \exp(+i\xi x) dx, \quad \phi(x) = \frac{1}{2\pi} \int_{-\infty}^{+\infty} \bar{\phi}(\xi) \exp(-i\xi x) d\xi \tag{2.4.5}$$

$$\tilde{\phi}(\eta) = \int_{-\infty}^{+\infty} \phi(y) \exp(+i\eta y) dy, \quad \phi(y) = \frac{1}{2\pi} \int_{-\infty}^{+\infty} \tilde{\phi}(\eta) \exp(-i\eta y) d\eta \tag{2.4.6}$$

Applying this triple integral transform with the quiescent and convergence conditions, the original differential equation (2.4.1) is transformed to the simple algebraic equation for the unknown function $\tilde{\bar{\phi}}^*$,

$$-\xi^2 \tilde{\bar{\phi}}^* - \eta^2 \tilde{\bar{\phi}}^* = (s/c)^2 \tilde{\bar{\phi}}^* - P \tag{2.4.7}$$

Then, the exact expression for $\bar{\bar{\phi}}^*$ in the transformed domain is given by

$$\bar{\bar{\phi}}^* = \frac{P}{\xi^2 + \eta^2 + (s/c)^2} \tag{2.4.8}$$

For the inversion, three inversion integrals must be carried out successively. The first one is the Fourier inversion integral with respect to the parameter η. This is reduced to the semi-infinite integral as

$$\bar{\phi}^* = \frac{1}{2\pi} \int_{-\infty}^{+\infty} \frac{P}{\xi^2 + \eta^2 + (s/c)^2} \exp(-i\eta y) d\eta = \frac{P}{\pi} \int_0^{\infty} \frac{\cos(\eta y)}{\eta^2 + \sqrt{\xi^2 + (s/c)^2}^2} d\eta \tag{2.4.9}$$

The integral on the far right side is easily evaluated by applying the formula (2.1.22). Then, the first Fourier inversion integral in Eq. (2.4.9) yields

$$\bar{\phi}^* = \frac{P}{2\sqrt{\xi^2 + (s/c)^2}} \exp\left\{ -|y|\sqrt{\xi^2 + (s/c)^2} \right\} \tag{2.4.10}$$

Secondly, we apply the inversion integral with respect to the parameter ξ,

$$\phi^* = \frac{1}{2\pi} \int_{-\infty}^{+\infty} \frac{P}{2\sqrt{\xi^2 + (s/c)^2}} \exp\left\{ -|y|\sqrt{\xi^2 + (s/c)^2} \right\} \exp(-i\xi x) d\xi \tag{2.4.11}$$

The above integral is also reduced to the semi-infinite integral, as

$$\phi^* = \frac{P}{2\pi} \int_0^{\infty} \frac{1}{\sqrt{\xi^2 + (s/c)^2}} \exp\left\{ -|y|\sqrt{\xi^2 + (s/c)^2} \right\} \cos(\xi x) d\xi \tag{2.4.12}$$

and we apply the integration formula (Erdélyi 1954, vol. I, pp. 17, 27)

$$\int_0^{\infty} \frac{1}{\sqrt{x^2 + a^2}} \exp\left(-c\sqrt{x^2 + a^2} \right) \cos(bx) dx = K_0\left(a\sqrt{b^2 + c^2} \right) \tag{2.4.13}$$

where $K_0(.)$ is the zeroth order modified Bessel function of the second kind. Then, Eq. (2.4.12) takes the compact form

$$\phi^* = \frac{P}{2\pi} K_0\left(\frac{s}{c}\sqrt{x^2 + y^2} \right) \tag{2.4.14}$$

The last inversion is the Laplace inversion. The Laplace inversion is symbolically expressed as

$$\phi = \frac{P}{2\pi} L^{-1}\left[K_0\left(\frac{s}{c}\sqrt{x^2+y^2}\right)\right] \tag{2.4.15}$$

We have the suitable inversion formula (Erdélyi 1954, vol. I, pp. 277, 8),

$$L^{-1}[K_0(as)] = \frac{H(t-a)}{\sqrt{t^2-a^2}} = \begin{cases} 0; & t<a \\ \frac{1}{\sqrt{t^2-a^2}}; & t>a \end{cases} \tag{2.4.16}$$

Applying this formula to Eq. (2.4.15), the simple expression for ϕ is obtained as

$$\phi = \frac{cP}{2\pi}\frac{H(ct-r)}{\sqrt{(ct)^2-r^2}} = \frac{cP}{2\pi}\begin{cases} 0; & ct<r \\ \frac{1}{\sqrt{(ct)^2-r^2}}; & ct>r \end{cases} \tag{2.4.17}$$

where $H(.)$ is Heaviside's unit step function and the radial distance r from the source is defined by

$$r = \sqrt{x^2+y^2} \tag{2.4.18}$$

Consequently, we have the exact expression for Green's function of the 2D wave equation as

$$\frac{\partial^2\phi}{\partial x^2} + \frac{\partial^2\phi}{\partial y^2} = \frac{1}{c^2}\frac{\partial^2\phi}{\partial t^2} - P\delta(x)\delta(y)\delta(t) \quad \Rightarrow \quad \phi = \frac{cP}{2\pi}\frac{H(ct-r)}{\sqrt{(ct)^2-r^2}} \tag{2.4.19}$$

2.5 2D Time-Harmonic Source

When the source is vibrating harmonically, the nonhomogeneous 2D wave equation is given by

$$\frac{\partial^2\phi}{\partial x^2} + \frac{\partial^2\phi}{\partial y^2} = \frac{1}{c^2}\frac{\partial^2\phi}{\partial t^2} - Q\delta(x)\delta(y)\exp(+i\omega t) \tag{2.5.1}$$

where Q and ω are the magnitude and the frequency of the source, respectively. We assume that its Green's function satisfies the convergence condition at infinity,

$$\phi|_{\sqrt{x^2+y^2}\to\infty} = \left.\frac{\partial\phi}{\partial x}\right|_{\sqrt{x^2+y^2}\to\infty} = \left.\frac{\partial\phi}{\partial y}\right|_{\sqrt{x^2+y^2}\to\infty} = 0 \tag{2.5.2}$$

As the standard technique for the time-harmonic response, the Green's function is assumed as the product

$$\phi(x, y, t) = \phi^{\#}(x, y)\exp(+i\omega t) \tag{2.5.3}$$

where $\phi^{\#}(x, y)$ is an amplitude function to be determined. Substituting this assumption into the nonhomogeneous wave equation (2.5.1), we have the reduced wave equation (so-called Helmholtz equation) for the amplitude function $\phi^{\#}(x, y)$,

$$\frac{\partial^2 \phi^{\#}}{\partial x^2} + \frac{\partial^2 \phi^{\#}}{\partial y^2} + (\omega/c)^2 \phi^{\#} = -Q\delta(x)\delta(y) \tag{2.5.4}$$

The convergence condition (2.5.2) is also rewritten for the amplitude function as

$$\phi^{\#}\big|_{\sqrt{x^2+y^2}\to\infty} = \frac{\partial \phi^{\#}}{\partial x}\bigg|_{\sqrt{x^2+y^2}\to\infty} = \frac{\partial \phi^{\#}}{\partial y}\bigg|_{\sqrt{x^2+y^2}\to\infty} = 0 \tag{2.5.5}$$

The double Fourier transform with respect to two space variables as defined by Eqs. (2.4.5) and (2.4.6) is applied to the nonhomogeneous Helmholtz equation (2.5.4),

$$\int_{-\infty}^{+\infty}\left[\int_{-\infty}^{+\infty}\left(\frac{\partial^2 \phi^{\#}}{\partial x^2} + \frac{\partial^2 \phi^{\#}}{\partial y^2} = -(\omega/c)^2 \phi^{\#} - Q\delta(x)\delta(y)\right)\exp(+i\xi x)dx\right]\exp(+i\eta y)dy \tag{2.5.6}$$

Defining the double transform of the amplitude function as

$$\bar{\bar{\phi}}^{\#} = \int_{-\infty}^{+\infty}\left[\int_{-\infty}^{+\infty}\phi^{\#}(x, y)\exp(+i\xi x)dx\right]\exp(+i\eta y)dy \tag{2.5.7}$$

Equation (2.5.6) is transformed into the algebraic equation for the unknown amplitude function $\bar{\bar{\phi}}^{\#}$,

$$-\xi^2\bar{\bar{\phi}}^{\#} - \eta^2\bar{\bar{\phi}}^{\#} = -(\omega/c)^2\bar{\bar{\phi}}^{\#} - Q \tag{2.5.8}$$

Then, the explicit expression for the amplitude function in the transformed domain is given by

$$\bar{\bar{\phi}}^{\#} = \frac{Q}{\xi^2 + \eta^2 - (\omega/c)^2} \tag{2.5.9}$$

As the first inversion, we apply the Fourier inversion integral with respect to the parameter η. Its formal inversion integral is simplified as

$$\bar{\phi}^{\#} = \frac{1}{2\pi}\int_{-\infty}^{+\infty} \frac{Q}{\xi^2 + \eta^2 - (\omega/c)^2} \exp(-i\eta y) d\eta = \frac{Q}{\pi}\int_{0}^{\infty} \frac{\cos(\eta y)}{\eta^2 + \sqrt{\xi^2 - (\omega/c)^2}^2} d\eta \tag{2.5.10}$$

The far right integral is evaluated with the aid of the formula (2.1.22) and yields

$$\bar{\phi}^{\#} = \frac{Q}{2\sqrt{\xi^2 - (\omega/c)^2}} \exp\left\{-|y|\sqrt{\xi^2 - (\omega/c)^2}\right\} \tag{2.5.11}$$

where the real part of the square root function $\sqrt{\xi^2 - (\omega/c)^2}$ must be positive in order to guarantee the convergence at infinity $|y| \to \infty$, i.e.

$$\mathrm{Re}\left(\sqrt{\xi^2 - (\omega/c)^2}\right) \geq 0 \tag{2.5.12}$$

The second Fourier inversion integral with respect to the parameter ξ is given by

$$\phi^{\#} = \frac{Q}{4\pi}\int_{-\infty}^{+\infty} \frac{1}{\sqrt{\xi^2 - (\omega/c)^2}} \exp\left\{-|y|\sqrt{\xi^2 - (\omega/c)^2}\right\} \exp(-i\xi x) d\xi \tag{2.5.13}$$

The integrand in the above inversion integral has two branch points at $\xi = \pm\omega/c$ which are on the integration path, and the square root function (radical) $\sqrt{\xi^2 - (\omega/c)^2}$ changes its sign around the branch point. Thus, we have to discuss the path deformation around the branch point as was done in the case of the 1D time-harmonic problem in Sect. 2.2. However, the singular point for the 1D problem was the simple pole, but that for the present 2D problem is the branch point for the square root function. Therefore, we have to discuss the introduction of the branch cut in the complex plane. As the detailed discussion for the introduction of branch cuts has been carried out in Sect. 1.3.2 in Chap. 1, we do not repeat, but, we cite here its result. If the reader needs more about the introduction of the branch cut, please return to Sect. 1.3 in Chap. 1.

In order to evaluate the inversion integral of Eq. (2.5.13), we define the integral I as

$$I = \frac{1}{2\pi}\int_{-\infty}^{+\infty} \frac{1}{\sqrt{\xi^2 - (\omega/c)^2}} \exp\left\{-i\xi x - |y|\sqrt{\xi^2 - (\omega/c)^2}\right\} d\xi \tag{2.5.14}$$

and consider the complex integral having the same integrand as that in the above equation,

$$\Phi = \frac{1}{2\pi}\int_C \frac{1}{\sqrt{\zeta^2 - (\omega/c)^2}} \exp\left\{-i\zeta x - |y|\sqrt{\zeta^2 - (\omega/c)^2}\right\} d\zeta \tag{2.5.15}$$

In the subsequent analysis, it should be understood that the real and imaginary parts of the complex variable ζ are ξ and η, i.e. $\zeta = \xi + i\eta$. The square root function $\sqrt{\zeta^2 - (\omega/c)^2}$ has two branch points at $\zeta = \pm\omega/c$, and the corresponding two branch cuts must be introduced in the complex ζ-plane, in order to make the radical be single-valued and to guarantee the convergence condition,

$$\mathrm{Re}\left(\sqrt{\zeta^2 - (\omega/c)^2}\right) \geq 0 \tag{2.5.16}$$

The discussion on the branch cut for the square root function in Sect. 1.3.2 shows two ways to consider the complex frequency: one is the complex frequency with a small **positive** imaginary part and the other with a small **negative** imaginary part. As the shape of the branch cut depends on the sign of the imaginary part, we evaluate the integral I for these two cases of the complex frequency, separately, and then determine the suitable imaginary part.

(1) Complex frequency with the negative imaginary part

If we add a small negative imaginary part to the frequency, the branch points are shifted from the real axis and then two branch cuts are introduced along the hyperbola in the second and the fourth quadrants in the complex ζ-plane as shown in Figs. 1.7 and 1.8 in Chap. 1. When the imaginary part of the frequency tends to zero as the limit, the branch cut lies on the real and imaginary axes as shown by the thick lines in Fig. 1.8a. Therefore, we take the closed loop C for our complex integral of Eq. (2.5.15) as shown in Fig. 2.5. We have two closed loops for C: one is the upper closed loop $C^{(+)}$ and the other is the lower loop $C^{(-)}$. The selection of the two loops depends on the convergence of the integrands at the infinity $|\zeta| \to \infty$. The lower loop is employed when the parameter (space variable) x is positive, and the upper loop is employed when $x<0$. The following evaluation is carried out for the case of the positive space variable, $x > 0$.

When $x > 0$, the loop for the complex integral is $C^{(-)}$ as shown in Fig. 2.5 where the integration path is the straight line $\overrightarrow{\mathrm{AOB}}$ on the real axis. In order to exclude the lower branch cut, the path along the branch cut is introduced. The path is composed of two lines, $\overline{\mathrm{CDE}}$ and $\overline{\mathrm{FGH}}$ along the branch cut, and a small circle $\overset{\frown}{\mathrm{EF}}$ around the branch point. It should be understood that the radius of the small circle vanishes as the limit. Two edges of the straight line $\overrightarrow{\mathrm{AOB}}$ are connected to the edges

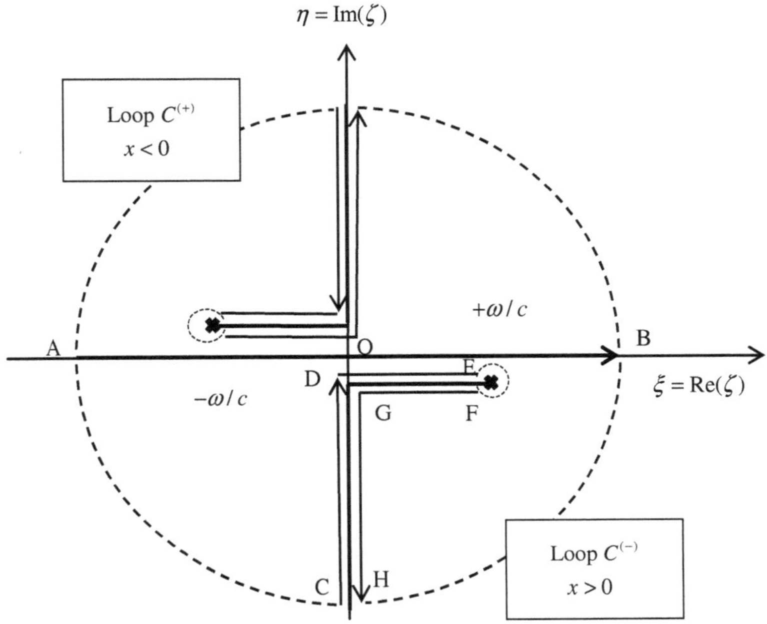

Fig. 2.5 Closed loop C for the complex integral Φ in Eq. (2.5.15) (the negative imaginary part of the complex frequency vanished)

of the path $\overline{\mathrm{CDE}}$ and $\overline{\mathrm{FGH}}$ through two quarter circles, $\overset{\frown}{\mathrm{AC}}$ and $\overset{\frown}{\mathrm{BH}}$, respectively. It is also understood that the radius of the quarter circle tends to infinity as the limit. Then, the closed loop $C^{(-)}$ is the sum of the integrals along these paths. In the clockwise direction, the closed loop is

$$C^{(-)}:\ \overrightarrow{\mathrm{AOB}}+\overset{\frown}{\mathrm{BH}}+\overrightarrow{\mathrm{HGF}}+\overset{\frown}{\mathrm{FE}}+\overrightarrow{\mathrm{EDC}}+\overset{\frown}{\mathrm{CA}} \tag{2.5.17}$$

The complex integral of Eq. (2.5.15) is decomposed into the integrals along the path segments as

$$\Phi=\frac{1}{2\pi}\left(\int_{\overrightarrow{\mathrm{AOB}}}+\int_{\overset{\frown}{\mathrm{BH}}}+\int_{\overrightarrow{\mathrm{HGF}}}+\int_{\overset{\frown}{\mathrm{FE}}}+\int_{\overrightarrow{\mathrm{EDC}}}+\int_{\overset{\frown}{\mathrm{CA}}}\right)\frac{1}{\sqrt{\zeta^2-(\omega/c)^2}}\exp\left\{-i\zeta x-|y|\sqrt{\zeta^2-(\omega/c)^2}\right\}d\zeta \tag{2.5.18}$$

The integral along the small circle $\overset{\frown}{\mathrm{FE}}$ around the branch point vanishes as its radius tends to zero. The two integrals along the large arcs $\overset{\frown}{\mathrm{BH}}$ and $\overset{\frown}{\mathrm{CA}}$ also vanish as the radius tends to infinity due to the convergence condition of Eq. (2.5.16). Thus, since no singular point is included in the closed loop and the total value of the complex integral vanishes, i.e. $\Phi = 0$, the integral along the real axis $\overrightarrow{\mathrm{AOB}}$ is converted to the integrals along the branch cut,

$$\begin{aligned}
&\frac{1}{2\pi}\int_{\overrightarrow{\mathrm{AOB}}}\frac{1}{\sqrt{\zeta^2-(\omega/c)^2}}\exp\left\{-i\zeta x-|y|\sqrt{\zeta^2-(\omega/c)^2}\right\}d\zeta \\
&=\frac{1}{2\pi}\left(\int_{\overrightarrow{\mathrm{CDE}}}+\int_{\overrightarrow{\mathrm{FGH}}}\right)\frac{1}{\sqrt{\zeta^2-(\omega/c)^2}}\exp\left\{-i\zeta x-|y|\sqrt{\zeta^2-(\omega/c)^2}\right\}d\zeta
\end{aligned} \tag{2.5.19}$$

where the directions of paths $\overrightarrow{\mathrm{CDE}}$ and $\overrightarrow{\mathrm{FGH}}$ are inverse of the paths $\overrightarrow{\mathrm{EDC}}$ and $\overrightarrow{\mathrm{HGF}}$, respectively. Equation (2.5.19) states that it is enough to consider the integrals along the branch cut to evaluate the integral on the real axis. The line paths $\overrightarrow{\mathrm{CDE}}$ and $\overrightarrow{\mathrm{FGH}}$ are further decomposed into the line segments along the real and imaginary axes. They are

$$\overrightarrow{\mathrm{CDE}}=\overrightarrow{\mathrm{CD}}+\overrightarrow{\mathrm{DE}},\ \overrightarrow{\mathrm{FGH}}=\overrightarrow{\mathrm{FG}}+\overrightarrow{\mathrm{GH}} \tag{2.5.20}$$

The argument and value of the square root function have already been determined in Sect. 1.3.2(2) in Chap. 1 and are tabulated in Table 2.1 with the integration variable along each line segment. Applying these results, the integral along each path is given by

$$\begin{aligned}
&\frac{1}{2\pi}\int_{\overrightarrow{\mathrm{AOB}}}\frac{1}{\sqrt{\zeta^2-(\omega/c)^2}}\exp\left\{-i\zeta x-|y|\sqrt{\zeta^2-(\omega/c)^2}\right\}d\zeta \\
&=\frac{1}{2\pi}\int_{-\infty}^{\infty}\frac{1}{\sqrt{\xi^2-(\omega/c)^2}}\exp\left\{-i\xi x-|y|\sqrt{\xi^2-(\omega/c)^2}\right\}d\xi
\end{aligned} \tag{2.5.21}$$

$$\begin{aligned}
&\frac{1}{2\pi}\int_{\overrightarrow{\mathrm{CD}}}\frac{1}{\sqrt{\zeta^2-(\omega/c)^2}}\exp\left\{-i\zeta x-|y|\sqrt{\zeta^2-(\omega/c)^2}\right\}d\zeta \\
&=\frac{1}{2\pi}\int_{0}^{\infty}\frac{1}{\sqrt{\eta^2+(\omega/c)^2}}\exp\left\{-\eta x-i|y|\sqrt{\eta^2+(\omega/c)^2}\right\}d\eta
\end{aligned} \tag{2.5.22}$$

Table 2.1 Value and argument of the square root function on the integration path in Fig. 2.5

Path	$\sqrt{\zeta^2-(\omega/c)^2}$	Variable ζ and its range
$\overrightarrow{\text{AOB}}$	$\sqrt{\xi^2-(\omega/c)^2}$	$\zeta=\xi$
		$-\infty<\xi<+\infty$
$\overrightarrow{\text{CD}}$	$+i\sqrt{\eta^2+(\omega/c)^2}$	$\zeta=-i\eta$
		$0<\eta<\infty$
$\overrightarrow{\text{DE}}$	$+i\sqrt{(\omega/c)^2-\xi^2}$	$\zeta=+\xi$
		$0<\xi<\omega/c$
$\overrightarrow{\text{FG}}$	$-i\sqrt{(\omega/c)^2-\xi^2}$	$\zeta=+\xi$
		$0<\xi<\omega/c$
$\overrightarrow{\text{GH}}$	$-i\sqrt{\eta^2+(\omega/c)^2}$	$\xi=-i\eta$
		$0<\eta<\infty$

ξ and η are positive real, except $\overrightarrow{\text{BOA}}$

$$\begin{aligned}&\frac{1}{2\pi}\int_{\overrightarrow{\text{DE}}}\frac{1}{\sqrt{\zeta^2-(\omega/c)^2}}\exp\left\{-i\zeta x-|y|\sqrt{\zeta^2-(\omega/c)^2}\right\}d\zeta\\&=-\frac{i}{2\pi}\int_0^{\omega/c}\frac{1}{\sqrt{(\omega/c)^2-\xi^2}}\exp\left\{-i\xi x-i|y|\sqrt{(\omega/c)^2-\xi^2}\right\}d\xi\end{aligned}\tag{2.5.23}$$

$$\begin{aligned}&\frac{1}{2\pi}\int_{\overrightarrow{\text{FG}}}\frac{1}{\sqrt{\zeta^2-(\omega/c)^2}}\exp\left\{-i\zeta x-|y|\sqrt{\zeta^2-(\omega/c)^2}\right\}d\zeta\\&=-\frac{i}{2\pi}\int_0^{\omega/c}\frac{1}{\sqrt{(\omega/c)^2-\xi^2}}\exp\left\{-i\xi x+i|y|\sqrt{(\omega/c)^2-\xi^2}\right\}d\xi\end{aligned}\tag{2.5.24}$$

$$\begin{aligned}&\frac{1}{2\pi}\int_{\overrightarrow{\text{GH}}}\frac{1}{\sqrt{\zeta^2-(\omega/c)^2}}\exp\left\{-i\zeta x-|y|\sqrt{\zeta^2-(\omega/c)^2}\right\}d\zeta\\&=\frac{1}{2\pi}\int_0^{\infty}\frac{1}{\sqrt{\eta^2+(\omega/c)^2}}\exp\left\{-\eta x+i|y|\sqrt{\eta^2+(\omega/c)^2}\right\}d\eta\end{aligned}\tag{2.5.25}$$

Substituting the above equations into Eq. (2.5.19) with the path decomposition of Eq. (2.5.20), we have for the integral I,

$$
\begin{aligned}
I &= \frac{1}{2\pi}\int_{-\infty}^{\infty}\frac{1}{\sqrt{\xi^2-(\omega/c)^2}}\exp\left\{-i\xi x-|y|\sqrt{\xi^2-(\omega/c)^2}\right\}d\xi \\
&= \frac{1}{\pi}\int_{0}^{\infty}\frac{1}{\sqrt{\eta^2+(\omega/c)^2}}\exp(-\eta x)\cos\left\{y\sqrt{\eta^2+(\omega/c)^2}\right\}d\eta \\
&\quad -\frac{1}{\pi}\int_{0}^{\omega/c}\frac{1}{\sqrt{(\omega/c)^2-\xi^2}}\sin(\xi x)\cos\left\{y\sqrt{(\omega/c)^2-\xi^2}\right\}d\xi \\
&\quad -\frac{i}{\pi}\int_{0}^{\omega/c}\frac{1}{\sqrt{(\omega/c)^2-\xi^2}}\cos(\xi x)\cos\left\{y\sqrt{(\omega/c)^2-\xi^2}\right\}d\xi
\end{aligned}
\tag{2.5.26}
$$

The Fourier inversion integral has just been converted to three real-valued integrals. Fortunately, we can evaluate these integrals with the aid of integration formulas. Two integration formulas (Gradshteyn and Ryzhik 1980, pp. 755, 6.677, No. 4)*

$$
\begin{aligned}
&\int_0^{\infty}\frac{1}{\sqrt{a^2+x^2}}\cos\left(b\sqrt{a^2+x^2}\right)\exp(-cx)dx-\int_0^{a}\frac{1}{\sqrt{a^2-x^2}}\cos\left(b\sqrt{a^2-x^2}\right)\sin(cx)dx \\
&= -\frac{\pi}{2}Y_0\left(a\sqrt{b^2+c^2}\right)
\end{aligned}
\tag{2.5.27}
$$

and (Erdélyi 1954, vol. I, pp. 28, 42)

$$
\int_0^{a}\frac{1}{\sqrt{a^2-x^2}}\cos\left(b\sqrt{a^2-x^2}\right)\cos(cx)dx=\frac{\pi}{2}J_0\left(a\sqrt{b^2+c^2}\right)
\tag{2.5.28}
$$

are applied to the right hand side of Eq. (2.5.26) and the Hankel function (Watson 1966, p. 73) is introduced. Then, Eq. (2.5.26) is simplified as

$$
\begin{aligned}
&\frac{1}{2\pi}\int_{-\infty}^{+\infty}\frac{1}{\sqrt{\xi^2-(\omega/c)^2}}\exp\left\{-i\xi x-|y|\sqrt{\xi^2-(\omega/c)^2}\right\}d\xi \\
&= -\frac{1}{2}Y_0\left(\frac{\omega}{c}\sqrt{x^2+y^2}\right)-\frac{i}{2}J_0\left(\frac{\omega}{c}\sqrt{x^2+y^2}\right) \\
&= -\frac{i}{2}H_0^{(2)}\left(\frac{\omega}{c}\sqrt{x^2+y^2}\right)
\end{aligned}
\tag{2.5.29}
$$

where $J_0(.)$ and $Y_0(.)$ are Bessel functions of the first and second kind, respectively, and Hankel function of the second kind is defined by $H_0^{(2)}(.) = J_0(.) - iY_0(.)$. Finally, we have the simple expression for the Fourier inversion integral I. That is

$$\frac{1}{2\pi}\int_{-\infty}^{+\infty} \frac{1}{\sqrt{\xi^2-(\omega/c)^2}} \exp\left\{-i\xi x - |y|\sqrt{\xi^2-(\omega/c)^2}\right\} d\xi = -\frac{i}{2} H_0^{(2)}\left(\frac{\omega}{c}\sqrt{x^2+y^2}\right) \quad (2.5.30)$$

It is easily understood that the above equation is also valid for the negative space variable, $x < 0$, since Eq. (2.5.30) is the even function of the space variable x. So, if we perform the complex integral with the upper closed loop $C^{(+)}$ in Fig. 2.5, we can obtain the same expression for the negative space variable. However, the result in Eq. (2.5.30) is valid only when the negative imaginary part of the complex frequency tends to zero.

(2) Complex frequency with the positive imaginary part

If we assume that the imaginary part of the complex frequency approaches to zero from the positive, the branch cuts corresponding to two branch points are introduced in the first and the third quadrants in the complex plane as was discussed in Sect. 1.3.2 in Chap. 1 and are shown in Fig. 1.11 as its zero limit. Thus, the

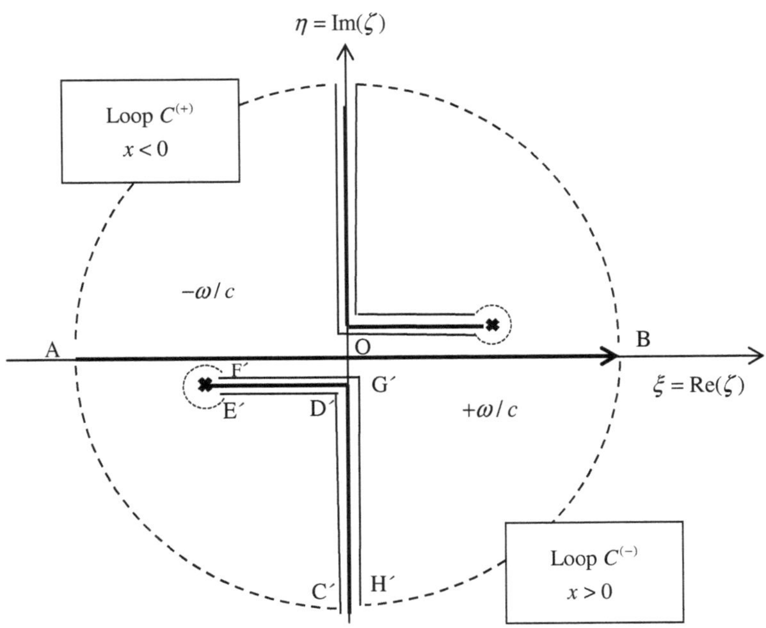

Fig. 2.6 Closed loop C for the complex integral Φ in Eq. (2.5.15) (the positive imaginary part of the complex frequency vanished)

Table 2.2 Value and argument of the root function along the integration path in Fig. 2.6

Path	$\sqrt{\zeta^2-(\omega/c)^2}$	Variable ζ and its range
$\overrightarrow{\mathrm{AOB}}$	$\sqrt{\xi^2-(\omega/c)^2}$	$\zeta=\xi$
		$-\infty<\xi<+\infty$
$\overrightarrow{\mathrm{C'D'}}$	$+i\sqrt{\eta^2+(\omega/c)^2}$	$\zeta=-i\eta$
		$0<\eta<\infty$
$\overrightarrow{\mathrm{D'E'}}$	$+i\sqrt{(\omega/c)^2-\xi^2}$	$\zeta=-\xi$
		$0<\xi<\omega/c$
$\overrightarrow{\mathrm{F'G'}}$	$-i\sqrt{(\omega/c)^2-\xi^2}$	$\zeta=-\xi$
		$0<\xi<\omega/c$
$\overrightarrow{\mathrm{G'H'}}$	$-i\sqrt{\eta^2+(\omega/c)^2}$	$\xi=-i\eta$
		$0<\eta<\infty$

ξ and η are positive real, except $\overrightarrow{\mathrm{BOA}}$

closed loop for the complex integral Φ is taken as $C^{(\pm)}$ in the complex ζ-plane as shown in Fig. 2.6 and the value and argument of the square root function are listed in Table 2.2. We can apply the same evaluation procedure as that in the previous subsection. The analysis is carried out for the case of the positive space variable, $x>0$. The integral on the real axis, i.e. the integral I, is given by

$$
\begin{aligned}
I &= \frac{1}{2\pi}\int_{\overrightarrow{\mathrm{AOB}}} \frac{1}{\sqrt{\zeta^2-(\omega/c)^2}}\exp\left\{-i\zeta x-|y|\sqrt{\zeta^2-(\omega/c)^2}\right\}d\zeta \\
&= \frac{1}{2\pi}\int_{-\infty}^{\infty} \frac{1}{\sqrt{\xi^2-(\omega/c)^2}}\exp\left\{-i\xi x-|y|\sqrt{\xi^2-(\omega/c)^2}\right\}d\xi
\end{aligned}
\tag{2.5.31}
$$

and the integrals along the branch cut are

$$
\begin{aligned}
&\frac{1}{2\pi}\int_{\overrightarrow{\mathrm{C'D'}}} \frac{1}{\sqrt{\zeta^2-(\omega/c)^2}}\exp\left\{-i\zeta x-|y|\sqrt{\zeta^2-(\omega/c)^2}\right\}d\zeta \\
&= \frac{1}{2\pi}\int_{0}^{\infty} \frac{1}{\sqrt{\eta^2+(\omega/c)^2}}\exp\left\{-\eta x-i|y|\sqrt{\eta^2+(\omega/c)^2}\right\}d\eta
\end{aligned}
\tag{2.5.32}
$$

$$
\begin{aligned}
&\frac{1}{2\pi}\int_{\overrightarrow{\mathrm{D'E'}}} \frac{1}{\sqrt{\zeta^2-(\omega/c)^2}}\exp\left\{-i\zeta x-|y|\sqrt{\zeta^2-(\omega/c)^2}\right\}d\zeta \\
&= \frac{i}{2\pi}\int_{0}^{\omega/c} \frac{1}{\sqrt{(\omega/c)^2-\xi^2}}\exp\left\{+i\xi x-i|y|\sqrt{(\omega/c)^2-\xi^2}\right\}d\xi
\end{aligned}
\tag{2.5.33}
$$

$$\begin{aligned}&\frac{1}{2\pi}\int_{\overrightarrow{F'G'}}\frac{1}{\sqrt{\zeta^2-(\omega/c)^2}}\exp\left\{-i\zeta x-|y|\sqrt{\zeta^2-(\omega/c)^2}\right\}d\zeta\\&=\frac{i}{2\pi}\int_0^{\omega/c}\frac{1}{\sqrt{(\omega/c)^2-\xi^2}}\exp\left\{+i\xi x+i|y|\sqrt{(\omega/c)^2-\xi^2}\right\}d\xi\end{aligned}\tag{2.5.34}$$

$$\begin{aligned}&\frac{1}{2\pi}\int_{\overrightarrow{G'H'}}\frac{1}{\sqrt{\zeta^2-(\omega/c)^2}}\exp\left\{-i\zeta x-|y|\sqrt{\zeta^2-(\omega/c)^2}\right\}d\zeta\\&=\frac{1}{2\pi}\int_0^{\infty}\frac{1}{\sqrt{\eta^2+(\omega/c)^2}}\exp\left\{-\eta x+i|y|\sqrt{\eta^2+(\omega/c)^2}\right\}d\eta\end{aligned}\tag{2.5.35}$$

Since no singular point is included in the closed loop, $\Phi = 0$. Thus, the integral I on the real axis is given by

$$\begin{aligned}I&=\frac{1}{2\pi}\int_{-\infty}^{\infty}\frac{1}{\sqrt{\xi^2-(\omega/c)^2}}\exp\left\{-i\xi x-|y|\sqrt{\xi^2-(\omega/c)^2}\right\}d\xi\\&=\frac{1}{\pi}\int_0^{\infty}\frac{1}{\sqrt{\eta^2+(\omega/c)^2}}\exp(-\eta x)\cos\left\{y\sqrt{\eta^2+(\omega/c)^2}\right\}d\eta\\&\quad-\frac{1}{\pi}\int_0^{\omega/c}\frac{1}{\sqrt{(\omega/c)^2-\xi^2}}\sin(\xi x)\cos\left\{y\sqrt{(\omega/c)^2-\xi^2}\right\}d\xi\\&\quad+\frac{i}{\pi}\int_0^{\omega/c}\frac{1}{\sqrt{(\omega/c)^2-\xi^2}}\cos(\xi x)\cos\left\{y\sqrt{(\omega/c)^2-\xi^2}\right\}d\xi\end{aligned}\tag{2.5.36}$$

The above equation can be simplified by applying the formulas (2.5.27) and (2.5.28). We have

$$\begin{aligned}&\frac{1}{2\pi}\int_{-\infty}^{\infty}\frac{1}{\sqrt{\zeta^2-(\omega/c)^2}}\exp\left\{-i\xi x-|y|\sqrt{\xi^2-(\omega/c)^2}\right\}d\xi\\&=+\frac{i}{2}H_0^{(1)}\left(\frac{\omega}{c}\sqrt{x^2+y^2}\right)\end{aligned}\tag{2.5.37}$$

where Hankel function of the first kind is defined by $H_0^{(1)}(.) = J_0(.) + iY_0(.)$.
*Derivation of formula (2.5.27).

We have the integration formula (Gradshteyn and Ryzhik 1980, pp. 755, 6.677),

$$\int_0^\infty Y_0\left(\alpha\sqrt{x^2+z^2}\right)\cos(\beta x)dx = \begin{cases} \dfrac{1}{\sqrt{\alpha^2-\beta^2}}\sin\left(z\sqrt{\alpha^2-\beta^2}\right); & \beta<\alpha \\ -\dfrac{1}{\sqrt{\beta^2-\alpha^2}}\exp\left(-z\sqrt{\beta^2-\alpha^2}\right); & \beta>\alpha \end{cases} \tag{2.5.38}$$

If we consider the above as the Fourier cosine transform with respect to the transform parameter β, its inverse cosine transform is given by

$$\begin{aligned} \frac{\pi}{2}Y_0\left(\alpha\sqrt{x^2+z^2}\right) &= \int_0^\alpha \frac{1}{\sqrt{\alpha^2-\beta^2}}\sin\left(z\sqrt{\alpha^2-\beta^2}\right)\cos(\beta x)d\beta \\ &\quad - \int_\alpha^\infty \frac{1}{\sqrt{\beta^2-\alpha^2}}\exp\left(-z\sqrt{\beta^2-\alpha^2}\right)\cos(\beta x)d\beta \end{aligned} \tag{2.5.39}$$

Making the change of variable, $u=\sqrt{\beta^2-\alpha^2}$, for the second integral in the right hand side, the formula (2.5.27) is obtained as

$$\begin{aligned} \frac{\pi}{2}Y_0\left(\alpha\sqrt{x^2+z^2}\right) &= \int_0^\alpha \frac{1}{\sqrt{\alpha^2-\beta^2}}\sin\left(z\sqrt{\alpha^2-\beta^2}\right)\cos(\beta x)d\beta \\ &\quad - \int_0^\infty \frac{1}{\sqrt{u^2+\alpha^2}}\exp(-zu)\cos\left(x\sqrt{u^2+\alpha^2}\right)du \end{aligned} \tag{2.5.40}$$

(3) Selection of the branch cut and the integration loop

The Fourier inversion integral I has just been evaluated in two ways of the branch cut and we get two expressions for the single integral I as

$$\begin{aligned} &\frac{1}{2\pi}\int_{-\infty}^{+\infty} \frac{1}{\sqrt{\xi^2-(\omega/c)^2}}\exp\left\{-i\xi x-|y|\sqrt{\xi^2-(\omega/c)^2}\right\}d\xi \\ &= \begin{cases} -\dfrac{i}{2}H_0^{(2)}(\omega r/c); & \omega\equiv\omega-i\varepsilon|_{\varepsilon\to 0} \\ +\dfrac{i}{2}H_0^{(1)}(\omega r/c); & \omega\equiv\omega+i\varepsilon|_{\varepsilon\to 0} \end{cases} \end{aligned} \tag{2.5.41}$$

where the radial distance r from the source point is defined by

$$r = \sqrt{x^2 + y^2} \tag{2.5.42}$$

In order to determine the correct evaluation, we multiply the time factor $\exp(+i\omega t)$ and replace the Hankel functions with their asymptotic forms (Watson 1966, p. 198) as

$$\left.\begin{aligned} H_0^{(1)}(z) &\sim \sqrt{\frac{2}{\pi z}}\exp\{-i(z-\pi/4)\} \\ H_0^{(2)}(z) &\sim \sqrt{\frac{2}{\pi z}}\exp\{+i(z-\pi/4)\} \end{aligned}\right\}; \quad z \to \infty \tag{2.5.43}$$

The asymptotic form of the integral I yields to

$$\begin{aligned} &\frac{\exp(+i\omega t)}{2\pi}\int_{-\infty}^{+\infty}\frac{1}{\sqrt{\xi^2-(\omega/c)^2}}\exp\left\{-i\xi x-|y|\sqrt{\xi^2-(\omega/c)^2}\right\}d\xi \\ &\underset{r\to\infty}{\approx}\begin{cases} -\dfrac{i}{2}\sqrt{\dfrac{2c}{\pi\omega r}}\exp(+\pi i/4)\exp\{+i\omega(t-r/c)\}; & \omega\equiv\omega-i\varepsilon|_{\varepsilon\to 0} \\ +\dfrac{i}{2}\sqrt{\dfrac{2c}{\pi\omega r}}\exp(-\pi i/4)\exp\{+i\omega(t+r/c)\}; & \omega\equiv\omega+i\varepsilon|_{\varepsilon\to 0} \end{cases} \end{aligned} \tag{2.5.44}$$

It is clear that the asymptotic form in the upper line shows the out-going wave from the source point $r = 0$, and that in the lower line does the in-coming wave from the infinity. Thus, the right evaluation for the integral I is

$$\frac{1}{2\pi}\int_{-\infty}^{+\infty}\frac{1}{\sqrt{\xi^2-(\omega/c)^2}}\exp\left\{-i\xi x-|y|\sqrt{\xi^2-(\omega/c)^2}\right\}d\xi = -\frac{i}{2}H_0^{(2)}(\omega r/c) \tag{2.5.45}$$

and the correct introduction of the branch cut and the integration loop are shown in Fig. 2.5 for our positive time factor $\exp(+i\omega t)$. However, if we employ the negative time factor $\exp(-i\omega t)$, the branch cut and the loop in Fig. 2.6 are the right choices, needless to say.

Anyway, we could evaluate the infinite integral of the Fourier inversion and obtained the amplitude function in the closed form as

$$\phi^{\#} = -\frac{iQ}{4}H_0^{(2)}(\omega r/c); \quad r = \sqrt{x^2 + y^2} \tag{2.5.46}$$

Thus, the 2D Green's function for the time-harmonic source is given by

$$\frac{\partial^2 \phi}{\partial x^2} + \frac{\partial^2 \phi}{\partial x^2} = \frac{1}{c^2}\frac{\partial^2 \phi}{\partial t^2} - Q\delta(x)\delta(y)\exp(+i\omega t) \Rightarrow \phi = -\frac{iQ}{4}H_0^{(2)}(\omega r/c)\exp(+i\omega t) \tag{2.5.47}$$

When the time factor is negative, we have

$$\frac{\partial^2 \phi}{\partial x^2} + \frac{\partial^2 \phi}{\partial x^2} = \frac{1}{c^2}\frac{\partial^2 \phi}{\partial t^2} - Q\delta(x)\delta(y)\exp(-i\omega t) \Rightarrow \phi = +\frac{iQ}{4}H_0^{(1)}(\omega r/c)\exp(-i\omega t) \tag{2.5.48}$$

(4) Convolution integral

The solution of Eq. (2.5.47) can be obtained directly by applying the convolution integral of the impulsive solution which is given by Eq. (2.4.19). Let us consider its convolution. The convolution integral for the Laplace transform is defined by

$$\phi(x, y, t) = \int_0^t \phi^{(implse)}(x, y, t')\exp\{+i\omega(t - t')\}dt' \tag{2.5.49}$$

where $\phi^{(implse)}$ is the solution for the impulsive source. We substitute the impulsive Green's function of Eq. (2.4.19) with the source magnitude Q into Eq. (2.5.49),

$$\phi(x, y, t) = \int_0^t \frac{cQ}{2\pi}\frac{H(ct' - r)}{\sqrt{(ct')^2 - r^2}}\exp\{+i\omega(t - t')\}dt' \tag{2.5.50}$$

and examine the supporting region for the step function. We have

$$\phi(x, y, t) = \frac{cQ}{2\pi}\exp(+i\omega t)H(t - r/c)\int_{r/c}^t \frac{1}{\sqrt{(ct')^2 - r^2}}\exp(-i\omega t')dt' \tag{2.5.51}$$

As we are considering the steady-state response, the time in the upper integration limit and that in the step function can be set to be infinite as

$$\phi(x, y, t) = \frac{cQ}{2\pi}\exp(+i\omega t)\lim_{t\to\infty}\left[H(t - r/c)\int_{r/c}^t \frac{1}{\sqrt{(ct')^2 - r^2}}\exp(-i\omega t')dt'\right] \tag{2.5.52}$$

Taking the limit for the time, we have

$$\phi(x, y, t) = \frac{Q}{2\pi} \exp(+i\omega t) \int_{r}^{\infty} \frac{1}{\sqrt{u^2 - r^2}} \exp(-i\omega u/c) du \tag{2.5.53}$$

This is just the integral representation for the Hankel function of the second kind (Watson 1966, p. 170),

$$H_0^{(2)}(x) = \frac{2i}{\pi} \int_{1}^{\infty} \frac{1}{\sqrt{u^2 - 1}} \exp(-ixu) du \tag{2.5.54}$$

Thus, we have

$$\phi(x, y, t) = -\frac{iQ}{4} \exp(+i\omega t) H_0^{(2)}(r\omega/c) \tag{2.5.55}$$

(5) Direct evaluation

In the former subsection (1) and (2), the Fourier inversion is carried out by applying the Cauchy complex integral. It was lengthy troublesome work. Without applying the Cauchy theorem, we can evaluate the integral I,

$$I = \frac{1}{2\pi} \int_{-\infty}^{+\infty} \frac{1}{\sqrt{\xi^2 - (\omega/c)^2}} \exp\left\{ -i\xi x - |y| \sqrt{\xi^2 - (\omega/c)^2} \right\} d\xi \tag{2.5.56}$$

Firstly, we reduce the infinite integral to the semi-infinite integral,

$$I = \frac{1}{\pi} \int_{0}^{+\infty} \frac{1}{\sqrt{\xi^2 - (\omega/c)^2}} \cos(\xi x) \exp\left\{ -|y| \sqrt{\xi^2 - (\omega/c)^2} \right\} d\xi \tag{2.5.57}$$

In the complex ξ-plane, the integration path for this integral is just on the real axis and two branch cuts for the radical $\sqrt{\xi^2 - (\omega/c)^2}$ are introduced on the real and imaginary axes as was discussed in Sect. 1.3.2 in Chap. 1. The slightly shifted branch cuts are shown in Fig. 2.7 where the integration path for the semi-infinite integral is OP. The integration path in the region $0 < \xi < \omega/c$ is slightly up from the

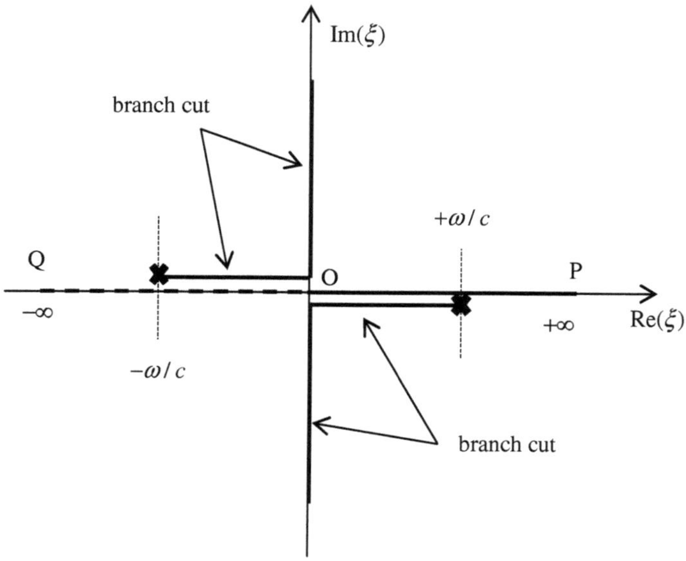

Fig. 2.7 Branch cuts and integration path for the inversion integral of Eq. (2.5.57)

lower branch cut. When the cut approaches to the real axis as was discussed in Sect. 1.3.2 (2), the argument of the square root function in the integrand yields to $+\pi/2$, i.e.

$$\sqrt{\xi^2-(\omega/c)^2}=+i\sqrt{(\omega/c)^2-\xi^2};\quad |\xi|<\omega/c \tag{2.5.58}$$

In the other positive region $\xi>\omega/c$ on the real axis, the root function is positive real. Then, the semi-infinite integral is decomposed into two integrals in the regions, $0<\xi<\omega/c$ and $\xi>\omega/c$, as

$$\begin{aligned} I=&\frac{1}{\pi}\int_0^{\omega/c}\frac{1}{(+i)\sqrt{(\omega/c)^2-\xi^2}}\cos(\xi x)\exp\left\{-i|y|\sqrt{(\omega/c)^2-\xi^2}\right\}d\xi\\ &+\frac{1}{\pi}\int_{\omega/c}^{+\infty}\frac{1}{\sqrt{\xi^2-(\omega/c)^2}}\cos(\xi x)\exp\left\{-|y|\sqrt{\xi^2-(\omega/c)^2}\right\}d\xi \end{aligned} \tag{2.5.59}$$

The above integral I is further decomposed into the real and imaginary parts,

$$
\begin{aligned}
I = & -\frac{1}{\pi}\int_{0}^{\omega/c}\frac{1}{\sqrt{(\omega/c)^2-\xi^2}}\cos(\xi x)\sin\left\{|y|\sqrt{(\omega/c)^2-\xi^2}\right\}d\xi \\
& +\frac{1}{\pi}\int_{\omega/c}^{+\infty}\frac{1}{\sqrt{\xi^2-(\omega/c)^2}}\cos(\xi x)\exp\left\{-|y|\sqrt{\xi^2-(\omega/c)^2}\right\}d\xi \\
& -\frac{i}{\pi}\int_{0}^{\omega/c}\frac{1}{\sqrt{(\omega/c)^2-\xi^2}}\cos(\xi x)\cos\left\{y\sqrt{(\omega/c)^2-\xi^2}\right\}d\xi
\end{aligned}
\tag{2.5.60}
$$

We make the change of variable for the first and second integrals in the above equation. The changes are defined by

$$
\eta=\sqrt{(\omega/c)^2-\xi^2},\quad \xi=\sqrt{(\omega/c)^2-\eta^2},\quad d\xi=-\frac{\eta d\eta}{\sqrt{(\omega/c)^2-\eta^2}}
\tag{2.5.61}
$$

for the first integral, and

$$
\eta=\sqrt{\xi^2-(\omega/c)^2},\quad \xi=\sqrt{\eta^2+(\omega/c)^2},\quad d\xi=\frac{\eta d\eta}{\sqrt{\eta^2+(\omega/c)^2}}
\tag{2.5.62}
$$

for the second integral. Equation (2.5.60) yields to

$$
\begin{aligned}
I = & -\frac{1}{\pi}\int_{0}^{\omega/c}\frac{1}{\sqrt{(\omega/c)^2-\eta^2}}\cos\left\{x\sqrt{(\omega/c)^2-\eta^2}\right\}\sin(|y|\eta)d\eta \\
& +\frac{1}{\pi}\int_{0}^{\infty}\frac{1}{\sqrt{\eta^2+(\omega/c)^2}}\cos\left\{x\sqrt{\eta^2+(\omega/c)^2}\right\}\exp(-|y|\eta)d\eta \\
& -\frac{i}{\pi}\int_{0}^{\omega/c}\frac{1}{\sqrt{(\omega/c)^2-\xi^2}}\cos(\xi x)\cos\left\{y\sqrt{(\omega/c)^2-\xi^2}\right\}d\xi
\end{aligned}
\tag{2.5.63}
$$

Then, we apply the formula of Eq. (2.5.27) to the first two integrals and the formula of Eq. (2.5.28) to the third integral. The integral I is exactly evaluated as

$$
I=-\frac{1}{2}Y_0\left(\frac{\omega}{c}\sqrt{x^2+y^2}\right)-\frac{i}{2}J_0\left(\frac{\omega}{c}\sqrt{x^2+y^2}\right)=-\frac{i}{2}H_0^{(2)}\left(\frac{\omega}{c}\sqrt{x^2+y^2}\right)
\tag{2.5.64}
$$

This is just the same as Eqs. (2.5.30) and (2.5.45) for the integral I.

2.6 3D Static Source

The static 3D Green's function for the Laplace equation is a particular solution of the nonhomogeneous differential equation,

$$\frac{\partial^2 \phi}{\partial x^2} + \frac{\partial^2 \phi}{\partial y^2} + \frac{\partial^2 \phi}{\partial z^2} = -S\delta(x)\delta(y)\delta(z) \tag{2.6.1}$$

where S is the magnitude of the source which is placed at the coordinate origin (0, 0, 0). The convergence condition at infinity,

$$\phi|_{\sqrt{x^2+y^2+z^2}\to\infty} = \left.\frac{\partial \phi}{\partial x}\right|_{\sqrt{x^2+y^2+z^2}\to\infty} = \left.\frac{\partial \phi}{\partial y}\right|_{\sqrt{x^2+y^2+z^2}\to\infty} = \left.\frac{\partial \phi}{\partial z}\right|_{\sqrt{x^2+y^2+z^2}\to\infty} = 0 \tag{2.6.2}$$

is applied to the Green's function ϕ.

The Green's function is obtained by the method of integral transform. For three space variables, the triple Fourier transform defined by

$$\bar{\phi}(\xi) = \int_{-\infty}^{+\infty} \phi(x)\exp(+i\xi x)dx, \quad \phi(x) = \frac{1}{2\pi}\int_{-\infty}^{+\infty} \bar{\phi}(\xi)\exp(-i\xi x)d\xi \tag{2.6.3}$$

$$\tilde{\phi}(\eta) = \int_{-\infty}^{+\infty} \phi(y)\exp(+i\eta y)dy, \quad \phi(y) = \frac{1}{2\pi}\int_{-\infty}^{+\infty} \tilde{\phi}(\eta)\exp(-i\eta y)d\eta \tag{2.6.4}$$

$$\hat{\phi}(\zeta) = \int_{-\infty}^{+\infty} \phi(z)\exp(+i\zeta z)dz, \quad \phi(z) = \frac{1}{2\pi}\int_{-\infty}^{+\infty} \hat{\phi}(\zeta)\exp(-i\zeta z)d\zeta \tag{2.6.5}$$

is applied to Eq. (2.6.1),

$$\int_{-\infty}^{+\infty}\left\langle \int_{-\infty}^{+\infty}\left[\int_{-\infty}^{+\infty}\left\{\frac{\partial^2 \phi}{\partial x^2} + \frac{\partial^2 \phi}{\partial y^2} + \frac{\partial^2 \phi}{\partial z^2} = -S\delta(x)\delta(y)\delta(z)\right\}\exp(+i\xi x)dx\right]\exp(+i\eta y)dy\right\rangle\exp(+i\zeta z)dz \tag{2.6.6}$$

Applying the convergence condition (2.6.2), the above Eq. (2.6.6) is transformed to the algebraic equation for the triple transformed function $\hat{\tilde{\bar{\phi}}}$,

$$-(\xi^2 + \eta^2 + \zeta^2)\hat{\tilde{\bar{\phi}}} = -S \tag{2.6.7}$$

The transformed function is determined explicitly,

$$\bar{\bar{\bar{\phi}}} = \frac{S}{\xi^2 + \eta^2 + \zeta^2} \tag{2.6.8}$$

The first Fourier inversion integral with respect to the parameter ζ is

$$\bar{\bar{\phi}} = \frac{1}{2\pi} \int_{-\infty}^{+\infty} \frac{S}{\xi^2 + \eta^2 + \zeta^2} \exp(-i\zeta z) d\zeta = \frac{S}{\pi} \int_0^{\infty} \frac{1}{\zeta^2 + \sqrt{\xi^2 + \eta^2}^2} \cos(\zeta z) d\zeta \tag{2.6.9}$$

The above integral is easily evaluated with the aid of the formula (2.1.22). It yields

$$\bar{\bar{\phi}} = \frac{S}{2\sqrt{\xi^2 + \eta^2}} \exp\left(-|z|\sqrt{\xi^2 + \eta^2}\right) \tag{2.6.10}$$

The second inversion integral with respect to the parameter η is reduced to the semi-infinite integral as

$$\begin{aligned} \bar{\phi} &= \frac{1}{2\pi} \int_{-\infty}^{+\infty} \frac{S}{2\sqrt{\xi^2 + \eta^2}} \exp\left(-|z|\sqrt{\xi^2 + \eta^2}\right) \exp(-i\eta y) d\eta \\ &= \frac{S}{2\pi} \int_0^{\infty} \frac{1}{\sqrt{\xi^2 + \eta^2}} \exp\left(-|z|\sqrt{\xi^2 + \eta^2}\right) \cos(\eta y) d\eta \end{aligned} \tag{2.6.11}$$

The latter semi-infinite integral can be evaluated by using the formula (2.4.13). It follows that

$$\bar{\phi} = \frac{S}{2\pi} K_0\left(|\xi|\sqrt{y^2 + z^2}\right) \tag{2.6.12}$$

The last inversion integral with respect to the parameter ξ is also reduced to the semi-infinite integral,

$$\begin{aligned} \phi &= \frac{S}{4\pi^2} \int_{-\infty}^{+\infty} K_0\left(|\xi|\sqrt{y^2 + z^2}\right) \exp(-i\xi x) d\xi \\ &= \frac{S}{2\pi^2} \int_0^{\infty} K_0\left(\xi\sqrt{y^2 + z^2}\right) \cos(\xi x) d\xi \end{aligned} \tag{2.6.13}$$

Fortunately, we have the suitable integration formula, (Erdélyi 1954, vol. I, pp. 49, 40)

$$\int_0^\infty K_0(a\xi)\cos(b\xi)d\xi = \frac{\pi}{2\sqrt{a^2+b^2}} \tag{2.6.14}$$

Applying this formula to Eq. (2.6.13), the last inversion integral is exactly evaluated as

$$\phi = \frac{S}{4\pi\sqrt{x^2+y^2+z^2}} \tag{2.6.15}$$

Consequently, we have the static 3D Green's function for the Laplace equation,

$$\frac{\partial^2\phi}{\partial x^2}+\frac{\partial^2\phi}{\partial y^2}+\frac{\partial^2\phi}{\partial z^2} = -S\delta(x)\delta(y)\delta(z) \quad \Rightarrow \quad \phi = \frac{S}{4\pi R} \tag{2.6.16}$$

where R is the 3D radial distance from the source,

$$R = \sqrt{x^2+y^2+z^2} \tag{2.6.17}$$

2.7 3D Impulsive Source

Green's function for the 3D wave equation is discussed. The wave equation with an impulsive point source located at the coordinate origin is given by

$$\frac{\partial^2\phi}{\partial x^2}+\frac{\partial^2\phi}{\partial y^2}+\frac{\partial^2\phi}{\partial z^2} = \frac{1}{c^2}\frac{\partial^2\phi}{\partial t^2} - P\delta(x)\delta(y)\delta(z)\delta(t) \tag{2.7.1}$$

where P is the magnitude of the source. The quiescent condition at an initial time,

$$\phi|_{t=0} = \left.\frac{\partial\phi}{\partial t}\right|_{t=0} = 0 \tag{2.7.2}$$

and the convergence condition at infinity,

$$\phi|_{\sqrt{x^2+y^2+z^2}\to\infty} = \left.\frac{\partial\phi}{\partial x}\right|_{\sqrt{x^2+y^2+z^2}\to\infty} = \left.\frac{\partial\phi}{\partial y}\right|_{\sqrt{x^2+y^2+z^2}\to\infty} = \left.\frac{\partial\phi}{\partial z}\right|_{\sqrt{x^2+y^2+z^2}\to\infty} = 0 \tag{2.7.3}$$

are also imposed.

Our dynamic Green's function has four variables: three space and one time variables. Laplace transform with respect to the time,

$$\phi^*(s) = \int_0^{+\infty} \phi(t)\exp(-st)dt \tag{2.7.4}$$

and the triple Fourier transform with respect to three space variables,

$$\bar{\phi}(\xi) = \int_{-\infty}^{+\infty} \phi(x)\exp(+i\xi x)dx, \quad \phi(x) = \frac{1}{2\pi}\int_{-\infty}^{+\infty} \bar{\phi}(\xi)\exp(-i\xi x)d\xi \tag{2.7.5}$$

$$\tilde{\phi}(\eta) = \int_{-\infty}^{+\infty} \phi(y)\exp(+i\eta y)dy, \quad \phi(y) = \frac{1}{2\pi}\int_{-\infty}^{+\infty} \tilde{\phi}(\eta)\exp(-i\eta y)d\eta \tag{2.7.6}$$

$$\hat{\phi}(\zeta) = \int_{-\infty}^{+\infty} \phi(z)\exp(+i\zeta z)dz, \quad \phi(z) = \frac{1}{2\pi}\int_{-\infty}^{+\infty} \hat{\phi}(\zeta)\exp(-i\zeta z)d\zeta \tag{2.7.7}$$

are applied to the nonhomogeneous wave equation (2.7.1). With the aid of the quiescent and convergence conditions, the wave equation is transformed to the simple algebraic equation for the multi-transformed unknown function $\hat{\tilde{\bar{\phi}}}^*$,

$$-\{\xi^2 + \eta^2 + \zeta^2 + (s/c)^2\}\hat{\tilde{\bar{\phi}}}^* = -P \tag{2.7.8}$$

The inversion starts from the transformed function,

$$\hat{\tilde{\bar{\phi}}}^* = \frac{P}{\xi^2 + \eta^2 + \zeta^2 + (s/c)^2} \tag{2.7.9}$$

As the first inversion, the Fourier inversion integral with respect to the parameter ζ

$$\begin{aligned} \tilde{\bar{\phi}}^* &= \frac{1}{2\pi}\int_{-\infty}^{+\infty} \frac{P}{\xi^2 + \eta^2 + \zeta^2 + (s/c)^2}\exp(-i\zeta z)dz \\ &= \frac{P}{\pi}\int_0^{\infty} \frac{\cos(\zeta z)}{\zeta^2 + \sqrt{\xi^2 + \eta^2 + (s/c)^2}^{\,2}}dz \end{aligned} \tag{2.7.10}$$

is carried out by applying the formula (2.1.22). It follows that

$$\tilde{\bar{\phi}}^* = \frac{P}{2\sqrt{\xi^2 + \eta^2 + (s/c)^2}} \exp\left\{-|z|\sqrt{\xi^2 + \eta^2 + (s/c)^2}\right\} \tag{2.7.11}$$

The second inversion is the integral with respect to the parameter η,

$$\begin{aligned}\bar{\phi}^* &= \frac{1}{2\pi}\int_{-\infty}^{+\infty} \frac{P}{2\sqrt{\xi^2 + \eta^2 + (s/c)^2}} \exp\left\{-|z|\sqrt{\xi^2 + \eta^2 + (s/c)^2}\right\} \exp(-i\eta y)d\eta \\ &= \frac{P}{2\pi}\int_0^{\infty} \frac{1}{\sqrt{\xi^2 + \eta^2 + (s/c)^2}} \exp\left\{-|z|\sqrt{\xi^2 + \eta^2 + (s/c)^2}\right\} \cos(\eta y)d\eta\end{aligned} \tag{2.7.12}$$

The latter semi-infinite integral is also evaluated by applying the formula (2.4.13). It yields

$$\bar{\phi}^* = \frac{P}{2\pi} K_0\left(\sqrt{\xi^2 + (s/c)^2}\sqrt{y^2 + z^2}\right) \tag{2.7.13}$$

The third inversion integral with respect to the parameter ξ is

$$\begin{aligned}\phi^* &= \frac{P}{2\pi}\frac{1}{2\pi}\int_{-\infty}^{+\infty} K_0\left(\sqrt{\xi^2 + (s/c)^2}\sqrt{y^2 + z^2}\right) \exp(-i\xi x)d\xi \\ &= \frac{P}{2\pi^2}\int_0^{\infty} K_0\left(\sqrt{\xi^2 + (s/c)^2}\sqrt{y^2 + z^2}\right) \cos(\xi x)d\xi\end{aligned} \tag{2.7.14}$$

The integration formula (Erdélyi 1954, vol. I, pp. 56, 43)

$$\int_0^{\infty} K_0\left(a\sqrt{\xi^2 + b^2}\right) \cos(c\xi)d\xi = \frac{\pi}{2\sqrt{a^2 + c^2}} \exp\left(-b\sqrt{a^2 + c^2}\right) \tag{2.7.15}$$

is very helpful for our task. Then, applying the above formula to the last integral in Eq. (2.7.14), we have

$$\phi^* = \frac{P}{4\pi\sqrt{x^2 + y^2 + z^2}} \exp\left(-\frac{s}{c}\sqrt{x^2 + y^2 + z^2}\right) \tag{2.7.16}$$

The last is the Laplace inversion. Its symbolical form is

$$\phi = \frac{P}{4\pi\sqrt{x^2+y^2+z^2}} L^{-1}\left[\exp\left(-\frac{s}{c}\sqrt{x^2+y^2+z^2}\right)\right] \tag{2.7.17}$$

The transform parameter "s" is included only in the argument of the exponential function. We remember the simple Laplace inversion formula for the delta function,

$$L^{-1}[\exp(-as)] = \delta(t-a) \tag{2.7.18}$$

Thus, applying this inversion formula, Eq. (2.7.17) is fully inverted as

$$\phi = \frac{P}{4\pi R}\delta(t - R/c) \tag{2.7.19}$$

where the radial distance R from the source is defined by Eq. (2.6.17). Finally, we have the 3D Green's function for the wave equation with the impulsive point source,

$$\frac{\partial^2\phi}{\partial x^2} + \frac{\partial^2\phi}{\partial y^2} + \frac{\partial^2\phi}{\partial z^2} = \frac{1}{c^2}\frac{\partial^2\phi}{\partial t^2} - P\delta(x)\delta(y)\delta(z)\delta(t) \Rightarrow \phi = \frac{P}{4\pi R}\delta(t-R/c) \tag{2.7.20}$$

This is the very simple expression in spite of the 3D nature!

2.8 3D Time-Harmonic Source

This section derives the 3D Green's function for a time-harmonic source. It is the convolution integral of the impulsive Green's function obtained in the previous section. The wave equation with the time-harmonic source is given by

$$\frac{\partial^2\phi}{\partial x^2} + \frac{\partial^2\phi}{\partial y^2} + \frac{\partial^2\phi}{\partial z^2} = \frac{1}{c^2}\frac{\partial^2\phi}{\partial t^2} - Q\delta(x)\delta(y)\delta(z)\exp(+i\omega t) \tag{2.8.1}$$

where Q is the magnitude of the source and ω the frequency of the time-harmonic vibration. The convergence condition at infinity,

$$\phi|_{\sqrt{x^2+y^2+z^2}\to\infty} = \left.\frac{\partial\phi}{\partial x}\right|_{\sqrt{x^2+y^2+z^2}\to\infty} = \left.\frac{\partial\phi}{\partial y}\right|_{\sqrt{x^2+y^2+z^2}\to\infty} = \left.\frac{\partial\phi}{\partial z}\right|_{\sqrt{x^2+y^2+z^2}\to\infty} = 0 \tag{2.8.2}$$

is also imposed.

The standard multiple integral transform technique is available for getting the time-harmonic Green's function. However, as shown in the case of 2D Green's function in Sect. 2.5 (4), we take a very simple way, i.e. the convolution integral of the impulsive Green's function.

In the previous section we get the Green's function for the impulsive source. Replacing the source magnitude with unit 1, the Green function is given by

$$\frac{\partial^2 \phi}{\partial x^2} + \frac{\partial^2 \phi}{\partial y^2} + \frac{\partial^2 \phi}{\partial z^2} = \frac{1}{c^2}\frac{\partial^2 \phi}{\partial t^2} - \delta(x)\delta(y)\delta(z)\delta(t) \Rightarrow \phi = \frac{1}{4\pi R}\delta(t - R/c) \quad (2.8.3)$$

Employing the time-harmonic source with the frequency ω and the magnitude Q

$$Q \exp\{+i\omega(t - t')\} \quad (2.8.4)$$

the convolution integral for the Laplace transform is given by

$$\phi = \int_0^t \frac{Q}{4\pi R}\delta(t' - R/c)\exp\{+i\omega(t - t')\}dt' \quad (2.8.5)$$

It is very easy to evaluate the above integral, since the integrand includes Dirac's delta function and we can apply the simple integration formula (1.2.3) in Sect. 1.2,

$$\int_a^b f(x)\delta(x - c)dx = \begin{cases} f(c); & a < c < b \\ 0; & c < a \text{ or } b < c \end{cases} \quad (2.8.6)$$

Then, we can evaluate the integral in Eq. (2.8.5) and have for ϕ

$$\phi = \frac{Q}{4\pi R}H(t - R/c)\exp\{+i\omega(t - R/c)\} \quad (2.8.7)$$

The step function ahead of the equation means that Eq. (2.8.7) is the transient response to the time-harmonic source and the disturbance starts from the wave arrival $t = R/c$. When sufficient long time has passed and the response becomes steady, the step function is meaningless. Then, we have the steady-state time-harmonic response as

$$\phi = \frac{Q}{4\pi R}\exp\{+i\omega(t - R/c)\} \quad (2.8.8)$$

Therefore, the 3D Green's function for the wave equation with the time-harmonic source is given by

Table 2.3 Green's function for Laplace and Wave equations

	Differential equation	Source	Green's function
1D	$\frac{\partial^2\phi}{\partial x^2} = \frac{1}{c^2}\frac{\partial^2\phi}{\partial t^2} - \text{Source}$	$P\delta(x)\delta(t)$	$\phi(x,t) = \frac{cP}{2}H(ct - \lvert x\rvert)$
		$Q\delta(x)\exp(\pm i\omega t)$	$\phi(x,t) = \mp\frac{iQ}{2(\omega/c)}\exp\{\pm i\omega(t - \lvert x\rvert/c)\}$
2D	$\frac{\partial^2\phi}{\partial x^2} + \frac{\partial^2\phi}{\partial y^2} = -\text{Source}$	$S\delta(x)\delta(y)$	$\phi(x,y) = -\frac{S}{4\pi}\log(r) + \text{arbitrary constant}; \quad r = \sqrt{x^2+y^2}$
	$\frac{\partial^2\phi}{\partial x^2} + \frac{\partial^2\phi}{\partial x^2} = \frac{1}{c^2}\frac{\partial^2\phi}{\partial t^2} - \text{Source}$	$P\delta(x)\delta(y)\delta(t)$	$\phi(x,y,t) = \frac{cP}{2\pi\sqrt{(ct)^2-r^2}}H(ct-r); \quad r = \sqrt{x^2+y^2}$
		$Q\delta(x)\delta(y)\exp(+i\omega t)$	$\phi(x,y,t) = -\frac{iQ}{4}H_0^{(2)}(\omega r/c)\exp(+i\omega t); \quad r = \sqrt{x^2+y^2}$
		$Q\delta(x)\delta(y)\exp(-i\omega t)$	$\phi(x,y,t) = +\frac{iQ}{4}H_0^{(1)}(\omega r/c)\exp(-i\omega t); \quad r = \sqrt{x^2+y^2}$
3D	$\frac{\partial^2\phi}{\partial x^2} + \frac{\partial^2\phi}{\partial y^2} + \frac{\partial^2\phi}{\partial z^2} = -\text{Source}$	$S\delta(x)\delta(y)\delta(z)$	$\phi(x,y,z) = \frac{S}{4\pi R}; \quad R = \sqrt{x^2+y^2+z^2}$
	$\frac{\partial^2\phi}{\partial x^2} + \frac{\partial^2\phi}{\partial y^2} + \frac{\partial^2\phi}{\partial z^2} = \frac{1}{c^2}\frac{\partial^2\phi}{\partial t^2} - \text{Source}$	$P\delta(x)\delta(y)\delta(z)\delta(t)$	$\phi(x,y,z,t) = \frac{P}{4\pi R}\delta(t - R/c); \quad R = \sqrt{x^2+y^2+z^2}$
		$Q\delta(x)\delta(y)\delta(z)\exp(\pm i\omega t)$	$\phi(x,y,z,t) = \frac{Q}{4\pi R}\exp\{\pm i\omega(t - R/c)\}; \quad R = \sqrt{x^2+y^2+z^2}$

$$\begin{aligned}\frac{\partial^2\phi}{\partial x^2}+\frac{\partial^2\phi}{\partial y^2}+\frac{\partial^2\phi}{\partial z^2}&=\frac{1}{c^2}\frac{\partial^2\phi}{\partial t^2}-Q\delta(x)\delta(y)\delta(z)\exp(+i\omega t)\\ \Rightarrow \phi&=\frac{Q}{4\pi R}\exp\{+i\omega(t-R/c)\}\end{aligned}\tag{2.8.9}$$

Exercise

(2.1) Using the impulsive response (2.1.27) for 1D wave equation, derive the time-harmonic Green's function through the convolution integral and compare it with the time-harmonic Green's function (2.2.27).

Appendix

See Table 2.3.

References

Erdélyi A (ed) (1954) Tables of integral transforms, vols I, II, McGraw-Hill, New York
Gradshteyn IS, Ryzhik IM (1980) In: Jefferey A (ed) Table of integrals, series, and products, 5th edn. Academic Press, San Diego
Watson GN (1966) A treatise on the theory of Bessel functions, Cambridge University Press, Cambridge

Chapter 3
Green's Dyadic for an Isotropic Elastic Solid

The present chapter shows how to derive an exact closed form solution, the so-called Green's dyadic, for elasticity equations. Introducing the Cartesian coordinate system $(x_i) \equiv (x, y, z)$, we employ the notation $\mathbf{u} \equiv (u_i)$ for the displacement, $\boldsymbol{\varepsilon} \equiv (\varepsilon_{ij})$ for the strain, $\boldsymbol{\sigma} \equiv (\sigma_{ij})$ for the stress, $\mathbf{B} \equiv (B_i)$ for the body force and ρ for the density. The governing equations for the deformation of an isotropic elastic solid are constituted by the following equations of motion,

$$\begin{aligned} \frac{\partial \sigma_{xx}}{\partial x} + \frac{\partial \sigma_{yx}}{\partial y} + \frac{\partial \sigma_{zx}}{\partial z} + \rho B_x &= \rho \frac{\partial^2 u_x}{\partial t^2} \\ \frac{\partial \sigma_{xy}}{\partial x} + \frac{\partial \sigma_{yy}}{\partial y} + \frac{\partial \sigma_{zy}}{\partial z} + \rho B_y &= \rho \frac{\partial^2 u_y}{\partial t^2} \\ \frac{\partial \sigma_{xz}}{\partial x} + \frac{\partial \sigma_{yz}}{\partial y} + \frac{\partial \sigma_{zz}}{\partial z} + \rho B_z &= \rho \frac{\partial^2 u_z}{\partial t^2} \end{aligned} \tag{3.1}$$

and by the stress-strain relation, the so-called Hooke's law,

$$\begin{bmatrix} \sigma_{xx} \\ \sigma_{yy} \\ \sigma_{zz} \\ \sigma_{xz} \\ \sigma_{yz} \\ \sigma_{xy} \end{bmatrix} = \begin{bmatrix} \lambda + 2\mu & \lambda & \lambda & 0 & 0 & 0 \\ & \lambda + 2\mu & \lambda & 0 & 0 & 0 \\ & & \lambda + 2\mu & 0 & 0 & 0 \\ & & & 2\mu & 0 & 0 \\ & \text{sym.} & & & 2\mu & 0 \\ & & & & & 2\mu \end{bmatrix} \begin{bmatrix} \varepsilon_{xx} \\ \varepsilon_{yy} \\ \varepsilon_{zz} \\ \varepsilon_{xz} \\ \varepsilon_{yz} \\ \varepsilon_{xy} \end{bmatrix} \tag{3.2}$$

Here (λ, μ) are Lame's constants which are expressed by Young's modulus E and Poisson ratio ν as

$$\lambda = \frac{\nu E}{(1+\nu)(1-2\nu)}, \quad \mu = \frac{E}{2(1+\nu)} \tag{3.3}$$

K. Watanabe, *Integral Transform Techniques for Green's Function*,
Lecture Notes in Applied and Computational Mechanics 76,
DOI 10.1007/978-3-319-17455-6_3

The strains in Eq. (3.2) are defined by the displacement gradient,

$$
\begin{aligned}
&\varepsilon_{xx} = \frac{\partial u_x}{\partial x}, \quad \varepsilon_{yy} = \frac{\partial u_y}{\partial y}, \quad \varepsilon_{zz} = \frac{\partial u_z}{\partial z}, \\
&\varepsilon_{xy} = \frac{1}{2}\left(\frac{\partial u_x}{\partial y} + \frac{\partial u_y}{\partial x}\right), \quad \varepsilon_{yz} = \frac{1}{2}\left(\frac{\partial u_y}{\partial z} + \frac{\partial u_z}{\partial y}\right), \quad \varepsilon_{zx} = \frac{1}{2}\left(\frac{\partial u_x}{\partial z} + \frac{\partial u_z}{\partial x}\right)
\end{aligned} \tag{3.4}
$$

It should be noticed that the symmetry relations for the stress and strain components are satisfied:

$$
\sigma_{ij} = \sigma_{ji}, \quad \varepsilon_{ij} = \varepsilon_{ji}; \quad i, j = x, y, z \tag{3.5}
$$

The elasticity equations constitute a set of coupled partial differential equations with 15 unknowns. In order to reduce the differential equations to more compact forms, the strain in Hooke's law is replaced with the displacement gradient and then the stress is substituted into the equations of motion. We then get a set of displacement equations, the so-called Navier equations, with only three unknown functions (displacement components),

$$
\begin{aligned}
(\gamma^2 - 1)\frac{\partial}{\partial x}\left(\frac{\partial u_x}{\partial x} + \frac{\partial u_y}{\partial y} + \frac{\partial u_z}{\partial z}\right) + \frac{\partial^2 u_x}{\partial x^2} + \frac{\partial^2 u_x}{\partial y^2} + \frac{\partial^2 u_x}{\partial z^2} &= \frac{1}{c_s^2}\frac{\partial^2 u_x}{\partial t^2} - \frac{1}{c_s^2}B_x \\
(\gamma^2 - 1)\frac{\partial}{\partial y}\left(\frac{\partial u_x}{\partial x} + \frac{\partial u_y}{\partial y} + \frac{\partial u_z}{\partial z}\right) + \frac{\partial^2 u_y}{\partial x^2} + \frac{\partial^2 u_y}{\partial y^2} + \frac{\partial^2 u_y}{\partial z^2} &= \frac{1}{c_s^2}\frac{\partial^2 u_y}{\partial t^2} - \frac{1}{c_s^2}B_y \\
(\gamma^2 - 1)\frac{\partial}{\partial z}\left(\frac{\partial u_x}{\partial x} + \frac{\partial u_y}{\partial y} + \frac{\partial u_z}{\partial z}\right) + \frac{\partial^2 u_z}{\partial x^2} + \frac{\partial^2 u_z}{\partial y^2} + \frac{\partial^2 u_z}{\partial z^2} &= \frac{1}{c_s^2}\frac{\partial^2 u_z}{\partial t^2} - \frac{1}{c_s^2}B_z
\end{aligned} \tag{3.6}
$$

where c_s and c_d are the velocities of shear and dilatational waves, respectively, and are defined by

$$
c_s = \sqrt{\frac{\mu}{\rho}}, \quad c_d = \sqrt{\frac{\lambda + 2\mu}{\rho}} \tag{3.7}
$$

The velocity ratio γ is defined and expressed by Poisson ratio v,

$$
\gamma = \frac{c_d}{c_s} = \sqrt{\frac{\lambda + 2\mu}{\mu}} = \sqrt{\frac{2(1 - v)}{1 - 2v}} \tag{3.8}
$$

Then, in this chapter we shall show a solution method for the coupled displacement Eqs. (3.6) with nonhomogeneous body force terms B_i. The solution method that we employ is the integral transforms: Fourier transforms with respect to the space variables and Laplace transform with respect the time variable.

A particular solution corresponding to a point body force is called a Green's function. However, each displacement component produced by one source component is called "Green's dyadic." Thus, the present chapter shows the solution method for the Green's dyadic.

3.1 2D Impulsive Source

We assume here the deformation of plane-strain. Take the 2D coordinate system (x, y) in an infinite elastic solid and neglect the anti-plane deformation produced by the anti-plane displacement u_z. The in-plane displacement components (u_x, u_y) are functions of two space variables (x, y) and the time t. As a wave source, we assume an impulsive point body force with magnitude P_i placed at the coordinate origin. Under these assumptions, the displacement equations (3.6) is reduced to the simpler form

$$\begin{aligned}(\gamma^2-1)\frac{\partial}{\partial x}\left(\frac{\partial u_x}{\partial x}+\frac{\partial u_y}{\partial y}\right)+\frac{\partial^2 u_x}{\partial x^2}+\frac{\partial^2 u_x}{\partial y^2}&=\frac{1}{c_s^2}\frac{\partial^2 u_x}{\partial t^2}-\frac{P_x}{c_s^2}\delta(x)\delta(y)\delta(t)\\(\gamma^2-1)\frac{\partial}{\partial y}\left(\frac{\partial u_x}{\partial x}+\frac{\partial u_y}{\partial y}\right)+\frac{\partial^2 u_y}{\partial x^2}+\frac{\partial^2 u_y}{\partial y^2}&=\frac{1}{c_s^2}\frac{\partial^2 u_y}{\partial t^2}-\frac{P_y}{c_s^2}\delta(x)\delta(y)\delta(t)\end{aligned}\tag{3.1.1}$$

Now, we consider these coupled differential equations. The solutions are two displacement components (u_x, u_y) corresponding to the nonhomogeneous body force P_i. Our solution strategy is very simple, namely, to transform the differential equations into the simultaneous algebraic equations in the transformed domain. We employ Laplace transform with respect to the time, defined by

$$u_j^* = \int_0^\infty u_j \exp(-st)dt \tag{3.1.2}$$

and the double Fourier transform with respect to the two space variables, defined by

$$\begin{aligned}\bar{u}_j &= \int_{-\infty}^{+\infty} u_j \exp(+i\xi x)dx, \qquad u_j = \frac{1}{2\pi}\int_{-\infty}^{+\infty} \bar{u}_j \exp(-i\xi x)d\xi\\ \tilde{u}_j &= \int_{-\infty}^{+\infty} u_j \exp(+i\eta y)dy, \qquad u_j = \frac{1}{2\pi}\int_{-\infty}^{+\infty} \tilde{u}_j \exp(-i\eta y)d\eta\end{aligned}\tag{3.1.3}$$

where the subscript j stands for x and y. The quiescent condition at an initial time,

$$u_j\big|_{t=0} = \frac{\partial u_j}{\partial t}\bigg|_{t=0} = 0 \tag{3.1.4}$$

and the convergence condition at infinity

$$u_j\big|_{\sqrt{x^2+y^2}\to\infty} = \frac{\partial u_j}{\partial x}\bigg|_{\sqrt{x^2+y^2}\to\infty} = \frac{\partial u_j}{\partial y}\bigg|_{\sqrt{x^2+y^2}\to\infty} = 0 \tag{3.1.5}$$

are assumed.

Applying the triple integral transform to the displacement equations (3.1.1), we have the algebraic equations for the transformed displacement components $\tilde{\bar{u}}_j^*$,

$$\begin{aligned} -i\xi(\gamma^2-1)\left(-i\xi\tilde{\bar{u}}_x^* - i\eta\tilde{\bar{u}}_y^*\right) - (\xi^2+\eta^2)\tilde{\bar{u}}_x^* &= (s/c_s)^2\tilde{\bar{u}}_x^* - P_x/c_s^2 \\ -i\eta(\gamma^2-1)\left(-i\xi\tilde{\bar{u}}_x^* - i\eta\tilde{\bar{u}}_y^*\right) - (\xi^2+\eta^2)\tilde{\bar{u}}_y^* &= (s/c_s)^2\tilde{\bar{u}}_y^* - P_y/c_s^2 \end{aligned} \tag{3.1.6}$$

Solving for the displacement, we have the exact expressions in the transformed domain (ξ, η, s),

$$\tilde{\bar{u}}_x^* = \frac{P_x}{c_s^2}\frac{1}{\alpha_s^2} - (\xi^2 P_x + \xi\eta P_y)\frac{1}{s^2}\left(\frac{1}{\alpha_d^2} - \frac{1}{\alpha_s^2}\right) \tag{3.1.7a}$$

$$\tilde{\bar{u}}_y^* = \frac{P_y}{c_s^2}\frac{1}{\alpha_s^2} - (\xi\eta P_x + \eta^2 P_y)\frac{1}{s^2}\left(\frac{1}{\alpha_d^2} - \frac{1}{\alpha_s^2}\right) \tag{3.1.7b}$$

where two radicals are defined by

$$\alpha_d = \sqrt{\xi^2+\eta^2+(s/c_d)^2}, \quad \alpha_s = \sqrt{\xi^2+\eta^2+(s/c_s)^2} \tag{3.1.8}$$

Our main task is to invert the transformed displacement to the original space (x, y, t). Examining the Eqs. (3.1.7), we learn that the inversion of the four transformed functions is enough for evaluating the displacement. These four fundamental functions are

$$\tilde{\bar{I}}_0^* = \frac{1}{c_s^2\alpha_s^2} \tag{3.1.9}$$

$$\tilde{\bar{I}}_{xx}^* = -\frac{\xi^2}{s^2}\left(\frac{1}{\alpha_d^2} - \frac{1}{\alpha_s^2}\right) \tag{3.1.10}$$

$$\bar{\bar{I}}_{xy}^{*} = -\frac{\xi\eta}{s^2}\left(\frac{1}{\alpha_d^2} - \frac{1}{\alpha_s^2}\right) \tag{3.1.11}$$

$$\bar{\bar{I}}_{yy}^{*} = -\frac{\eta^2}{s^2}\left(\frac{1}{\alpha_d^2} - \frac{1}{\alpha_s^2}\right) \tag{3.1.12}$$

If we could have the four inversions, the displacement is expressed as

$$u_x = P_x(I_0 + I_{xx}) + P_y I_{xy}, \quad u_y = P_x I_{xy} + P_y(I_0 + I_{yy}) \tag{3.1.13}$$

Then, we shall consider the inversion of the four fundamentals, successively.
(1) Inversion of $\bar{\bar{I}}_0^*$
Rewriting $\bar{\bar{I}}_0^*$ as

$$\bar{\bar{I}}_0^* = \frac{1}{c_s^2}\frac{1}{\xi^2 + \eta^2 + (s/c_s)^2}, \tag{3.1.14}$$

we can find that this equation is the same as Eq. (2.4.8) for the transformed Green's function $\bar{\bar{\phi}}^*$ with the replacement $P \rightarrow 1/c_s^2$ and $c \rightarrow c_s$. Then, we apply the same mathematics as that in Sect. 2.4 and have the inversion,

$$I_0(x, y, t) = \frac{1}{2\pi c_s}\frac{H(c_c t - r)}{\sqrt{(c_s t)^2 - r^2}}. \tag{3.1.15}$$

where $H(.)$ is Heaviside's unit step function.
(2) Inversion of $\bar{\bar{I}}_{xx}^*$
As we have no previous result for the second inversion $\bar{\bar{I}}_{xx}^*$, the inversion integrals are carried out successively. The Fourier inversion integral with respect to the parameter η is given by

$$\begin{aligned}\bar{I}_{xx}^* &= -\frac{\xi^2}{2\pi s^2}\int_{-\infty}^{\infty}\left(\frac{1}{\alpha_d^2} - \frac{1}{\alpha_s^2}\right)\exp(-i\eta y)d\eta \\ &= -\frac{\xi^2}{\pi s^2}\int_0^{\infty}\left\{\frac{1}{\xi^2 + \eta^2 + (s/c_d)^2} - \frac{1}{\xi^2 + \eta^2 + (s/c_s)^2}\right\}\cos(\eta y)d\eta\end{aligned} \tag{3.1.16}$$

The semi-infinite integral is evaluated with the use of the formula (2.1.22) and is arranged so that each term has the same radical,

$$\bar{I}_{xx}^{*} = -\frac{1}{2s^2}\left\{\sqrt{\xi^2 + (s/c_d)^2} - \frac{(s/c_d)^2}{\sqrt{\xi^2 + (s/c_d)^2}}\right\}\exp\left\{-|y|\sqrt{\xi^2 + (s/c_d)^2}\right\}$$
$$+\frac{1}{2s^2}\left\{\sqrt{\xi^2 + (s/c_s)^2} - \frac{(s/c_s)^2}{\sqrt{\xi^2 + (s/c_s)^2}}\right\}\exp\left\{-|y|\sqrt{\xi^2 + (s/c_s)^2}\right\} \tag{3.1.17}$$

The next Fourier inversion integral with respect to the parameter ξ is reduced to the semi-infinite integral as

$$I_{xx}^{*} = -\frac{1}{2\pi s^2}\int_0^\infty \left\{\sqrt{\xi^2 + (s/c_d)^2} - \frac{(s/c_d)^2}{\sqrt{\xi^2 + (s/c_d)^2}}\right\}\exp\left\{-|y|\sqrt{\xi^2 + (s/c_d)^2}\right\}\cos(\xi x)d\xi$$
$$+\frac{1}{2\pi s^2}\int_0^\infty \left\{\sqrt{\xi^2 + (s/c_s)^2} - \frac{(s/c_s)^2}{\sqrt{\xi^2 + (s/c_s)^2}}\right\}\exp\left\{-|y|\sqrt{\xi^2 + (s/c_s)^2}\right\}\cos(\xi x)d\xi \tag{3.1.18}$$

We have already had the integration formula (2.4.13). Here, it is recited again,

$$\int_0^\infty \frac{1}{\sqrt{x^2 + a^2}}\exp\left(-c\sqrt{x^2 + a^2}\right)\cos(bx)dx = K_0\left(a\sqrt{b^2 + c^2}\right) \tag{3.1.19}$$

where $K_0(.)$ is the zeroth order modified Bessel function of the second kind. This formula is applicable only for the second term in the bracket; another formula is necessary for the first term. Differentiating the formula (3.1.19) twice with respect to the parameter "c," we have the new formula for our use,

$$\int_0^\infty \sqrt{x^2 + a^2}\exp\left(-c\sqrt{x^2 + a^2}\right)\cos(bx)dx$$
$$= \frac{a^2c^2}{b^2 + c^2}K_0\left(a\sqrt{b^2 + c^2}\right) - \frac{a}{\sqrt{b^2 + c^2}}\frac{b^2 - c^2}{b^2 + c^2}K_1\left(a\sqrt{b^2 + c^2}\right) \tag{3.1.20}$$

Applying these two formulas to the integrals in Eq. (3.1.18), we have

$$\begin{aligned} I_{xx}^{*} = &+ \frac{1}{2\pi c_d^2} \frac{x^2}{r^2} K_0(rs/c_d) + \frac{1}{2\pi s r c_d} \frac{x^2 - y^2}{r^2} K_1(rs/c_d) \\ &- \frac{1}{2\pi c_s^2} \frac{x^2}{r^2} K_0(rs/c_s) - \frac{1}{2\pi s r c_s} \frac{x^2 - y^2}{r^2} K_1(rs/c_s) \end{aligned} \tag{3.1.21}$$

where the radial distance from the source is defined by

$$r = \sqrt{x^2 + y^2} \tag{3.1.22}$$

The last inversion is the Laplace inversion. Its inversion is expressed symbolically as

$$\begin{aligned} I_{xx} = &+ \frac{1}{2\pi c_d^2} \frac{x^2}{r^2} L^{-1}[K_0(rs/c_d)] + \frac{1}{2\pi r c_d} \frac{x^2 - y^2}{r^2} L^{-1}\left[\frac{1}{s} K_1(rs/c_d)\right] \\ &- \frac{1}{2\pi c_s^2} \frac{x^2}{r^2} L^{-1}[K_0(rs/c_s)] - \frac{1}{2\pi r c_s} \frac{x^2 - y^2}{r^2} L^{-1}\left[\frac{1}{s} K_1(rs/c_s)\right] \end{aligned} \tag{3.1.23}$$

We have already used the inversion formula (2.4.16),

$$L^{-1}[K_0(as)] = \frac{H(t-a)}{\sqrt{t^2 - a^2}} \tag{3.1.24}$$

One more inversion formula that includes the modified Bessel function of the second kind is that (Erdélyi 1954, vol. I, pp. 277, 11),

$$L^{-1}\left[\frac{1}{s} K_1(as)\right] = H(t-a) \frac{\sqrt{t^2 - a^2}}{a} \tag{3.1.25}$$

Applying these two inversion formulas to Eq. (3.1.23), we have the final form for I_{xx},

$$\begin{aligned} I_{xx}(x,y,t) = &+ \frac{H(t - r/c_d)}{2\pi c_d} \left\{ \frac{x^2}{r^2} \frac{1}{\sqrt{(c_d t)^2 - r^2}} + \frac{x^2 - y^2}{r^4} \sqrt{(c_d t)^2 - r^2} \right\} \\ &- \frac{H(t - r/c_s)}{2\pi c_s} \left\{ \frac{x^2}{r^2} \frac{1}{\sqrt{(c_s t)^2 - r^2}} + \frac{x^2 - y^2}{r^4} \sqrt{(c_s t)^2 - r^2} \right\} \end{aligned} \tag{3.1.26}$$

(3) Inversion of $\bar{\bar{I}}^*_{yy}$

The same inversion procedure as that for $\bar{\bar{I}}^*_{xx}$ is applied to the inversion of $\bar{\bar{I}}^*_{yy}$, and its final form is

$$I_{yy}(x,y,t) = +\frac{H(t-r/c_d)}{2\pi c_d}\left\{\frac{y^2}{r^2}\frac{1}{\sqrt{(c_d t)^2 - r^2}} + \frac{y^2-x^2}{r^4}\sqrt{(c_d t)^2 - r^2}\right\}$$
$$-\frac{H(t-r/c_s)}{2\pi c_s}\left\{\frac{y^2}{r^2}\frac{1}{\sqrt{(c_s t)^2 - r^2}} + \frac{y^2-x^2}{r^4}\sqrt{(c_s t)^2 - r^2}\right\} \quad (3.1.27)$$

The difference between I_{xx} in Eq. (3.1.26) and I_{yy} in Eq. (3.1.27) should be noticed. The difference is in the space variables only. If we replace x with y in I_{xx} of Eq. (3.1.26), we could have I_{yy} of Eq. (3.1.27). This is easily anticipated from the comparison of the transformed forms in Eqs. (3.1.10) and (3.1.12). In these equations, each transform parameter corresponds to the original space variable. If we exchange ξ with η in Eq. (3.1.10), it yields Eq. (3.1.12). Similarly, in the original space we can exchange the space variables.

(4) Inversion of $\bar{\bar{I}}^*_{xy}$

The inversion technique for $\bar{\bar{I}}^*_{xy}$ is essentially the same as that for the former two cases, but the integration formulas are slightly different. The Fourier inversion integral with respect to the parameter η is reduced to the semi-infinite integral

$$\bar{I}^*_{xy} = -\frac{1}{2\pi s^2}\int_{-\infty}^{\infty}\xi\eta\left(\frac{1}{\alpha_d^2}-\frac{1}{\alpha_s^2}\right)\exp(-i\eta y)d\eta = \frac{i\xi}{\pi s^2}\int_0^{\infty}\left(\frac{\eta}{\alpha_d^2}-\frac{\eta}{\alpha_s^2}\right)\sin(\eta y)d\eta$$
$$(3.1.28)$$

The integration formula (Erdélyi 1954, vol. I, pp. 65, 15)

$$\int_0^{\infty}\frac{x}{x^2+a^2}\sin(bx)dx = \frac{\pi}{2}\mathrm{sgn}(b)\exp(-a|b|), \quad (3.1.29)$$

is applied to Eq. (3.1.28). It then follows that

$$\bar{I}^*_{xy} = \frac{i\xi}{2s^2}\mathrm{sgn}(y)\left[\exp\left\{-|y|\sqrt{\xi^2+(s/c_d)^2}\right\} - \exp\left\{-|y|\sqrt{\xi^2+(s/c_s)^2}\right\}\right] \quad (3.1.30)$$

where sgn(.) is the sign function defined by

$$\operatorname{sgn}(x) = \begin{cases} +1\,; & x > 0 \\ -1\,; & x < 0 \end{cases} \tag{3.1.31}$$

The next Fourier inversion integral with respect to the parameter ξ

$$I_{xy}^* = \frac{\operatorname{sgn}(y)}{2s^2}\frac{1}{2\pi}\int_{-\infty}^{+\infty}(+i\xi)\left[\exp\left\{-|y|\sqrt{\xi^2+(s/c_d)^2}\right\} - \exp\left\{-|y|\sqrt{\xi^2+(s/c_s)^2}\right\}\right]e^{-i\xi x}d\xi \tag{3.1.32}$$

is also reduced to the semi-infinite integral

$$I_{xy}^* = \frac{\operatorname{sgn}(y)}{2\pi s^2}\int_0^{\infty}\xi\left[\exp\left\{-|y|\sqrt{\xi^2+(s/c_d)^2}\right\} - \exp\left\{-|y|\sqrt{\xi^2+(s/c_s)^2}\right\}\right]\sin(\xi x)d\xi \tag{3.1.33}$$

Applying the integration formula (Erdélyi 1954, vol. I, pp. 75, 35)

$$\int_0^{\infty} x\exp\left(-b\sqrt{x^2+a^2}\right)\sin(cx)dx = \frac{a^2bc}{b^2+c^2}K_2\left(a\sqrt{b^2+c^2}\right) \tag{3.1.34}$$

to the integral in Eq. (3.1.33), we have for I_{xy}^*

$$I_{xy}^* = \frac{xy}{2\pi r^2}\left\{\frac{1}{c_d^2}K_2(rs/c_d) - \frac{1}{c_s^2}K_2(rs/c_s)\right\}. \tag{3.1.35}$$

Lastly, the Laplace inversion formula (Erdélyi 1954, vol. I, pp. 277, 12)

$$L^{-1}[K_2(bs)] = H(t-b)\left(\frac{1}{\sqrt{t^2-b^2}} + \frac{2}{b^2}\sqrt{t^2-b^2}\right) \tag{3.1.36}$$

is applied to Eq. (3.1.35). The final form of I_{xy} is given by

$$\begin{aligned} I_{xy} = &+\frac{xy}{2\pi r^4}\left\{\frac{(r/c_d)^2}{\sqrt{t^2-(r/c_d)^2}} + 2\sqrt{t^2-(r/c_d)^2}\right\}H(t-r/c_d) \\ &-\frac{xy}{2\pi r^4}\left\{\frac{(r/c_s)^2}{\sqrt{t^2-(r/c_s)^2}} + 2\sqrt{t^2-(r/c_s)^2}\right\}H(t-r/c_s). \end{aligned} \tag{3.1.37}$$

(5) Green's dyadic

We have obtained the exact expressions for the three fundamentals, I_{xx}, I_{xy} and I_{yy}, and the unified expression for these is given by

$$\begin{aligned} I_{ij}(x,y,t) = &+\frac{H(t-r/c_d)}{2\pi c_d}\left\{\frac{x_i x_j}{r^2}\frac{1}{\sqrt{(c_d t^2)-r^2}}+\left(\frac{2x_i x_j}{r^2}-\delta_{ij}\right)\frac{\sqrt{(c_d t^2)-r^2}}{r^2}\right\} \\ &-\frac{H(t-r/c_s)}{2\pi c_s}\left\{\frac{x_i x_j}{r^2}\frac{1}{\sqrt{(c_s t)^2-r^2}}+\left(\frac{2x_i x_j}{r^2}-\delta_{ij}\right)\frac{\sqrt{(c_s t)^2-r^2}}{r^2}\right\} \end{aligned} \tag{3.1.38}$$

where the subscripts i and j stand for x and y, and it should be understood that $x_x \equiv x$, $x_y \equiv y$. Further, δ_{ij} is Kronecker's delta defined by

$$\delta_{ij} = \begin{cases} 1; & i=j \\ 0; & i \neq j \end{cases} \tag{3.1.39}$$

Thus, we have just obtained the exact expressions for the four fundamental functions. Substituting Eqs. (3.1.15) and (3.1.38) into Eq. (3.1.13), the displacement in the actual space (x, y, t) is expressed as

$$\begin{aligned} u_i(x,y,t) = \sum_{j=x,y} P_j \Bigg[&\frac{H(t-r/c_d)}{2\pi c_d}\left\{\frac{x_i x_j}{r^2}\frac{1}{\sqrt{(c_d t)^2-r^2}}+\left(\frac{2x_i x_j}{r^2}-\delta_{ij}\right)\frac{\sqrt{(c_d t)^2-r^2}}{r^2}\right\} \\ &-\frac{H(t-r/c_s)}{2\pi c_s}\left\{\left(\frac{x_i x_j}{r^2}-\delta_{ij}\right)\frac{1}{\sqrt{(c_s t)^2-r^2}}+\left(\frac{2x_i x_j}{r^2}-\delta_{ij}\right)\frac{\sqrt{(c_s t)^2-r^2}}{r^2}\right\}\Bigg] \end{aligned} \tag{3.1.40}$$

In this equation, the operation of Heaviside's unit step function determines the disturbed region. $H(t - r/c_d)$ shows a circular cylindrical region disturbed by the dilatational wave which expands with velocity c_d, while $H(t - r/c_s)$ does that by the shear wave with c_s. These two waves are the basic disturbances in the 2D dynamic deformation and are shown in Fig. 3.1.

Now, we rewrite the above Eq. (3.1.40) as

$$u_i(x,y,t) = \sum_{j=x,y} P_j G_{ij}(x,y,t) \tag{3.1.41}$$

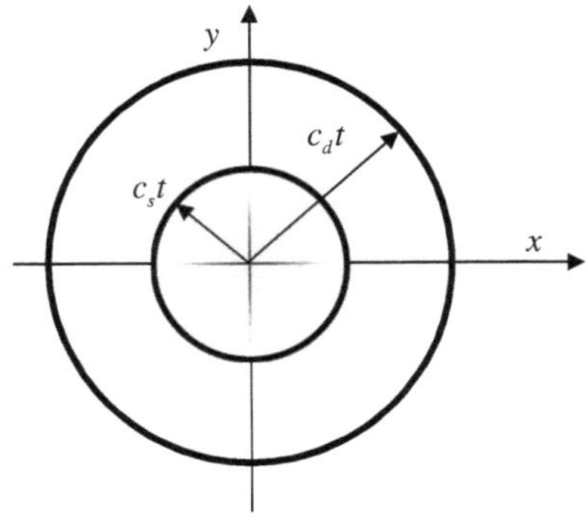

Fig. 3.1 Radiation of P and SV waves from a point source

The function $G_{ij}(x, y, t)$ is called "Green's dyadic" and it expresses the displacement component in the i-axis direction due to the unit body force in the j-axis direction. The explicit form of the dyadic G_{ij} is given by

$$G_{ij}(x,y,t) = \frac{H(t-r/c_d)}{2\pi c_d}\left\{\frac{x_i x_j}{r^2}\frac{1}{\sqrt{(c_d t)^2 - r^2}} + \left(\frac{2x_i x_j}{r^2} - \delta_{ij}\right)\frac{\sqrt{(c_d t)^2 - r^2}}{r^2}\right\}$$
$$- \frac{H(t-r/c_s)}{2\pi c_s}\left\{\left(\frac{x_i x_j}{r^2} - \delta_{ij}\right)\frac{1}{\sqrt{(c_s t)^2 - r^2}} + \left(\frac{2x_i x_j}{r^2} - \delta_{ij}\right)\frac{\sqrt{(c_s t)^2 - r^2}}{r^2}\right\} \quad (3.1.42)$$

3.2 2D Time-Harmonic Source

When the source is harmonically vibrating with frequency ω, the non-homogeneous body force term in the displacement equation (3.1.1) is replaced with the time-harmonic source, i.e.

$$(\gamma^2 - 1)\frac{\partial}{\partial x}\left(\frac{\partial u_x}{\partial x} + \frac{\partial u_y}{\partial y}\right) + \frac{\partial^2 u_x}{\partial x^2} + \frac{\partial^2 u_x}{\partial y^2} = \frac{1}{c_s^2}\frac{\partial^2 u_x}{\partial t^2} - \frac{Q_x}{c_s^2}\delta(x)\delta(y)\exp(i\omega t) \quad (3.2.1a)$$

$$(\gamma^2 - 1)\frac{\partial}{\partial y}\left(\frac{\partial u_x}{\partial x} + \frac{\partial u_y}{\partial y}\right) + \frac{\partial^2 u_y}{\partial x^2} + \frac{\partial^2 u_y}{\partial y^2} = \frac{1}{c_s^2}\frac{\partial^2 u_y}{\partial t^2} - \frac{Q_y}{c_s^2}\delta(x)\delta(y)\exp(i\omega t) \quad (3.2.1b)$$

where Q_i is the magnitude of the source component.

It is possible to apply the integral transform method for obtaining the time-harmonic Green's dyadic. However, we do not employ the method here since the cumbersome complex integration must be discussed as that for the time-harmonic Green's function in Sect. 2.5. We employ a simpler way, i.e. the convolution integral. The time-harmonic response can be obtained by the convolution integral of the impulsive response as

$$u_i(x,y,t) = \lim_{t\to\infty} \int_0^t u_i^{(impulse)}(x,y,t') \exp\{+i\omega(t-t')\}dt' \tag{3.2.2}$$

where $u_i^{(impulse)}(x,y,t')$ is the impulsive response given by Eq. (3.1.41) with (3.1.42), where the source magnitude P_i for the impulsive solution is replaced with Q_i. In this convolution integral, the time-harmonic function, exp($+i\omega t$), is excluded from the limit, but we keep the upper limit of the integral be infinite, since the steady-state response takes place long time after the initial disturbance. Then, the convolution integral for the time-harmonic response takes the form

$$u_i(x,y,t) = \exp(+i\omega t) \int_0^{t\to\infty} u_i^{(impulse)}(x,y,t') \exp(-i\omega t')dt' \tag{3.2.3}$$

Substituting Eqs. (3.1.41) with (3.1.42) into Eq. (3.2.3), we have

$$u_i(x,y,t) = \exp(+i\omega t)\sum_{j=x,y} Q_j \left[\frac{1}{2\pi c_d} \left\{ \begin{array}{l} \frac{x_i x_j}{r^2} \int_{r/c_d}^{\infty} \frac{\exp(-i\omega t')}{\sqrt{(c_d t')^2 - r^2}} dt' \\ + \left(\frac{2x_i x_j}{r^2} - \delta_{ij}\right) \frac{1}{r^2} \int_{r/c_d}^{\infty} \sqrt{(c_d t)^2 - r^2} \exp(-i\omega t')dt' \end{array} \right\} \right.$$
$$\left. - \frac{1}{2\pi c_s} \left\{ \begin{array}{l} \left(\frac{x_i x_j}{r^2} - \delta_{ij}\right) \int_{r/c_s}^{\infty} \frac{\exp(-i\omega t')}{\sqrt{(c_s t')^2 - r^2}} dt' \\ + \left(\frac{2x_i x_j}{r^2} - \delta_{ij}\right) \frac{1}{r^2} \int_{r/c_s}^{\infty} \sqrt{(c_s t)^2 - r^2} \exp(-i\omega t')dt' \end{array} \right\} \right] \tag{3.2.4}$$

Two integrals in the above equation are the integral representations of Hankel function of the second kind (Watson 1966, p. 169). They are

$$\begin{aligned} \int_a^{\infty} \frac{\exp(-ixz)}{\sqrt{x^2 - a^2}} dx &= -\frac{i\pi}{2} H_0^{(2)}(az) \\ \int_a^{\infty} \sqrt{x^2 - a^2} \exp(-ixz)dx &= +\frac{i\pi a}{2z} H_1^{(2)}(az) \end{aligned} \tag{3.2.5}$$

Applying these formulas to the integrals in Eq. (3.2.4), the time-harmonic response is expressed as

$$u_i(x,y,t) = \sum_{j=x,y} Q_j g_{ij}(x,y)\exp(+i\omega t) \tag{3.2.6}$$

where $g_{ij}(x, y)$ is the Green's dyadic for the time-harmonic response and is given by

$$\begin{aligned} g_{ij}(x,y) = -\frac{i}{4}\Bigg[&\frac{1}{c_d^2}\left\{\frac{x_ix_j}{r^2}H_0^{(2)}(r\omega/c_d) + \left(\delta_{ij} - \frac{2x_ix_j}{r^2}\right)\frac{c_d}{r\omega}H_1^{(2)}(r\omega/c_d)\right\} \\ &- \frac{1}{c_s^2}\left\{\left(\frac{x_ix_j}{r^2} - \delta_{ij}\right)H_0^{(2)}(r\omega/c_s) + \left(\delta_{ij} - \frac{2x_ix_j}{r^2}\right)\frac{c_s}{r\omega}H_1^{(2)}(r\omega/c_s)\right\}\Bigg] \end{aligned} \tag{3.2.7}$$

Then, Eq. (3.2.6) with Eq. (3.2.7) gives the particular solution of the displacement equation (3.2.1).

If the frequency is negative, $\omega \to -\omega(=\omega e^{-\pi i})$, and the time factor is $\exp(-i\omega t)$, we apply the formula (Watson 1966, p. 75)

$$H_n^{(2)}(xe^{-\pi i}) = (-1)^{n+1}H_n^{(1)}(x) \tag{3.2.8}$$

to the Hankel function in Eq. (3.2.7), and then the displacement is expressed as

$$u_i(x,y,t) = \sum_{j=x,y} Q_j g_{ij}(x,y)\exp(-i\omega t) \tag{3.2.9}$$

and its dyadic g_{ij} (x, y) is given by

$$\begin{aligned} g_{ij}(x,y) = +\frac{i}{4}\Bigg[&\frac{1}{c_d^2}\left\{\frac{x_ix_j}{r^2}H_0^{(1)}(r\omega/c_d) + \left(\delta_{ij} - \frac{2x_ix_j}{r^2}\right)\frac{c_d}{r\omega}H_1^{(1)}(r\omega/c_d)\right\} \\ &- \frac{1}{c_s^2}\left\{\left(\frac{x_ix_j}{r^2} - \delta_{ij}\right)H_0^{(1)}(r\omega/c_s) + \left(\delta_{ij} - \frac{2x_ix_j}{r^2}\right)\frac{c_s}{r\omega}H_1^{(1)}(r\omega/c_s)\right\}\Bigg] \end{aligned} \tag{3.2.10}$$

where $H_n^{(1)}(.)$ is the n-th order Hankel function of the first kind and defined by

$$H_n^{(1)}(x) = J_n(x) + iY_n(x) \tag{3.2.11}$$

3.3 2D Static Source

If we consider to derive the static solution from the time-harmonic solution, we have to take the limit $\omega \to 0$ in the dyadic given by Eq. (3.2.7) or (3.2.10). However, the limit does not exist since the static displacement in the 2D plane deformation does not vanish at infinity. We have to start from the original differential equation and solve it. The static source is a non-time-dependent point body force and the displacement equation which has no inertia term is given by

$$\begin{aligned}(\gamma^2-1)\frac{\partial}{\partial x}\left(\frac{\partial u_x}{\partial x}+\frac{\partial u_y}{\partial y}\right)+\frac{\partial^2 u_x}{\partial x^2}+\frac{\partial^2 u_x}{\partial y^2}&=-\frac{S_x}{c_s^2}\delta(x)\delta(y)\\(\gamma^2-1)\frac{\partial}{\partial y}\left(\frac{\partial u_x}{\partial x}+\frac{\partial u_y}{\partial y}\right)+\frac{\partial^2 u_y}{\partial x^2}+\frac{\partial^2 u_y}{\partial y^2}&=-\frac{S_y}{c_s^2}\delta(x)\delta(y)\end{aligned} \tag{3.3.1}$$

where the static body force is placed at the coordinate origin and its magnitude is S_i. For these coupled differential equations, the convergence condition at infinity

$$u_j\Big|_{\sqrt{x^2+y^2}\to\infty}=\frac{\partial u_j}{\partial x}\Big|_{\sqrt{x^2+y^2}\to\infty}=\frac{\partial u_j}{\partial y}\Big|_{\sqrt{x^2+y^2}\to\infty}=0 \tag{3.3.2}$$

is assumed in order to apply the Fourier transform (but, the final solution does not satisfy the first condition in the above equation). We should understand that the convergence condition guarantees the application of the Fourier transform.

In order to obtain the static Green's dyadic, we apply the double Fourier transform with respect to two space variables. Applying the double Fourier transform defined by Eq. (3.1.3) to the displacement equations (3.3.1), we have the algebraic equations for the transformed displacement components $\tilde{\tilde{u}}_j$,

$$\begin{aligned}-i\xi(\gamma^2-1)(-i\xi\tilde{\tilde{u}}_x-i\eta\tilde{\tilde{u}}_y)-(\xi^2+\eta^2)\tilde{\tilde{u}}_x&=-S_x/c_s^2\\-i\eta(\gamma^2-1)(-i\xi\tilde{\tilde{u}}_x-i\eta\tilde{\tilde{u}}_y)-(\xi^2+\eta^2)\tilde{\tilde{u}}_y&=-S_y/c_s^2\end{aligned} \tag{3.3.3}$$

and their solutions are

$$\begin{aligned}\tilde{\tilde{u}}_x&=\frac{S_x}{c_s^2}\left\{\frac{1}{\xi^2+\eta^2}-\frac{\gamma^2-1}{\gamma^2}\frac{\xi^2}{(\xi^2+\eta^2)^2}\right\}-\frac{S_y}{c_s^2}\frac{\gamma^2-1}{\gamma^2}\frac{\xi\eta}{(\xi^2+\eta^2)^2}\\\tilde{\tilde{u}}_y&=-\frac{S_x}{c_s^2}\frac{\gamma^2-1}{\gamma^2}\frac{\xi\eta}{(\xi^2+\eta^2)^2}+\frac{S_y}{c_s^2}\left\{\frac{1}{\xi^2+\eta^2}-\frac{\gamma^2-1}{\gamma^2}\frac{\eta^2}{(\xi^2+\eta^2)^2}\right\}\end{aligned} \tag{3.3.4}$$

Examining the above equations, we learn that four inversions are needed for the full Fourier inversion since the displacement is given by

$$\begin{aligned}\tilde{\tilde{u}}_x&=\frac{S_x}{c_s^2}\left(\tilde{\tilde{I}}_0-\frac{\gamma^2-1}{\gamma^2}\tilde{\tilde{I}}_{xx}\right)-\frac{S_y}{c_s^2}\frac{\gamma^2-1}{\gamma^2}\tilde{\tilde{I}}_{xy}\\\tilde{\tilde{u}}_y&=-\frac{S_x}{c_s^2}\frac{\gamma^2-1}{\gamma^2}\tilde{\tilde{I}}_{xy}+\frac{S_y}{c_s^2}\left(\tilde{\tilde{I}}_0-\frac{\gamma^2-1}{\gamma^2}\tilde{\tilde{I}}_{yy}\right)\end{aligned} \tag{3.3.5}$$

where the four fundamentals to be inverted are

$$\bar{\bar{I}}_0 = \frac{1}{\xi^2 + \eta^2}, \quad \bar{\bar{I}}_{xx} = \frac{\xi^2}{(\xi^2 + \eta^2)^2}, \quad \bar{\bar{I}}_{yy} = \frac{\eta^2}{(\xi^2 + \eta^2)^2}, \quad \bar{\bar{I}}_{xy} = \frac{\xi\eta}{(\xi^2 + \eta^2)^2} \tag{3.3.6}$$

The inversion for each fundamental is carried out in the following subsections.

(1) Inversion of $\bar{\bar{I}}_0$

We have already inverted this function in Sect. 2.3. The result is given by Eq. (2.3.21), i.e.

$$\bar{\bar{\phi}} = \frac{S}{\xi^2 + \eta^2} \quad \Rightarrow \quad \phi = -\frac{S}{4\pi}\log(x^2 + y^2) + \text{const.} \tag{3.3.7}$$

Applying this result, we have the inversion for $\bar{\bar{I}}_0$ as

$$I_0 = -\frac{1}{4\pi}\log(x^2 + y^2) + \text{const.} \tag{3.3.8}$$

(2) Inversion of $\bar{\bar{I}}_{xx}$

The Fourier inversion integral with respect to the parameter η is reduced to the semi-infinite integral,

$$\bar{I}_{xx} = \frac{1}{2\pi}\int_{-\infty}^{+\infty} \frac{\xi^2}{(\xi^2 + \eta^2)^2}\exp(-i\eta y)d\eta = \frac{\xi^2}{\pi}\int_0^{\infty} \frac{1}{(\xi^2 + \eta^2)^2}\cos(\eta y)d\eta \tag{3.3.9}$$

The integration formula (Gradshteyn and Ryzhik 1980, pp. 449, 3.729 1)

$$\int_0^{\infty} \frac{1}{(x^2 + b^2)^2}\cos(ax)dx = \frac{\pi}{4b^3}(1 + ab)\exp(-ab); \quad a > 0, \quad b > 0 \tag{3.3.10}$$

is applied to the far right integral in Eq. (3.3.9). We have for $\bar{I}_{xx}$

$$\bar{I}_{xx} = \frac{\xi^2}{\pi}\int_0^{\infty} \frac{1}{(\xi^2 + \eta^2)^2}\cos(\eta y)d\eta = \left(\frac{1}{4|\xi|} + \frac{|y|}{4}\right)\exp(-|\xi||y|) \tag{3.3.11}$$

The last Fourier inversion with respect to the parameter ξ is given by the sum of two integrals,

$$I_{xx} = \frac{1}{2\pi}\int_{-\infty}^{+\infty}\left(\frac{1}{4|\xi|}+\frac{|y|}{4}\right)\exp(-|\xi||y|)\exp(-i\xi x)d\xi$$
$$= \frac{1}{4\pi}\int_{0}^{\infty}\frac{1}{\xi}\exp(-\xi|y|)\cos(\xi x)d\xi + \frac{|y|}{4\pi}\int_{0}^{\infty}\exp(-\xi|y|)\cos(\xi x)d\xi \tag{3.3.12}$$

Inspecting the two integrals in the last equation, the second integral can be evaluated by applying the formula (2.3.16) in Chap. 2. The first integral, however, has no formula since its integrand has the first order singularity at $\xi = 0$ and it is impossible to evaluate the integral in this form. We extract the first integral

$$I_{xx}^{(1)} = \frac{1}{4\pi}\int_{0}^{\infty}\frac{1}{\xi}\exp(-\xi|y|)\cos(\xi x)d\xi \tag{3.3.13}$$

and consider its derivative with respect to each space variable as

$$\frac{\partial I_{xx}^{(1)}}{\partial x} = -\frac{1}{4\pi}\int_{0}^{\infty}\exp(-\xi|y|)\sin(\xi x)d\xi$$
$$\frac{\partial I_{xx}^{(1)}}{\partial |y|} = -\frac{1}{4\pi}\int_{0}^{\infty}\exp(-\xi|y|)\cos(\xi x)d\xi \tag{3.3.14}$$

These two integrals are easily evaluated by the formulas (2.3.16) to yield

$$\frac{\partial I_{xx}^{(1)}}{\partial x} = -\frac{1}{4\pi}\frac{x}{x^2+y^2}, \quad \frac{\partial I_{xx}^{(1)}}{\partial |y|} = -\frac{1}{4\pi}\frac{|y|}{x^2+y^2} \tag{3.3.15}$$

Then, we return the two derivatives to the original one by the integration with respect to each space variable. Two expressions for the single $I_{xx}^{(1)}$ are obtained as

$$I_{xx}^{(1)} = -\frac{1}{4\pi}\int\frac{x}{x^2+y^2}dx = -\frac{1}{8\pi}\log(x^2+y^2)+C_1(y)$$
$$I_{xx}^{(1)} = -\frac{1}{4\pi}\int\frac{|y|}{x^2+y^2}d|y| = -\frac{1}{8\pi}\log(x^2+y^2)+C_2(x) \tag{3.3.16}$$

These two equations must be equal. This condition is satisfied by setting $C_1(y) = C_2(x)$ and thus the two functions must be a pure constant,

$$C_1(y) = C_2(x) = \text{const.} \tag{3.3.17}$$

Then, we have for $I_{xx}^{(1)}$,

$$I_{xx}^{(1)} = -\frac{1}{8\pi}\log(x^2 + y^2) + \text{const.} \tag{3.3.18}$$

The second integral in Eq. (3.3.12) is easily evaluated by the formula (2.3.16) and it yields

$$I_{xx}^{(2)} = \frac{|y|}{4\pi}\int_0^\infty \exp(-\xi|y|)\cos(\xi x)d\xi = \frac{1}{4\pi}\frac{y^2}{x^2 + y^2} \tag{3.3.19}$$

Finally, substituting Eqs. (3.3.18) and (3.3.19) into (3.3.12), we have the inversion

$$I_{xx} = -\frac{1}{8\pi}\log(x^2 + y^2) + \frac{1}{4\pi}\frac{y^2}{x^2 + y^2} + \text{const.} \tag{3.3.20}$$

Further, since the constant in the above equation is arbitrary, we can rewrite I_{xx} as

$$I_{xx} = -\frac{1}{8\pi}\log(x^2 + y^2) - \frac{1}{4\pi}\frac{x^2}{x^2 + y^2} + \text{const.} \tag{3.3.21}$$

(3) Inversion of $\tilde{\tilde{I}}_{yy}$

Rewriting the transformed equation in Eq. (3.3.5), we learn that this inversion can be decomposed into the sum of the former two functions, as

$$\tilde{\tilde{I}}_{yy} = \frac{1}{\xi^2 + \eta^2} - \frac{\xi^2}{(\xi^2 + \eta^2)^2} = \tilde{\tilde{I}}_0 - \tilde{\tilde{I}}_{xx} \tag{3.3.22}$$

Then, we can apply the former results, Eqs. (3.3.8) and (3.3.21), to the above equation,

$$\begin{aligned} I_{yy} &= I_0 - I_{xx} \\ &= -\frac{1}{4\pi}\log(x^2 + y^2) + \text{const.} + \frac{1}{8\pi}\log(x^2 + y^2) - \frac{1}{4\pi}\frac{y^2}{x^2 + y^2} + \text{const.} \end{aligned} \tag{3.3.23}$$

The final form of I_{yy} is given by

$$I_{yy} = -\frac{1}{8\pi}\log(x^2 + y^2) - \frac{1}{4\pi}\frac{y^2}{x^2 + y^2} + \text{const.} \tag{3.3.24}$$

This can be derived from I_{xx} in Eq. (3.3.21) by the exchange of variables. Changing x with y in Eq. (3.3.21), we have the inversion for I_{yy}.

(4) Inversion of $\bar{\bar{I}}_{xy}$

The Fourier inversion integral with respect to the parameter η is reduced to the semi-infinite integral,

$$\bar{I}_{xy} = \frac{1}{2\pi}\int\limits_{-\infty}^{+\infty} \frac{\xi\eta}{(\xi^2 + \eta^2)^2}\exp(-i\eta y)d\eta = -\frac{i\xi}{\pi}\int\limits_{0}^{\infty} \frac{\eta}{(\xi^2 + \eta^2)^2}\sin(\eta y)d\eta \tag{3.3.25}$$

The integration formula (Gradshteyn and Ryzhik 1980, pp. 449, 3.729 2),

$$\int\limits_{0}^{\infty} \frac{x}{(x^2 + b^2)^2}\sin(ax)dx = \frac{\pi a}{4b}\exp(-ab) \tag{3.3.26}$$

is applied to the last integral in Eq. (3.3.25). We have

$$\bar{I}_{xy} = -\frac{iy}{4}\operatorname{sgn}(\xi)\exp(-|\xi||y|) \tag{3.3.27}$$

The next Fourier inversion integral with respect to the parameter ξ is reduced to the simple integral,

$$\begin{aligned} I_{xy} &= -\frac{iy}{4}\frac{1}{2\pi}\int\limits_{-\infty}^{+\infty} \operatorname{sgn}(\xi)\exp(-|\xi||y|)\exp(-i\xi x)d\xi \\ &= -\frac{y}{4\pi}\int\limits_{0}^{\infty} \exp(-\xi|y|)\sin(\xi x)d\xi \end{aligned} \tag{3.3.28}$$

Applying the formula (2.3.16), we have for I_{xy},

$$I_{xy} = -\frac{1}{4\pi}\frac{xy}{x^2 + y^2} \tag{3.3.29}$$

(5) 2D Kelvin's solution

The four inversions in Eq. (3.3.5) have thus been completed. Since the displacement in the original space is given by

$$u_x = \frac{S_x}{c_s^2}\left(I_0 - \frac{\gamma^2-1}{\gamma^2}I_{xx}\right) - \frac{S_y}{c_s^2}\frac{\gamma^2-1}{\gamma^2}I_{xy}$$
$$u_y = -\frac{S_x}{c_s^2}\frac{\gamma^2-1}{\gamma^2}I_{xy} + \frac{S_y}{c_s^2}\left(I_0 - \frac{\gamma^2-1}{\gamma^2}I_{yy}\right) \tag{3.3.30}$$

we substitute I_0 and I_{ij} given by Eqs. (3.3.8), (3.3.21), (3.3.24) and (3.3.29) into the above Eq. (3.3.30). The final form for the displacement is given by

$$u_x = \frac{S_x}{4\pi c_d^2}\left\{-(\gamma^2+1)\log(r) + (\gamma^2-1)\left(\frac{x}{r}\right)^2\right\} + \frac{S_y}{4\pi c_d^2}(\gamma^2-1)\frac{xy}{r^2}$$
$$u_y = \frac{S_x}{4\pi c_d^2}(\gamma^2-1)\frac{xy}{r^2} + \frac{S_y}{4\pi c_d^2}\left\{-(\gamma^2+1)\log(r) + (\gamma^2-1)\left(\frac{y}{r}\right)^2\right\} \tag{3.3.31}$$

where $c_d = \gamma c_s$ defined by Eq. (3.8) is used. The constant term is omitted since it gives a simple rigid motion with no strain. The reader should notice that the constant term in the displacement breaks the applicability of the Fourier transform; thus the last inversion integral with respect to the parameter ξ has the singular point.

The expression for the displacement is rewritten in terms of the dyadic,

$$u_x = S_x g_{xx}^{(static)}(x,y) + S_y g_{xy}^{(static)}(x,y)$$
$$u_y = S_x g_{xy}^{(static)}(x,y) + S_y g_{yy}^{(static)}(x,y) \tag{3.3.32}$$

where the dyadic for the static source is given by

$$g_{xx}^{(static)}(x,y) = \frac{1}{4\pi c_d^2}\left\{-(\gamma^2+1)\log(r) + (\gamma^2-1)\left(\frac{x}{r}\right)^2\right\}$$
$$g_{xy}^{(static)}(x,y) = \frac{1}{4\pi c_d^2}(\gamma^2-1)\frac{xy}{r^2} \tag{3.3.33}$$
$$g_{yy}^{(static)}(x,y) = \frac{1}{4\pi c_d^2}\left\{-(\gamma^2+1)\log(r) + (\gamma^2-1)\left(\frac{y}{r}\right)^2\right\}$$

This static Green's dyadic for the plane deformation is called the "two dimensional Kelvin's solution."

3.4 3D Impulsive Source

An impulsive point source placed at the coordinate origin is expressed by the body force

$$\begin{pmatrix} B_x \\ B_y \\ B_z \end{pmatrix} = \begin{pmatrix} P_x \\ P_y \\ P_z \end{pmatrix} \delta(x)\delta(y)\delta(z)\delta(t), \tag{3.4.1}$$

where P_i is the magnitude in the i-direction. The displacement equation with this impulsive source is given by

$$\begin{aligned}
(\gamma^2-1)\frac{\partial}{\partial x}\left(\frac{\partial u_x}{\partial x}+\frac{\partial u_y}{\partial y}+\frac{\partial u_z}{\partial z}\right)+\frac{\partial^2 u_x}{\partial x^2}+\frac{\partial^2 u_x}{\partial y^2}+\frac{\partial^2 u_x}{\partial z^2}=\frac{1}{c_s^2}\frac{\partial^2 u_x}{\partial t^2}-\frac{P_x}{c_s^2}\delta(x)\delta(y)\delta(z)\delta(t)\\
(\gamma^2-1)\frac{\partial}{\partial y}\left(\frac{\partial u_x}{\partial x}+\frac{\partial u_y}{\partial y}+\frac{\partial u_z}{\partial z}\right)+\frac{\partial^2 u_y}{\partial x^2}+\frac{\partial^2 u_y}{\partial y^2}+\frac{\partial^2 u_y}{\partial z^2}=\frac{1}{c_s^2}\frac{\partial^2 u_y}{\partial t^2}-\frac{P_y}{c_s^2}\delta(x)\delta(y)\delta(z)\delta(t)\\
(\gamma^2-1)\frac{\partial}{\partial z}\left(\frac{\partial u_x}{\partial x}+\frac{\partial u_y}{\partial y}+\frac{\partial u_z}{\partial z}\right)+\frac{\partial^2 u_z}{\partial x^2}+\frac{\partial^2 u_z}{\partial y^2}+\frac{\partial^2 u_z}{\partial z^2}=\frac{1}{c_s^2}\frac{\partial^2 u_z}{\partial t^2}-\frac{P_z}{c_s^2}\delta(x)\delta(y)\delta(z)\delta(t)
\end{aligned} \tag{3.4.2}$$

where c_s is the shear wave velocity and the velocity ratio γ is defined by Eq. (3.8). On these displacement equations, we impose the quiescent condition at an initial time,

$$u_i|_{t=0} = \frac{\partial u_i}{\partial t}|_{t=0} = 0 \tag{3.4.3}$$

and the convergence condition at infinity,

$$u_i|_{\sqrt{x^2+y^2+z^2}\to\infty} = \frac{\partial u_i}{\partial x}|_{\sqrt{x^2+y^2+z^2}\to\infty} = \frac{\partial u_i}{\partial y}|_{\sqrt{x^2+y^2+z^2}\to\infty} = \frac{\partial u_i}{\partial z}|_{\sqrt{x^2+y^2+z^2}\to\infty} = 0 \tag{3.4.4}$$

Laplace transform with respect to the time variable,

$$u_j^* = \int_0^\infty u_j \exp(-st)dt \tag{3.4.5}$$

and the triple Fourier transform with respect to three space variables,

$$\begin{aligned}
\bar{u}_j &= \int_{-\infty}^{+\infty} u_j \exp(+i\xi x)dx, \qquad u_j = \frac{1}{2\pi}\int_{-\infty}^{+\infty} \bar{u}_j \exp(-i\xi x)d\xi \\
\tilde{u}_j &= \int_{-\infty}^{+\infty} u_j \exp(+i\eta y)dy, \qquad u_j = \frac{1}{2\pi}\int_{-\infty}^{+\infty} \tilde{u}_j \exp(-i\eta y)d\eta \\
\hat{u}_j &= \int_{-\infty}^{+\infty} u_j \exp(+i\zeta z)dz, \qquad u_j = \frac{1}{2\pi}\int_{-\infty}^{+\infty} \hat{u}_j \exp(-i\zeta z)d\zeta
\end{aligned} \tag{3.4.6}$$

are applied to the displacement equations (3.4.2). The simple algebraic equations for the transformed displacement are

$$\begin{aligned}
(\gamma^2-1)\left(-\xi^2\hat{\tilde{\bar{u}}}_x^* - \xi\eta\hat{\tilde{\bar{u}}}_y^* - \xi\zeta\hat{\tilde{\bar{u}}}_z^*\right) - (\xi^2+\eta^2+\zeta^2)\hat{\tilde{\bar{u}}}_x^* &= \left(\frac{s}{c_s}\right)^2 \hat{\tilde{\bar{u}}}_x^* - \frac{P_x}{c_s^2} \\
(\gamma^2-1)\left(-\xi\eta\hat{\tilde{\bar{u}}}_x^* - \eta^2\hat{\tilde{\bar{u}}}_y^* - \eta\zeta\hat{\tilde{\bar{u}}}_z^*\right) - (\xi^2+\eta^2+\zeta^2)\hat{\tilde{\bar{u}}}_y^* &= \left(\frac{s}{c_s}\right)^2 \hat{\tilde{\bar{u}}}_y^* - \frac{P_y}{c_s^2} \\
(\gamma^2-1)\left(-\xi\zeta\hat{\tilde{\bar{u}}}_x^* - \eta\zeta\hat{\tilde{\bar{u}}}_y^* - \zeta^2\hat{\tilde{\bar{u}}}_z^*\right) - (\xi^2+\eta^2+\zeta^2)\hat{\tilde{\bar{u}}}_z^* &= \left(\frac{s}{c_s}\right)^2 \hat{\tilde{\bar{u}}}_z^* - \frac{P_z}{c_s^2}
\end{aligned} \tag{3.4.7}$$

and the displacement components in the transformed domain are given by

$$\begin{aligned}
\hat{\tilde{\bar{u}}}_x^* &= \frac{P_x}{c_s^2\beta_s^2} - (\xi^2P_x + \xi\eta P_y + \xi\zeta P_z)\frac{1}{s^2}\left(\frac{1}{\beta_d^2} - \frac{1}{\beta_s^2}\right) \\
\hat{\tilde{\bar{u}}}_y^* &= \frac{P_y}{c_s^2\beta_s^2} - (\xi\eta P_x + \eta^2P_y + \eta\zeta P_z)\frac{1}{s^2}\left(\frac{1}{\beta_d^2} - \frac{1}{\beta_s^2}\right) \\
\hat{\tilde{\bar{u}}}_z^* &= \frac{P_z}{c_s^2\beta_s^2} - (\xi\zeta P_x + \eta\zeta P_y + \zeta^2P_z)\frac{1}{s^2}\left(\frac{1}{\beta_d^2} - \frac{1}{\beta_d^2}\right)
\end{aligned} \tag{3.4.8}$$

where the radicals are defined by

$$\beta_d = \sqrt{\xi^2+\eta^2+\zeta^2+(s/c_d)^2}, \quad \beta_s = \sqrt{\xi^2+\eta^2+\zeta^2+(s/c_s)^2} \tag{3.4.9}$$

In order to explore the convenient way for the inversion of the transformed displacement, we inspect the expression of Eq. (3.4.8). It is found that seven inversion formulas are necessary.

They are

$$\hat{\bar{\bar{I}}}^*_0(\xi,\eta,\varsigma,s)=\frac{1}{c_s^2\beta_s^2},\tag{3.4.10}$$

$$\begin{aligned}&\hat{\bar{\bar{I}}}^*_{11}(\xi,\eta,\zeta,s)=\frac{\xi^2}{s^2}\left(\frac{1}{\beta_d^2}-\frac{1}{\beta_s^2}\right),\quad \hat{\bar{\bar{I}}}^*_{12}(\xi,\eta,\zeta,s)=\frac{\eta^2}{s^2}\left(\frac{1}{\beta_d^2}-\frac{1}{\beta_s^2}\right),\\&\hat{\bar{\bar{I}}}^*_{13}(\xi,\eta,\zeta,s)=\frac{\zeta^2}{s^2}\left(\frac{1}{\beta_d^2}-\frac{1}{\beta_s^2}\right)\end{aligned}\tag{3.4.11}$$

$$\begin{aligned}&\hat{\bar{\bar{I}}}^*_{21}(\xi,\eta,\zeta,s)=\frac{\xi\eta}{s^2}\left(\frac{1}{\beta_d^2}-\frac{1}{\beta_s^2}\right),\quad \hat{\bar{\bar{I}}}^*_{22}(\xi,\eta,\zeta,s)=\frac{\eta\zeta}{s^2}\left(\frac{1}{\beta_d^2}-\frac{1}{\beta_s^2}\right),\\&\hat{\bar{\bar{I}}}^*_{23}(\xi,\eta,\zeta,s)=\frac{\xi\zeta}{s^2}\left(\frac{1}{\beta_d^2}-\frac{1}{\beta_s^2}\right)\end{aligned}\tag{3.4.12}$$

The first inversion for $\hat{\bar{\bar{I}}}^*_0$ is the same as that of the 3D wave equation in Sect. 2.7. Applying that result to our inversion, we have

$$I_0(x,y,z,t)=\frac{1}{4\pi c_s^2R}\delta(t-R/c_s)\tag{3.4.13}$$

where the 3D radial distance is defined by

$$R=\sqrt{x^2+y^2+z^2}\tag{3.4.14}$$

Subsequently, we consider the inversion of $\hat{\bar{\bar{I}}}^*_{1j}$ in Eq. (3.4.11). The three Fourier transforms are the same in the form of definition integral and thus the exchange of two space variables is equivalent to the exchange of two integration parameters. That is

$$\begin{aligned}&\hat{\bar{\bar{I}}}^*_{12}(\underline{\xi},\underline{\eta},\zeta,s)=\hat{\bar{\bar{I}}}^*_{11}(\underline{\eta},\underline{\xi},\zeta,s)\quad\Rightarrow\quad I_{12}(\underline{x},\underline{y},z,t)=I_{11}(\underline{y},\underline{x},z,t)\\&\hat{\bar{\bar{I}}}^*_{13}(\underline{\xi},\eta,\underline{\zeta},s)=\hat{\bar{\bar{I}}}^*_{11}(\underline{\zeta},\eta,\underline{\xi},s)\quad\Rightarrow\quad I_{13}(\underline{x},y,\underline{z},t)=I_{11}(\underline{z},y,\underline{x},t)\end{aligned}\tag{3.4.15}$$

Due to this exchangeability, if we could obtain the inversion for $\hat{\bar{\bar{I}}}^*_{11}$ as $F(x, y, z, t)$, the inversion of $\hat{\bar{\bar{I}}}^*_{12}$ is given by the exchange of the space variables as $F(\underline{y}, \underline{x}, z, t)$ and the inversion of $\hat{\bar{\bar{I}}}^{*}_{13}$ is also given by the variable exchange as $F(\underline{z}, y, \underline{x}, t)$. Thus, we can draw a schematic inversion diagram as

$$\begin{array}{ccc} \hat{\bar{\bar{I}}}^{*}_{11}(\xi,\eta,\zeta,s) = \dfrac{\xi^2}{s^2}\left(\dfrac{1}{\beta_d^2} - \dfrac{1}{\beta_s^2}\right) & \Rightarrow & I_{11}(x,y,z,t) = F(\underline{x},\underline{y},z,t) \\ \downarrow \text{exchange } \xi \text{ with } \eta & & \downarrow \text{exchange } x \text{ with } y \\ \hat{\bar{\bar{I}}}^{*}_{12}(\xi,\eta,\zeta,s) = \dfrac{\eta^2}{s^2}\left(\dfrac{1}{\beta_d^2} - \dfrac{1}{\beta_s^2}\right) & \Rightarrow & I_{12}(x,y,z,t) = F(\underline{y},\underline{x},z,t) \end{array} \tag{3.4.16}$$

Similarly, the third inversion $\hat{\bar{\bar{I}}}^{*}_{13}$ is also obtained by the exchange of the space variables as

$$\begin{array}{ccc} \hat{\bar{\bar{I}}}^{*}_{11}(\xi,\eta,\zeta,s) = \dfrac{\xi^2}{s^2}\left(\dfrac{1}{\beta_d^2} - \dfrac{1}{\beta_s^2}\right) & \Rightarrow & I_{11}(x,y,z,t) = F(\underline{x},y,\underline{z},t) \\ \downarrow \text{exchange } \xi \text{ with } \zeta & & \downarrow \text{exchange } x \text{ with } z \\ \hat{\bar{\bar{I}}}^{*}_{13}(\xi,\eta,\zeta,s) = \dfrac{\zeta^2}{s^2}\left(\dfrac{1}{\beta_d^2} - \dfrac{1}{\beta_s^2}\right) & \Rightarrow & I_{13}(x,y,z,t) = F(\underline{z},y,\underline{x},t) \end{array} \tag{3.4.17}$$

Then, the necessary inversion formula is reduced to only one, that is

$$\hat{\bar{\bar{I}}}^{*}_{11}(\xi,\eta,\zeta,s) = \frac{\xi^2}{s^2}\left(\frac{1}{\beta_d^2} - \frac{1}{\beta_s^2}\right) \tag{3.4.18}$$

As for the second group of the inversion in Eq. (3.4.12), we learn that the necessary inversion formula is

$$\hat{\bar{\bar{I}}}^{*}_{21}(\xi,\eta,\zeta,s) = \frac{\xi\eta}{s^2}\left(\frac{1}{\beta_d^2} - \frac{1}{\beta_s^2}\right) \tag{3.4.19}$$

If we could get the inversion as $G(x, y, z, t)$, the other two are given by the exchange of the space variables. They are

$$\begin{array}{ccc} \hat{\bar{\bar{I}}}^{*}_{21}(\xi,\eta,\zeta,s) = \dfrac{\xi\eta}{s^2}\left(\dfrac{1}{\beta_d^2} - \dfrac{1}{\beta_s^2}\right) & \Rightarrow & I_{21}(x,y,z,t) = G(\underline{x},y,\underline{z},t) \\ \downarrow \text{exchange } \xi \text{ with } \zeta & & \downarrow \text{exchange } x \text{ with } z \\ \hat{\bar{\bar{I}}}^{*}_{22}(\xi,\eta,\zeta,s) = \dfrac{\eta\zeta}{s^2}\left(\dfrac{1}{\beta_d^2} - \dfrac{1}{\beta_s^2}\right) & \Rightarrow & I_{22}(x,y,z,t) = G(\underline{z},y,\underline{x},t) \end{array} \tag{3.4.20}$$

and

$$\begin{array}{lcl} \hat{\bar{\bar{I}}}^{*}_{21}(\xi,\eta,\zeta,s)=\dfrac{\xi\eta}{s^2}\left(\dfrac{1}{\beta_d^2}-\dfrac{1}{\beta_s^2}\right) & \Rightarrow & I_{21}(x,y,z,t)=G(x,\underline{y},\underline{z},t) \\ \downarrow \text{exchange } \eta \text{ with } \zeta & & \downarrow \text{exchange } y \text{ with } z \\ \hat{\bar{\bar{I}}}^{*}_{23}(\xi,\eta,\zeta,s)=\dfrac{\xi\zeta}{s^2}\left(\dfrac{1}{\beta_d^2}-\dfrac{1}{\beta_s^2}\right) & \Rightarrow & I_{23}(x,y,z,t)=G(x,\underline{z},\underline{y},t) \end{array} \tag{3.4.21}$$

Consequently, it is sufficient to develop the inversion formulas only for $\hat{\bar{\bar{I}}}^{*}_{11}$ and $\hat{\bar{\bar{I}}}^{*}_{21}$.

(1) Inversion of $\hat{\bar{\bar{I}}}^{*}_{11}$

The formal Fourier inversion integral with respect to the parameter ζ is given by

$$\begin{aligned} \bar{\bar{I}}^{*}_{11}(\xi,\eta,z,s) &= \frac{1}{2\pi}\int\limits_{-\infty}^{+\infty}\frac{\xi^2}{s^2}\left(\frac{1}{\beta_d^2}-\frac{1}{\beta_s^2}\right)\exp(-i\zeta z)d\zeta \\ &= \frac{\xi^2}{\pi s^2}\int\limits_{0}^{\infty}\left(\frac{1}{\zeta^2+\alpha_d^2}-\frac{1}{\zeta^2+\alpha_s^2}\right)\cos(\zeta z)d\zeta \end{aligned} \tag{3.4.22}$$

where

$$\alpha_j=\sqrt{\xi^2+\eta^2+(s/c_j)^2};\quad j=d,s \tag{3.4.23}$$

Applying the simple integration formula (2.1.22) to the integral in Eq. (3.4.22), we have

$$\bar{\bar{I}}^{*}_{11}(\xi,\eta,z,s)=\frac{\xi^2}{2s^2}\left\{\frac{1}{\alpha_d}\exp(-\alpha_d|z|)-\frac{1}{\alpha_s}\exp(-\alpha_s|z|)\right\} \tag{3.4.24}$$

The second inversion integral with respect to the parameter η is given by

$$\begin{aligned} \bar{I}^{*}_{11}(\xi,y,z,s) &= \frac{1}{2\pi}\int\limits_{-\infty}^{+\infty}\frac{\xi^2}{2s^2}\left\{\frac{1}{\alpha_d}\exp(-\alpha_d|z|)-\frac{1}{\alpha_s}\exp(-\alpha_s|z|)\right\}\exp(-i\eta y)d\eta \\ &= \frac{\xi^2}{2\pi s^2}\int\limits_{0}^{\infty}\left[\frac{\exp\left\{-|z|\sqrt{\xi^2+\eta^2+(s/c_d)^2}\right\}}{\sqrt{\xi^2+\eta^2+(s/c_d)^2}}-\frac{\exp\left\{-|z|\sqrt{\xi^2+\eta^2+(s/c_s)^2}\right\}}{\sqrt{\xi^2+\eta^2+(s/c_s)^2}}\right]\cos(\eta y)d\eta \end{aligned} \tag{3.4.25}$$

The integration formula (2.4.13) in Chap. 2,

$$\int_0^\infty \frac{1}{\sqrt{x^2+a^2}} \exp\left(-c\sqrt{x^2+a^2}\right)\cos(bx)dx = K_0\left(a\sqrt{b^2+c^2}\right) \tag{3.4.26}$$

is applied to the integral in Eq. (3.4.25). We have for $\bar{I}_{11}^*$

$$\bar{I}_{11}^*(\xi,y,z,s) = \frac{\xi^2}{2\pi s^2}\left\{K_0\left(\bar{r}\sqrt{\xi^2+(s/c_d)^2}\right) - K_0\left(\bar{r}\sqrt{\xi^2+(s/c_s)^2}\right)\right\} \tag{3.4.27}$$

where $K_0(.)$ is the zeroth order modified Bessel function of the second kind and

$$\bar{r} = \sqrt{y^2+z^2} \tag{3.4.28}$$

The last Fourier inversion integral with respect to the parameter ξ is given by

$$\begin{aligned} I_{11}^*(x,y,z,s) &= \frac{1}{2\pi}\int_{-\infty}^{+\infty} \frac{\xi^2}{2\pi s^2}\left\{K_0\left(\bar{r}\sqrt{\xi^2+(s/c_d)^2}\right) - K_0\left(\bar{r}\sqrt{\xi^2+(s/c_s)^2}\right)\right\}\exp(-i\xi x)d\xi \\ &= \frac{1}{2\pi^2 s^2}\int_0^\infty \xi^2\left\{K_0\left(\bar{r}\sqrt{\xi^2+(s/c_d)^2}\right) - K_0\left(\bar{r}\sqrt{\xi^2+(s/c_s)^2}\right)\right\}\cos(\xi x)d\xi \end{aligned} \tag{3.4.29}$$

The last integral, which includes the modified Bessel function, has not yet been tabulated in reference books. However, we have one formula which resembles our integral, i.e.x

$$I = \int_0^\infty K_0\left(a\sqrt{x^2+c^2}\right)\cos(bx)dx = \frac{\pi}{2}\frac{\exp\left(-c\sqrt{a^2+b^2}\right)}{\sqrt{a^2+b^2}} \tag{3.4.30*}$$

*Notice: the above equation is just the Fourier cosine inversion of Eq. (3.4.26)!

If we differentiate this integration formula with respect to the parameter "b" twice, it yields to our necessary formula,

$$-\frac{\partial^2 I}{\partial b^2} = \int_0^\infty x^2 K_0\left(a\sqrt{x^2+c^2}\right)\cos(bx)dx = -\frac{\pi}{2}\frac{\partial^2}{\partial b^2}\left\{\frac{\exp\left(-c\sqrt{a^2+b^2}\right)}{\sqrt{a^2+b^2}}\right\} \tag{3.4.31}$$

That is,

$$\int_0^\infty x^2 K_0\left(a\sqrt{x^2+c^2}\right)\cos(bx)dx$$
$$=\frac{\pi}{2}\frac{1}{a^2+b^2}\left\{\left(\frac{1}{\sqrt{a^2+b^2}}+c\right)\left(1-\frac{3b^2}{a^2+b^2}\right)-\frac{b^2c^2}{\sqrt{a^2+b^2}}\right\}\exp\left(-c\sqrt{a^2+b^2}\right) \tag{3.4.32}$$

Applying this new formula (3.4.32) to the integral in Eq. (3.4.29), we have

$$\begin{aligned} I_{11}^*(x,y,z,s) = \frac{1}{4\pi R^2}\Bigg[&\left\{\left(1-\frac{3x^2}{R^2}\right)\frac{1}{Rs^2}+\left(1-\frac{3x^2}{R^2}\right)\frac{1}{c_d s}-\frac{x^2}{c_d^2 R}\right\}\exp\{-s(R/c_d)\} \\ &-\left\{\left(1-\frac{3x^2}{R^2}\right)\frac{1}{Rs^2}+\left(1-\frac{3x^2}{R^2}\right)\frac{1}{c_s s}-\frac{x^2}{c_s^2 R}\right\}\exp\{-s(R/c_s)\}\Bigg] \end{aligned} \tag{3.4.33}$$

where the 3D radial distance R is defined by Eq. (3.4.14).

The last Laplace inversion is carried out by applying the inversion formulas,

$$L^{-1}[\exp(-as)] = \delta(t-a) \tag{3.4.34}$$

$$L^{-1}\left[\frac{1}{s}\exp(-as)\right] = H(t-a) = \begin{cases} 1; & t>a \\ 0; & t<a \end{cases} \tag{3.4.35}$$

$$L^{-1}\left[\frac{1}{s^2}\exp(-as)\right] = H(t-a)(t-a) = \begin{cases} t-a; & t>a \\ 0; & t<a \end{cases} \tag{3.4.36}$$

where $\delta(.)$ and $H(.)$ are Dirac's delta and Heaviside's unit step functions defined in Sect. 1.2. Finally, the application of Laplace inversion formulas leads to the explicit expression for the inversion:

$$\begin{aligned} I_{11}(x,y,z,t) = \frac{1}{4\pi R}\Bigg\{&-\frac{x^2}{c_d^2R^2}\delta(t-R/c_d)+\frac{x^2}{c_s^2R^2}\delta(t-R/c_s) \\ &+\left(1-\frac{3x^2}{R^2}\right)\frac{t}{R^2}H(t-R/c_d)H(R/c_s-t)\Bigg\} \end{aligned} \tag{3.4.37}$$

The other inversions for $\bar{\hat{\bar{I}}}{}^*_{12}$ and $\bar{\hat{\bar{I}}}{}^*_{13}$ can be obtained by the simple exchange of the space variables as shown in Eqs. (3.4.16) and (3.4.17). They are

$$I_{12}(x,y,z,t) = \frac{1}{4\pi R}\left\{-\frac{y^2}{c_d^2R^2}\delta(t-R/c_d) + \frac{y^2}{c_s^2R^2}\delta(t-R/c_s) + \left(1-\frac{3y^2}{R^2}\right)\frac{t}{R^2}H(t-R/c_d)H(R/c_s-t)\right\} \tag{3.4.38}$$

$$I_{13}(x,y,z,t) = \frac{1}{4\pi R}\left\{-\frac{z^2}{c_d^2R^2}\delta(t-R/c_d) + \frac{z^2}{c_s^2R^2}\delta(t-R/c_s) + \left(1-\frac{3z^2}{R^2}\right)\frac{t}{R^2}H(t-R/c_d)H(R/c_s-t)\right\} \tag{3.4.39}$$

(2) Inversion of $\bar{\hat{\bar{I}}}{}^*_{21}$

The Fourier inversion integral with respect to the parameter ζ

$$\begin{aligned}\tilde{\bar{I}}{}^*_{21}(\xi,\eta,z,s) &= \frac{1}{2\pi}\int_{-\infty}^{+\infty}\frac{\xi\eta}{s^2}\left(\frac{1}{\beta_d^2}-\frac{1}{\beta_s^2}\right)\exp(-i\zeta z)d\zeta \\ &= \frac{\xi\eta}{\pi s^2}\int_0^{\infty}\left(\frac{1}{\zeta^2+\alpha_d^2}-\frac{1}{\zeta^2+\alpha_s^2}\right)\cos(\zeta z)d\zeta\end{aligned} \tag{3.4.40}$$

is evaluated by applying the formula (2.1.22) in Chap. 2. It is

$$\tilde{\bar{I}}{}^*_{21}(\xi,\eta,z,s) = \frac{\xi\eta}{2s^2}\left\{\frac{1}{\alpha_d}\exp(-\alpha_d|z|) - \frac{1}{\alpha_s}\exp(-\alpha_s|z|)\right\} \tag{3.4.41}$$

The second inversion integral with respect to the parameter η is given by

$$\begin{aligned}\bar{I}{}^*_{21}(\xi,y,z,s) &= \frac{1}{2\pi}\int_{-\infty}^{+\infty}\frac{\xi\eta}{2s^2}\left\{\frac{1}{\alpha_d}\exp(-\alpha_d|z|)-\frac{1}{\alpha_s}\exp(-\alpha_s|z|)\right\}\exp(-i\eta y)d\eta \\ &= \frac{-i\xi}{2\pi s^2}\int_0^{\infty}\left\{\frac{\eta}{\sqrt{\eta^2+\gamma_d^2}}\exp\left(-|z|\sqrt{\eta^2+\gamma_d^2}\right)-\frac{\eta}{\sqrt{\eta^2+\gamma_s^2}}\exp\left(-|z|\sqrt{\eta^2+\gamma_s^2}\right)\right\}\sin(\eta y)d\eta\end{aligned} \tag{3.4.42}$$

where

$$\gamma_j = \sqrt{\xi^2 + (s/c_j)^2}; \quad j = d, s \tag{3.4.43}$$

We apply the integration formula (Erdélyi 1954, vol. I, pp. 75, 36),

$$\int_0^\infty \frac{x}{\sqrt{x^2+a^2}} \exp\left(-c\sqrt{x^2+a^2}\right) \sin(bx)dx = \frac{ab}{\sqrt{b^2+c^2}} K_1\left(a\sqrt{b^2+c^2}\right) \tag{3.4.44}$$

to the integral in Eq. (3.4.42). It follows that

$$\begin{aligned}\bar{I}^*_{21}(\xi, y, z, s) = \frac{y}{2\pi\bar{r}s^2}\Big\{ & -i\xi\sqrt{\xi^2+(s/c_d)^2}K_1\left(\bar{r}\sqrt{\xi^2+(s/c_d)^2}\right) \\ & + i\xi\sqrt{\xi^2+(s/c_s)^2}K_1\left(\bar{r}\sqrt{\xi^2+(s/c_s)^2}\right)\Big\}\end{aligned} \tag{3.4.45}$$

where $\bar{r}$ has already been defined by Eq. (3.4.28). The last Fourier inversion integral with respect to the parameter ξ is given by

$$\begin{aligned}I^*_{21}(\xi, y, z, s) = & \frac{y}{2\pi\bar{r}s^2}\frac{1}{2\pi}\int_{-\infty}^{+\infty}\Big\{-i\xi\sqrt{\xi^2+(s/c_d)^2}K_1\left(\bar{r}\sqrt{\xi^2+(s/c_d)^2}\right) \\ & + i\xi\sqrt{\xi^2+(s/c_s)^2}K_1\left(\bar{r}\sqrt{\xi^2+(s/c_s)^2}\right)\Big\}\exp(-i\xi x)d\xi \\ = & -\frac{y}{2\pi^2\bar{r}s^2}\int_0^\infty\Big\{\xi\sqrt{\xi^2+(s/c_d)^2}K_1\left(\bar{r}\sqrt{\xi^2+(s/c_d)^2}\right) \\ & - \xi\sqrt{\xi^2+(s/c_s)^2}K_1\left(\bar{r}\sqrt{\xi^2+(s/c_s)^2}\right)\Big\}\sin(\xi x)d\xi\end{aligned} \tag{3.4.46}$$

We have the formula (Erdélyi 1954, vol. I, pp. 113, 45)

$$\begin{aligned}& \int_0^\infty x\sqrt{x^2+b^2}K_1\left(a\sqrt{x^2+b^2}\right)\sin(cx)dx \\ & = \frac{\pi}{2}\frac{ab^2c}{(a^2+c^2)^{3/2}}\left\{1+\frac{3}{b\sqrt{a^2+c^2}}+\frac{3}{b^2(a^2+c^2)}\right\}\exp\left(-b\sqrt{a^2+c^2}\right)\end{aligned} \tag{3.4.47}$$

The integrals in Eq. (3.4.46) are evaluated and we have

$$I_{21}^{*}(x,y,z,s) = -\frac{xy}{4\pi R^3}\left[\frac{1}{c_d^2}\left\{1+\frac{3c_d}{sR}+\frac{3c_d^2}{s^2R^2}\right\}\exp\{-s(R/c_d)\}\right. \\ \left. -\frac{1}{c_s^2}\left\{1+\frac{3c_s}{sR}+\frac{3c_s^2}{s^2R^2}\right\}\exp\{-s(R/c_s)\}\right] \tag{3.4.48}$$

The Laplace inversion is carried out by applying the formulas (3.4.34)–(3.4.36). Then, the final form of $I_{21}(x, y, z, t)$ is given by

$$I_{21}(x,y,z,t) = -\frac{xy}{4\pi R^3}\left\{\frac{1}{c_d^2}\delta(t-R/c_d)-\frac{1}{c_s^2}\delta(t-R/c_s)+\frac{3t}{R^2}H(t-R/c_d)H(R/c_s-t)\right\} \tag{3.4.49}$$

We have just obtained the exact expressions for the two principal parts: I_{11} and I_{21}. Other two inversions can be obtained by the exchange of the space variables as shown in Eqs. (3.4.20) and (3.4.21). They are

$$I_{22}(x,y,z,t) = -\frac{yz}{4\pi R^3}\left\{\frac{1}{c_d^2}\delta(t-R/c_d)-\frac{1}{c_s^2}\delta(t-R/c_s)+\frac{3t}{R^2}H(t-R/c_d)H(R/c_s-t)\right\} \tag{3.4.50}$$

$$I_{23}(x,y,z,t) = -\frac{xz}{4\pi R^3}\left\{\frac{1}{c_d^2}\delta(t-R/c_d)-\frac{1}{c_s^2}\delta(t-R/c_s)+\frac{3t}{R^2}H(t-R/c_d)H(R/c_s-t)\right\} \tag{3.4.51}$$

(3) Green's dyadic

We have just obtained the exact inversion of the fundamentals. Inspecting Eq. (3.4.8), we learn that the displacement components can be expressed in terms of $I_{ij}(x, y, z, t)$ as

$$\begin{aligned} u_x &= P_x\{I_0(x,y,z,t)-I_{11}(x,y,z,t)\}+P_y\{-I_{21}(x,y,z,t)\}+P_z\{-I_{23}(x,y,z,t)\} \\ u_y &= P_x\{-I_{21}(x,y,z,t)\}+P_y\{I_0(x,y,z,t)-I_{12}(x,y,z,t)\}+P_z\{-I_{22}(x,y,z,t)\} \\ u_z &= P_x\{-I_{23}(x,y,z,t)\}+P_y\{-I_{22}(x,y,z,t)\}+P_z\{I_0(x,y,z,t)-I_{13}(x,y,z,t)\} \end{aligned} \tag{3.4.52}$$

Substituting Eqs. (3.4.13), (3.4.37–3.4.39) and (3.4.49–3.4.51) into Eq. (3.4.52), we have the following explicit expressions for the displacement components:

$$
\begin{aligned}
u_x = {} & \frac{P_x}{4\pi R}\left\{\begin{array}{l} \dfrac{x^2}{c_d^2R^2}\delta(t-R/c_d)+\dfrac{1}{c_s^2}\left(1-\dfrac{x^2}{R^2}\right)\delta(t-R/c_s) \\ -\dfrac{t}{R^2}\left(1-\dfrac{3x^2}{R^2}\right)H(t-R/c_d)H(R/c_s-t) \end{array}\right\} \\
& +\frac{P_y}{4\pi R}\left\{\frac{xy}{c_d^2R^2}\delta(t-R/c_d)-\frac{xy}{c_s^2R^2}\delta(t-R/c_s)+\frac{3xyt}{R^4}H(t-R/c_d)H(R/c_s-t)\right\} \\
& +\frac{P_z}{4\pi R}\left\{\frac{xz}{c_d^2R^2}\delta(t-R/c_d)-\frac{xz}{c_s^2R^2}\delta(t-R/c_s)+\frac{3xzt}{R^4}H(t-R/c_d)H(R/c_s-t)\right\}
\end{aligned}
\tag{3.4.53a}
$$

$$
\begin{aligned}
u_y = {} & \frac{P_x}{4\pi R}\left\{\frac{xy}{c_d^2R^2}\delta(t-R/c_d)-\frac{xy}{c_s^2R^2}\delta(t-R/c_s)+\frac{3xyt}{R^4}H(t-R/c_d)H(R/c_s-t)\right\} \\
& +\frac{P_y}{4\pi R}\left\{\begin{array}{l} \dfrac{y^2}{c_d^2R^2}\delta(t-R/c_d)+\dfrac{1}{c_s^2}\left(1-\dfrac{y^2}{R^2}\right)\delta(t-R/c_s) \\ -\dfrac{t}{R^2}\left(1-\dfrac{3y^2}{R^2}\right)H(t-R/c_d)H(R/c_s-t) \end{array}\right\} \\
& +\frac{P_z}{4\pi R}\left\{\frac{yz}{c_d^2R^2}\delta(t-R/c_d)-\frac{yz}{c_s^2R^2}\delta(t-R/c_s)+\frac{3yzt}{R^4}H(t-R/c_d)H(R/c_s-t)\right\}
\end{aligned}
\tag{3.4.53b}
$$

$$
\begin{aligned}
u_z = {} & \frac{P_x}{4\pi R}\left\{\frac{xz}{c_d^2R^2}\delta(t-R/c_d)-\frac{xz}{c_s^2R^2}\delta(t-R/c_s)+\frac{3xzt}{R^4}H(t-R/c_d)H(R/c_s-t)\right\} \\
& +\frac{P_y}{4\pi R}\left\{\frac{yz}{c_d^2R^2}\delta(t-R/c_d)-\frac{yz}{c_s^2R^2}\delta(t-R/c_s)+\frac{3yzt}{R^4}H(t-R/c_d)H(R/c_s-t)\right\} \\
& +\frac{P_z}{4\pi R}\left\{\begin{array}{l} \dfrac{z^2}{c_d^2R^2}\delta(t-R/c_d)+\dfrac{1}{c_s^2}\left(1-\dfrac{z^2}{R^2}\right)\delta(t-R/c_s) \\ -\dfrac{t}{R^2}\left(1-\dfrac{3z^2}{R^2}\right)H(t-R/c_d)H(R/c_s-t) \end{array}\right\}
\end{aligned}
\tag{3.4.53c}
$$

In terms of the Green's dyadic, we can rewrite the displacement components as

$$
u_i = \sum_{j=x,y,z} P_j G_{ij}(x,y,z,t); \quad i = x,y,z \tag{3.4.54}
$$

where the dyadic $G_{ij}(x, y, z, t)$ is given by

$$G_{ij}(x,y,z,t) = \frac{1}{4\pi R}\left\{\begin{array}{l} \dfrac{x_i x_j}{c_d^2 R^2}\delta(t - R/c_d) + \dfrac{1}{c_s^2}\left(\delta_{ij} - \dfrac{x_i x_j}{R^2}\right)\delta(t - R/c_s) \\ -\dfrac{t}{R^2}\left(\delta_{ij} - \dfrac{3x_i x_j}{R^2}\right)H(t - R/c_d)H(R/c_s - t) \end{array}\right\} \tag{3.4.55}$$

It is clear that there are two spherical waves. Their fronts are given by $c_d t = R$ and $c_s t = R$ for dilatational and shear waves, respectively. We also find that the displacement has the singularity of the delta function (i.e. the first order singularity) at their wave fronts.

3.5 3D Time-Harmonic Source

When the source is a time-harmonic vibration, the body force with magnitude Q_i is given by

$$\begin{pmatrix} B_x \\ B_y \\ B_z \end{pmatrix} = \begin{pmatrix} Q_x \\ Q_y \\ Q_z \end{pmatrix} \delta(x)\delta(y)\delta(z)\exp(+i\omega t), \tag{3.5.1}$$

and the displacement equation is

$$\begin{aligned} &(\gamma^2 - 1)\frac{\partial}{\partial x}\left(\frac{\partial u_x}{\partial x} + \frac{\partial u_y}{\partial y} + \frac{\partial u_z}{\partial z}\right) + \frac{\partial^2 u_x}{\partial x^2} + \frac{\partial^2 u_x}{\partial y^2} + \frac{\partial^2 u_x}{\partial z^2} \\ &= \frac{1}{c_s^2}\frac{\partial^2 u_x}{\partial t^2} - \frac{Q_x}{c_s^2}\delta(x)\delta(y)\delta(z)\exp(i\omega t) \\ &(\gamma^2 - 1)\frac{\partial}{\partial y}\left(\frac{\partial u_x}{\partial x} + \frac{\partial u_y}{\partial y} + \frac{\partial u_z}{\partial z}\right) + \frac{\partial^2 u_y}{\partial x^2} + \frac{\partial^2 u_y}{\partial y^2} + \frac{\partial^2 u_y}{\partial z^2} \\ &= \frac{1}{c_s^2}\frac{\partial^2 u_y}{\partial t^2} - \frac{Q_y}{c_s^2}\delta(x)\delta(y)\delta(z)\exp(i\omega t) \\ &(\gamma^2 - 1)\frac{\partial}{\partial z}\left(\frac{\partial u_x}{\partial x} + \frac{\partial u_y}{\partial y} + \frac{\partial u_z}{\partial z}\right) + \frac{\partial^2 u_z}{\partial x^2} + \frac{\partial^2 u_z}{\partial y^2} + \frac{\partial^2 u_z}{\partial z^2} \\ &= \frac{1}{c_s^2}\frac{\partial^2 u_z}{\partial t^2} - \frac{Q_z}{c_s^2}\delta(x)\delta(y)\delta(z)\exp(i\omega t) \end{aligned} \tag{3.5.2}$$

In this section we do not employ the method of the integral transform for solving the displacement equations. Instead, we take a short-cut. That is the convolution integral with the impulsive Green's dyadicobtained in the previous section. When we express the displacement in terms of the time-harmonic Green's dyadic,

$$u_i = \sum_{j=x,y,z} Q_j g_{ij}(x,y,z,t); \quad i = x,y,z \tag{3.5.3}$$

the dyadic for the time-harmonic response can be derived by the convolution integral of the impulsive dyadic given by Eq. (3.4.55). The convolution integral is evaluated as follows

$$\begin{aligned} g_{ij}(x,y,z,t) &= \lim_{t\to\infty} \int_0^t G_{ij}(x,y,z,t') \exp\{+i\omega(t-t')\} dt' \\ &= \exp(+i\omega t) \int_0^{t\to\infty} G_{ij}(x,y,z,t') \exp(-i\omega t') dt' \end{aligned} \tag{3.5.4}$$

Substituting Eq. (3.4.55) into the last integral in the above equation, we can easily perform the integration and obtain the time-harmonic Green's dyadic g_{ij} as

$$\begin{aligned} & g_{ij}(x,y,z,t) \\ &= \frac{1}{4\pi R}\left\{\frac{x_i x_j}{c_d^2 R^2} + \frac{1}{(\omega R)^2}\left(1 + i\frac{\omega R}{c_d}\right)\left(\delta_{ij} - \frac{3x_i x_j}{R^2}\right)\right\} \exp\{i\omega(t - R/c_d)\} \\ &\quad + \frac{1}{4\pi R}\left\{\frac{1}{c_s^2}\left(\delta_{ij} - \frac{x_i x_j}{R^2}\right) - \frac{1}{(\omega R)^2}\left(1 + i\frac{\omega R}{c_s}\right)\left(\delta_{ij} - \frac{3x_i x_j}{R^2}\right)\right\} \exp\{i\omega(t - R/c_s)\} \end{aligned} \tag{3.5.5}$$

3.6 3D Static Source

When a static point force is placed at the coordinate origin, the body force with magnitude S_i is given by

$$\begin{pmatrix} B_x \\ B_y \\ B_z \end{pmatrix} = \begin{pmatrix} S_x \\ S_y \\ S_z \end{pmatrix} \delta(x)\delta(y)\delta(z) \tag{3.6.1}$$

The static deformation can be considered as the time-harmonic response with zero-frequency. Taking the limit $\omega \rightarrow 0$ in Eq. (3.5.4), the displacement can be expressed in terms of the static dyadic as

$$u_i = \sum_{j=x,y,z} S_j G_{ij}^{(K)}(x,y,z); \quad i = x,y,z \tag{3.6.2}$$

where the source magnitude Q_i is replaced with S_i and the static dyadic $G_{ij}^{(K)}(x,y,z)$ is derived from the limit and is given by

$$G_{ij}^{(K)}(x,y,z) = \lim_{\omega\to 0} g_{ij}(x,y,z,t) = \frac{1}{8\pi c_d^2 R}\left\{(\gamma^2+1)\delta_{ij} + (\gamma^2-1)\frac{x_i x_j}{R^2}\right\} \tag{3.6.3}$$

Since the velocity ratio can be replaced with the simple function of Poisson ratio v as

$$\gamma^2 = \frac{2(1-v)}{1-2v} \tag{3.6.4}$$

the dyadic can also be rewritten as

$$G_{ij}^{(K)}(x,y,z) = \frac{1}{8\pi(1-2v)c_d^2 R}\left\{(3-4v)\delta_{ij} + \frac{x_i x_j}{R^2}\right\} \tag{3.6.5}$$

This dyadic is called "Kelvin's solution" for the static deformation.

3.7 Torsional Source

This section shows two impulsive Green's functions for the axisymmetric torsion problem. The first one is corresponding to a circular ring force, and the second to a point torque. We take the cylindrical coordinate system (r, θ, z) in an infinite elastic solid and assume that the symmetry axis of the deformation/torsion is the z-axis. The ring force lies on the $z = 0$ plane and its symmetry axis is also the z-axis. On the other hand, the point torque is place on the coordinate origin. Both produce the axisymmetric torsional deformation.

Due to the axisymmetric nature of the deformation, the non-vanishing displacement component is the circumferential displacement u_θ only, and the governing equation for the axisymmetric torsional deformation is given by

$$\frac{\partial^2 u_\theta}{\partial r^2} + \frac{1}{r}\frac{\partial u_\theta}{\partial r} - \frac{u_\theta}{r^2} + \frac{\partial^2 u_\theta}{\partial z^2} - \frac{1}{c_s^2}\frac{\partial^2 u_\theta}{\partial t^2} = -\frac{1}{c_s^2}B_\theta(r,z,t) \tag{3.7.1}$$

where c_s is the shear (torsional) wave velocity and B_θ is the circumferential body force which represents the ring or point source. When the source is an impulsive ring force with radius a, the body force is defined by

$$B_\theta(r,z,t) = Q\frac{\delta(r-a)}{r}\delta(z)\delta(t) \tag{3.7.2}$$

where $\delta(\cdot)$ is Dirac's delta function and Q is the magnitude of the source. On the other hand, when the source is an impulsive point torque, the body force is defined by the limit of the ring force as

$$B_\theta(r,z,t) = \frac{T}{2\pi\rho}\left[\lim_{a\to 0}\frac{\delta(r-a)}{ar}\right]\delta(z)\delta(t) \tag{3.7.3}$$

where T is the magnitude of the torque. As we have no suitable mathematical formula for the limit in the square bracket, the limiting form for the point torque source is retained. In this section, we shall discuss Green's functions for these two sources, separately.

3.7.1 Ring Source

In order to solve the differential equation (3.7.1) with the source (3.7.2), we apply the triple integral transform which is defined in Chap. 1: Hankel transform with respect to the radial variable r, Fourier transform with respect to the axial variable z and Laplace transform with respect to the time t. The triple transform is expressed as

$$\int_0^\infty \exp(-st)dt \int_{-\infty}^\infty \exp(+i\zeta z)dz \int_0^\infty \left\{ \begin{aligned} &\frac{\partial^2 u_\theta}{\partial r^2} + \frac{1}{r}\frac{\partial u_\theta}{\partial r} - \frac{u_\theta}{r^2} + \frac{\partial^2 u_\theta}{\partial z^2} - \frac{1}{c_s^2}\frac{\partial^2 u_\theta}{\partial t^2} \\ &= -\frac{1}{c_s^2}Q\frac{\delta(r-a)}{r}\delta(z)\delta(t) \end{aligned} \right\} rJ_1(\xi r)dr \tag{3.7.4}$$

The convergence condition at infinity.

$$u_\theta|_{\sqrt{r^2+z^2}\to\infty} = \left.\frac{\partial u_\theta}{\partial r}\right|_{\sqrt{r^2+z^2}\to\infty} = \left.\frac{\partial u_\theta}{\partial z}\right|_{\sqrt{r^2+z^2}\to\infty} = 0 \tag{3.7.5}$$

and the quiescent condition at the initial time

$$u_\theta\big|_{t=0} = \left.\frac{\partial u_\theta}{\partial t}\right|_{t=0} = 0 \tag{3.7.6}$$

are employed, and the transformed displacement is defined as

$$\hat{\tilde{u}}_\theta^* = \int_0^\infty \exp(-st)dt \int_{-\infty}^\infty \exp(+i\zeta z)dz \int_0^\infty r u_\theta J_1(\xi r)dr \tag{3.7.7}$$

The application of the triple integral transform yields to the simple algebraic equation for the transformed displacement,

$$-\left\{\xi^2 + \zeta^2 + (s/c_s)^2\right\}\hat{\tilde{u}}_\theta^* = -\frac{Q}{c_s^2}J_1(\xi a) \tag{3.7.8}$$

Then, we consider the inversion of the displacement,

$$\hat{\tilde{u}}_\theta^* = \frac{Q}{c_s^2}\frac{J_1(\xi a)}{\xi^2 + \zeta^2 + (s/c_s)^2} \tag{3.7.9}$$

As the first step, the Fourier inversion integral defined by Eq. (1.1.21c) in Chap. 1 is applied and it is reduced to the semi-infinite integral as

$$\tilde{u}_\theta^* = \frac{Q}{\pi c_s^2}J_1(\xi a)\int_0^\infty \frac{1}{\xi^2 + \zeta^2 + (s/c_s)^2}\cos(\zeta z)d\zeta \tag{3.7.10}$$

The simple integration formula (2.1.22) in Chap. 2 is applied to the above integral. It follows that

$$\tilde{u}_\theta^* = \frac{Q}{2c_s^2}\frac{J_1(\xi a)}{\sqrt{\xi^2 + (s/c_s)^2}}\exp\left\{-|z|\sqrt{\xi^2 + (s/c_s)^2}\right\} \tag{3.7.11}$$

The second step is to apply the Hankel inversion with $n = 1$ defined by Eq. (1.1.19) in Chap. 1. The Hankel inversion integral is given by

$$u_\theta^* = \frac{Q}{2c_s^2}\int_0^\infty \frac{\xi}{\sqrt{\xi^2 + (s/c_s)^2}}\exp\left\{-|z|\sqrt{\xi^2 + (s/c_s)^2}\right\}J_1(\xi a)J_1(\xi r)d\xi \tag{3.7.12}$$

The product of two Bessel functions is replaced with the integral form of the single Bessel function (Watson 1966, p. 361),

$$J_1(\xi a)J_1(\xi r) = \frac{1}{\pi}\int_0^{\pi} J_0\left(\xi\sqrt{r^2 + a^2 - 2ar\cos\varphi}\right)\cos\varphi d\varphi \tag{3.7.13}$$

and the order of integration is exchanged. Equation (3.7.12) yields

$$u_\theta^* = \frac{Q}{2\pi c_s^2}\int_0^{\pi}\cos\varphi d\varphi\int_0^{\infty}\frac{\xi}{\sqrt{\xi^2 + (s/c_s)^2}}\exp\left\{-|z|\sqrt{\xi^2 + (s/c_s)^2}\right\}J_0(\xi Z)d\xi \tag{3.7.14}$$

where

$$Z = \sqrt{r^2 + a^2 - 2ar\cos\varphi} \tag{3.7.15}$$

Fortunately, we have the nice integration formula (Erdélyi 1954, vol. II, pp. 9, 24) for the inner integral, i.e.

$$\begin{aligned}&\int_0^{\infty}\frac{\xi}{\sqrt{\xi^2 + (s/c_s)^2}}\exp\left\{-|z|\sqrt{\xi^2 + (s/c_s)^2}\right\}J_0(\xi Z)d\xi\\ &= \frac{1}{\sqrt{Z^2 + z^2}}\exp\left\{-(s/c_s)\sqrt{Z^2 + z^2}\right\}\end{aligned} \tag{3.7.16}$$

Applying this integration formula to Eq. (3.7.14), and recalling the definition of Z, we have the single finite integral for the Laplace transformed displacement,

$$u_\theta^* = \frac{Q}{2\pi c_s^2}\int_0^{\pi}\frac{\exp\left\{-(s/c_s)\sqrt{r^2 + a^2 - 2ar\cos\varphi + z^2}\right\}}{\sqrt{r^2 + a^2 - 2ar\cos\varphi + z^2}}\cos\varphi d\varphi \tag{3.7.17}$$

Before going to the Laplace inversion, the integral is simplified by the change of variable, $\varphi \to u$,

$$u = \frac{1}{c_s}\sqrt{r^2 + a^2 - 2ar\cos\varphi + z^2} \tag{3.7.18}$$

Equation (3.7.17) is simplified as

$$u_\theta^* = \frac{Q}{4\pi arc_s} \int_{R_1/c_s}^{R_2/c_s} \left\{ \sqrt{\frac{R_2^2 - (c_s u)^2}{(c_s u)^2 - R_1^2}} - \sqrt{\frac{(c_s u)^2 - R_1^2}{R_2^2 - (c_s u)^2}} \right\} \exp(-su)du \tag{3.7.19}$$

where

$$R_1 = \sqrt{(r-a)^2 + z^2}, \quad R_2 = \sqrt{(r+a)^2 + z^2} \tag{3.7.20}$$

Examining the integrand, we learn that the Laplace transform parameter s is included only in the argument of the exponential function. So, the simple Laplace inversion formula for the delta function

$$L^{-1}[\exp(-as)] = \delta(t-a) \tag{3.7.21}$$

is applied to Eq. (3.7.19). It yields

$$u_\theta = \frac{Q}{4\pi arc_s} \int_{R_1/c_s}^{R_2/c_s} \left\{ \sqrt{\frac{R_2^2 - (c_s u)^2}{(c_s u)^2 - R_1^2}} - \sqrt{\frac{(c_s u)^2 - R_1^2}{R_2^2 - (c_s u)^2}} \right\} \delta(t-u)du \tag{3.7.22}$$

In order to evaluate the integral, we examine the supporting region for the delta function and applied the simple formula (1.2.3) in Chap. 1,

$$\int_a^b f(x)\delta(x-c)dx = \begin{cases} f(c); & a < c < b \\ 0; & c < a \text{ or } b < c \end{cases} \tag{3.7.23}$$

We have the simple form for the torsional displacement, i.e. Green's function for the torsional deformation,

$$u_\theta = \frac{Q}{4\pi c_s ar} H(R_2 - c_s t)H(c_s t - R_1) \left\{ \sqrt{\frac{R_2^2 - (c_s t)^2}{(c_s t)^2 - R_1^2}} - \sqrt{\frac{(c_s t)^2 - R_1^2}{R_2^2 - (c_s t)^2}} \right\} \tag{3.7.24}$$

where $H(.)$ is Heaviside's unit step function defined in Sect. 1.2.

3.7.2 Point Torque Source

This subsection considers the Green's function corresponding to the point torque source defined by Eq. (3.7.3). As was employed in the previous subsection, we also apply the triple integral transform to the governing Eq. (3.7.1) with the point torque source,

$$\int_0^\infty \exp(-st)dt \int_{-\infty}^\infty \exp(+i\zeta z)dz \int_0^\infty \left\{ \begin{array}{l} \frac{\partial^2 u_\theta}{\partial r^2} + \frac{1}{r}\frac{\partial u_\theta}{\partial r} - \frac{u_\theta}{r^2} + \frac{\partial^2 u_\theta}{\partial z^2} - \frac{1}{c_s^2}\frac{\partial^2 u_\theta}{\partial t^2} \\ = -\frac{T}{2\pi\mu}\left[\lim_{a\to 0}\frac{\delta(r-a)}{ar}\right]\delta(z)\delta(t) \end{array} \right\} rJ_1(\xi r)dr \tag{3.7.25}$$

The limit in the source term is taken after the Hankel transform. Its procedure is

$$\int_0^\infty \lim_{a\to 0}\frac{\delta(r-a)}{ar} rJ_1(\xi r)dr = \lim_{a\to 0}\int_0^\infty \frac{\delta(r-a)}{ar} rJ_1(\xi r)dr = \lim_{a\to 0}\frac{J_1(\xi a)}{a} = \frac{\xi}{2} \tag{3.7.26}$$

The transformed governing equation yields to the simple algebraic equation,

$$-\left\{\xi^2 + \zeta^2 + (s/c_s)^2\right\}\hat{\tilde{u}}_\theta^* = -\frac{T}{4\pi\mu}\xi \tag{3.7.27}$$

Thus, the triple transformed displacement is given by

$$\hat{\tilde{u}}_\theta^* = \frac{T}{4\pi\mu}\frac{\xi}{\xi^2 + \zeta^2 + (s/c_s)^2} \tag{3.7.28}$$

Now, we consider the inversion. As was done for the inversion integral of Eq. (3.7.10), the Fourier inversion integral with respect to the parameter ζ is easily evaluated as

$$\tilde{u}_\theta^* = \frac{T}{8\pi\mu}\frac{\xi}{\sqrt{\xi^2 + (s/c_s)^2}}\exp\left\{-|z|\sqrt{\xi^2 + (s/c_s)^2}\right\} \tag{3.7.29}$$

The second inversion, i.e. Hankel inversion integral, is separated into two terms as

$$\begin{aligned} u_\theta^* &= \frac{T}{8\pi\mu}\int_0^\infty \frac{\xi^2}{\sqrt{\xi^2 + (s/c_s)^2}}\exp\left\{-|z|\sqrt{\xi^2 + (s/c_s)^2}\right\}J_1(\xi r)d\xi \\ &= \frac{T}{8\pi\mu}\int_0^\infty \left\{\sqrt{\xi^2 + (s/c_s)^2} - \frac{(s/c_s)^2}{\sqrt{\xi^2 + (s/c_s)^2}}\right\}\exp\left\{-|z|\sqrt{\xi^2 + (s/c_s)^2}\right\}J_1(\xi r)d\xi \end{aligned} \tag{3.7.30}$$

The second term in the last line can be evaluated by applying the formula (Erdélyi 1954, vol. II, pp. 19, 10)

$$\int_0^\infty \frac{1}{\sqrt{\xi^2+\beta^2}} \exp\left\{-\alpha\sqrt{\xi^2+\beta^2}\right\} J_1(\xi y) d\xi = \frac{1}{\beta y}\left\{\exp(-\alpha\beta) - \exp\left(-\beta\sqrt{\alpha^2+y^2}\right)\right\} \tag{3.7.31}$$

As to the first term in the Hankel inversion integral (3.7.30), we derive a suitable integration formula from the formula (3.7.31). Differentiating Eq. (3.7.31) twice with respect to the parameter α, we have

$$\int_0^\infty \sqrt{\xi^2+\beta^2} \exp\left\{-\alpha\sqrt{\xi^2+\beta^2}\right\} J_1(\xi y) d\xi = \frac{\beta}{y}\exp(-\alpha\beta) + \frac{1}{\sqrt{\alpha^2+y^2}}\left(\frac{y}{\alpha^2+y^2} - \frac{\alpha^2\beta}{y\sqrt{\alpha^2+y^2}}\right)\exp\left(-\beta\sqrt{\alpha^2+y^2}\right) \tag{3.7.32}$$

Applying these two formulas to Eq. (3.7.30), we have the Laplace transformed displacement as

$$u_\theta^* = \frac{T}{8\pi\mu}\frac{r}{R^2}\left(\frac{s}{c_s} + \frac{1}{R}\right)\exp(-sR/c_s) \tag{3.7.33}$$

where R is the distance from the source point (coordinate origin),

$$R = \sqrt{r^2+z^2} \tag{3.7.34}$$

The last Laplace inversion is very simple. Fortunately, we have two simple inversion formulas,

$$L^{-1}\{\exp(-as)\} = \delta(t-a), \quad L^{-1}\{s\exp(-as)\} = \delta'(t-a) \tag{3.7.35}$$

where $\delta'(.)$ is the derivative of the delta function as

$$\delta'(t) = \frac{d\delta(t)}{dt} \tag{3.7.36}$$

Applying these two inversion formulas to Eq. (3.7.33), we have the final form of the torsional displacement as

$$u_\theta = \frac{T}{8\pi\mu}\frac{r}{R^2}\left\{\frac{1}{c_s}\delta'(t - R/c_s) + \frac{1}{R}\delta(t - R/c_s)\right\} \tag{3.7.37}$$

The above equation shows only one spherical wave and its front is $R = c_s t$. It should be noticed that disturbance is only on the wave front and no disturbance its inside, $R < c_s t$, due to the nature of the delta function.

Exercises

(3.1) From the unified expression for the Green's dyadic (3.1.42), derive the explicit expressions for G_{xx}, G_{xy}, G_{yx}, G_{yy} and show that $G_{xy} = G_{yx}$.

(3.2) Show that the explicit expression of the Green's dyadic (3.4.55) is the same as the corresponding one in the displacement equation (3.4.53).

(3.3) When the body force B_θ in the governing equation (3.7.1) is a suddenly applied point torque,

$$B_\theta(r, z, t) = \frac{T_0}{2\pi\rho}\left[\lim_{a\to 0}\frac{\delta(r - a)}{ar}\right]\delta(z)H(t) \tag{3.7.39}$$

where $H(t)$ is Heaviside's unit step function and T_0 is the magnitude of the torque, show that the corresponding Green's function is given by

$$u_\theta = \frac{T_0}{8\pi\mu}\frac{r}{R^2}\left\{\frac{1}{c_s}\delta(t - R/c_s) + \frac{1}{R}H(t - R/c_s)\right\} \tag{3.7.40}$$

Appendix

See Table 3.1.

Table 3.1 Green's dyadic for elastodynamic equations

	Displacement equations	Source B_i	Green's dyadic
2D	2D inplane deformation (plane strain) $(\gamma^2-1)\frac{\partial}{\partial x}\left(\frac{\partial u_x}{\partial x}+\frac{\partial u_y}{\partial y}\right)+\frac{\partial^2 u_x}{\partial x^2}+\frac{\partial^2 u_x}{\partial y^2}=\frac{1}{c_s^2}\frac{\partial^2 u_x}{\partial t^2}-\frac{1}{c_s^2}B_x(x,y,t)$ $(\gamma^2-1)\frac{\partial}{\partial y}\left(\frac{\partial u_x}{\partial x}+\frac{\partial u_y}{\partial y}\right)+\frac{\partial^2 u_y}{\partial x^2}+\frac{\partial^2 u_y}{\partial y^2}=\frac{1}{c_s^2}\frac{\partial^2 u_y}{\partial t^2}-\frac{1}{c_s^2}B_y(x,y,t)$	$P_i\delta(x)\delta(y)\delta(t)$	$u_i(x,y,t)=\sum_{j=x,y}P_jG_{ij}(x,y,t);\quad i,j=x,y$ $G_{ij}(x,y,t)=\frac{H(t-r/c_d)}{2\pi c_d}\left\{\frac{x_ix_j}{r^2}\frac{1}{\sqrt{(c_dt)^2-r^2}}+\left(\frac{2x_ix_j}{r^2}-\delta_{ij}\right)\frac{\sqrt{(c_dt)^2-r^2}}{r^2}\right\}$ $-\frac{H(t-r/c_s)}{2\pi c_s}\left\{\left(\frac{x_ix_j}{r^2}-\delta_{ij}\right)\frac{1}{\sqrt{(c_st)^2-r^2}}+\left(\frac{2x_ix_j}{r^2}-\delta_{ij}\right)\frac{\sqrt{(c_st)^2-r^2}}{r^2}\right\}$
		$Q_i\delta(x)\delta(y)\exp(+i\omega t)$	$u_i(x,y,t)=\sum_{j=x,y}Q_jg_{ij}(x,y)\exp(+i\omega t);\quad i,j=x,y$ $g_{ij}(x,y)=-\frac{i}{4}\left[\frac{1}{c_d^2}\left\{\frac{x_ix_j}{r^2}H_0^{(2)}(r\omega/c_d)+\left(\delta_{ij}-\frac{2x_ix_j}{r^2}\right)\frac{c_d}{r\omega}H_1^{(2)}(r\omega/c_d)\right\}\right.$ $\left.-\frac{1}{c_s^2}\left\{\left(\frac{x_ix_j}{r^2}-\delta_{ij}\right)H_0^{(2)}(r\omega/c_s)+\left(\delta_{ij}-\frac{2x_ix_j}{r^2}\right)\frac{c_s}{r\omega}H_1^{(2)}(r\omega/c_s)\right\}\right]$
		$S_i\delta(x)\delta(y)$	$u_i=\sum_{j=x,y}S_jg_{ij}^{(static)}(x,y);\quad i,j=x,y$ $g_{xx}^{(static)}(x,y)=\frac{1}{4\pi c_d^2}\left\{-(\gamma^2+1)\log(r)+(\gamma^2-1)\left(\frac{x}{r}\right)^2\right\}$ $g_{xy}^{(static)}(x,y)=\frac{1}{4\pi c_d^2}(\gamma^2-1)\frac{xy}{r^2}$ $g_{yy}^{(static)}(x,y)=\frac{1}{4\pi c_d^2}\left\{-(\gamma^2+1)\log(r)+(\gamma^2-1)\left(\frac{y}{r}\right)^2\right\}$
3D	$(\gamma^2-1)\frac{\partial}{\partial x}\left(\frac{\partial u_x}{\partial x}+\frac{\partial u_y}{\partial y}+\frac{\partial u_z}{\partial z}\right)+\frac{\partial^2 u_x}{\partial x^2}+\frac{\partial^2 u_x}{\partial y^2}+\frac{\partial^2 u_x}{\partial z^2}=\frac{1}{c_s^2}\frac{\partial^2 u_x}{\partial t^2}-\frac{1}{c_s^2}B_x(x,y,z,t)$ $(\gamma^2-1)\frac{\partial}{\partial y}\left(\frac{\partial u_x}{\partial x}+\frac{\partial u_y}{\partial y}+\frac{\partial u_z}{\partial z}\right)+\frac{\partial^2 u_y}{\partial x^2}+\frac{\partial^2 u_y}{\partial y^2}+\frac{\partial^2 u_y}{\partial z^2}=\frac{1}{c_s^2}\frac{\partial^2 u_y}{\partial t^2}-\frac{1}{c_s^2}B_y(x,y,z,t)$ $(\gamma^2-1)\frac{\partial}{\partial z}\left(\frac{\partial u_x}{\partial x}+\frac{\partial u_y}{\partial y}+\frac{\partial u_z}{\partial z}\right)+\frac{\partial^2 u_z}{\partial x^2}+\frac{\partial^2 u_z}{\partial y^2}+\frac{\partial^2 u_z}{\partial z^2}=\frac{1}{c_s^2}\frac{\partial^2 u_z}{\partial t^2}-\frac{1}{c_s^2}B_z(x,y,z,t)$	$P_i\delta(x)\delta(y)\delta(z)\delta(t)$	$u_i=\sum_{j=x,y,z}P_jG_{ij}(x,y,z,t);\quad i,j=x,y,z$ $G_{ij}(x,y,z,t)=\frac{1}{4\pi R}\left\{\begin{array}{l}\frac{x_ix_j}{c_d^2R^2}\delta(t-R/c_d)+\frac{1}{c_s^2}\left(\delta_{ij}-\frac{x_ix_j}{R^2}\right)\delta(t-R/c_s)\\-\frac{t}{R^2}\left(\delta_{ij}-\frac{3x_ix_j}{R^2}\right)H(t-R/c_d)H(R/c_s-t)\end{array}\right\}$
		$Q_i\delta(x)\delta(y)\delta(z)$ $\exp(+i\omega t)$	$u_i=\sum_{j=x,y,z}Q_jg_{ij}(x,y,z,t);\quad i,j=x,y,z$ $g_{ij}(x,y,z,t)=\frac{1}{4\pi R}\left\{\frac{x_ix_j}{c_d^2R^2}+\frac{1}{(\omega R)^2}\left(1+i\frac{\omega R}{c_d}\right)\left(\delta_{ij}-\frac{3x_ix_j}{R^2}\right)\right\}\exp\{i\omega(t-R/c_d)\}$ $+\frac{1}{4\pi R}\left\{\frac{1}{c_s^2}\left(\delta_{ij}-\frac{x_ix_j}{R^2}\right)-\frac{1}{(\omega R)^2}\left(1+i\frac{\omega R}{c_s}\right)\left(\delta_{ij}-\frac{3x_ix_j}{R^2}\right)\right\}\exp\{i\omega(t-R/c_s)\}$

(continued)

Table 3.1 (continued)

Displacement equations	Source B_i	Green's dyadic
	$S_i\delta(x)\delta(y)\delta(z)$	$u_i = \sum_{j=x,y,z} S_j G_{ij}^{(K)}(x,y,z); \quad i,j = x,y,z$ $G_{ij}^{(K)}(x,y,z) = \frac{1}{8\pi(1-2\nu)c_d^2 R}\left\{(3-4\nu)\delta_{ij} + \frac{x_i x_j}{R^2}\right\}$
Torsional ring source $\frac{\partial^2 u_\theta}{\partial r^2} + \frac{1}{r}\frac{\partial u_\theta}{\partial r} - \frac{u_\theta}{r^2} + \frac{\partial^2 u_\theta}{\partial z^2} - \frac{1}{c_s^2}\frac{\partial^2 u_\theta}{\partial t^2} = -\frac{1}{c_s^2} Q \frac{\delta(r-a)}{r}\delta(z)\delta(t)$		$u_\theta = \frac{Q}{4\pi c_s a r} H(R_2 - c_s t)H(c_s t - R_1)\left\{\sqrt{\frac{R_2^2 - (c_s t)^2}{(c_s t)^2 - R_1^2}} - \sqrt{\frac{(c_s t)^2 - R_1^2}{R_2^2 - (c_s t)^2}}\right\}$ $R_1 = \sqrt{(r-a)^2 + z^2}, \quad R_2 = \sqrt{(r+a)^2 + z^2}$
Point torque source $\frac{\partial^2 u_\theta}{\partial r^2} + \frac{1}{r}\frac{\partial u_\theta}{\partial r} - \frac{u_\theta}{r^2} + \frac{\partial^2 u_\theta}{\partial z^2} - \frac{1}{c_s^2}\frac{\partial^2 u_\theta}{\partial t^2} = -\frac{T}{2\pi\mu}\left[\lim_{a\to 0}\frac{\delta(r-a)}{ar}\right]\delta(z)\delta(t)$		$u_\theta = \frac{T}{8\pi\mu}\frac{r}{R^2}\left\{\frac{1}{c_s}\delta'(t - R/c_s) + \frac{1}{R}\delta(t - R/c_s)\right\}$ $\delta\prime(t) = \frac{d\delta(t)}{dt}, \ R = \sqrt{r^2 + z^2}$
Suddenly applied point torque $\frac{\partial^2 u_\theta}{\partial r^2} + \frac{1}{r}\frac{\partial u_\theta}{\partial r} - \frac{u_\theta}{r^2} + \frac{\partial^2 u_\theta}{\partial z^2} - \frac{1}{c_s^2}\frac{\partial^2 u_\theta}{\partial t^2} = -\frac{T_0}{2\pi\mu}\left[\lim_{a\to 0}\frac{\delta(r-a)}{ar}\right]\delta(z)H(t)$		$u_\theta = \frac{T_0}{8\pi\mu}\frac{r}{R^2}\left\{\frac{1}{c_s}\delta(t - R/c_s) + \frac{1}{R}H(t - R/c_s)\right\}$ $R = \sqrt{r^2 + z^2}$

References

Erdélyi A (ed) (1954) Tables of integral transforms, vol I and II. MacGraw-Hill, New-York
Gradshteyn IS, Ryzhik IM (Jefferey A (ed)) (1980) Table of integrals, series, and products, 5th edn. Academic Press, San Diego
Watson GN (1966) A treatise on the theory of bessel functions. Cambridge University Press, Cambridge (1966)

Chapter 4
Acoustic Wave in a Uniform Flow

Traditionally, when we use the word "wave," we think about acoustic waves or water waves. The acoustic wave is much more familiar to our daily life and we experience many wave phenomena such as reflection, refraction, diffraction, the Doppler effects, etc. The governing equations for the acoustic wave are rigorously derived from the fluid equations and Green's function for the acoustic wave in a flowing fluid is discussed by applying the method of integral transform.

4.1 Compressive Viscous Fluid

Motions and disturbances in fluids such as water, oil and gas are governed by four groups of equations:

(1) Equations of motion,

$$\begin{aligned}
\frac{\partial \sigma_{xx}}{\partial x}+\frac{\partial \sigma_{yx}}{\partial y}+\frac{\partial \sigma_{zx}}{\partial z}+\rho B_x &= \rho\left(\frac{\partial v_x}{\partial t}+v_x\frac{\partial v_x}{\partial x}+v_y\frac{\partial v_x}{\partial y}+v_z\frac{\partial v_x}{\partial z}\right)\\
\frac{\partial \sigma_{xy}}{\partial x}+\frac{\partial \sigma_{yy}}{\partial y}+\frac{\partial \sigma_{zy}}{\partial z}+\rho B_y &= \rho\left(\frac{\partial v_y}{\partial t}+v_x\frac{\partial v_y}{\partial x}+v_y\frac{\partial v_y}{\partial y}+v_z\frac{\partial v_y}{\partial z}\right)\\
\frac{\partial \sigma_{xz}}{\partial x}+\frac{\partial \sigma_{yz}}{\partial y}+\frac{\partial \sigma_{zz}}{\partial z}+\rho B_z &= \rho\left(\frac{\partial v_z}{\partial t}+v_x\frac{\partial v_z}{\partial x}+v_y\frac{\partial v_z}{\partial y}+v_z\frac{\partial v_z}{\partial z}\right)
\end{aligned} \tag{4.1.1}$$

(2) Constitutive equations for the linear Newtonian fluid,

$$\begin{aligned}
\sigma_{xx} &= -p-\frac{2\mu}{3}\left(\frac{\partial v_x}{\partial x}+\frac{\partial v_y}{\partial y}+\frac{\partial v_z}{\partial z}\right)+2\mu\frac{\partial v_x}{\partial x}\\
\sigma_{yy} &= -p-\frac{2\mu}{3}\left(\frac{\partial v_x}{\partial x}+\frac{\partial v_y}{\partial y}+\frac{\partial v_z}{\partial z}\right)+2\mu\frac{\partial v_y}{\partial y}\\
\sigma_{zz} &= -p-\frac{2\mu}{3}\left(\frac{\partial v_x}{\partial x}+\frac{\partial v_y}{\partial y}+\frac{\partial v_z}{\partial z}\right)+2\mu\frac{\partial v_z}{\partial z}\\
\sigma_{xy} &= \mu\left(\frac{\partial v_x}{\partial y}+\frac{\partial v_y}{\partial x}\right),\quad \sigma_{yz}=\mu\left(\frac{\partial v_y}{\partial z}+\frac{\partial v_z}{\partial y}\right),\quad \sigma_{zx}=\mu\left(\frac{\partial v_x}{\partial z}+\frac{\partial v_z}{\partial x}\right)
\end{aligned} \tag{4.1.2}$$

K. Watanabe, *Integral Transform Techniques for Green's Function*,
Lecture Notes in Applied and Computational Mechanics 76,
DOI 10.1007/978-3-319-17455-6_4

(3) Continuity equation

$$\frac{\partial \rho}{\partial t}+\frac{\partial(\rho v_x)}{\partial x}+\frac{\partial(\rho v_y)}{\partial y}+\frac{\partial(\rho v_z)}{\partial z}=0 \tag{4.1.3}$$

(4) Equation of state for the acoustic medium (adiabatic change)

$$\frac{p}{p_0}=\left(\frac{\rho}{\rho_0}\right)^{\kappa} \tag{4.1.4}$$

In above equations, the viscosity μ is a known/given constant that specifies the nature of the fluid. The body force B_i is assumed to be a source and its functional form must be specified. The stress σ_{ij}, the particle velocity v_i, the hydro-static pressure p, and the density ρ are unknown functions to be determined. The subscript "0" stands for the quantities at a reference state. The constant parameter κ is the ratio of two specific heats. Thus, the governing equation for the linear fluid is a set of strongly coupled partial differential equations with 11 unknowns.

Substituting the constitutive Eq. (4.1.2) into the equations of motion (4.1.1), we obtain the well-known Navier-Stokes equations,

$$\begin{aligned}&-\frac{\partial p}{\partial x}+\frac{\mu}{3}\frac{\partial}{\partial x}\left(\frac{\partial v_x}{\partial x}+\frac{\partial v_y}{\partial y}+\frac{\partial v_z}{\partial z}\right)+\mu\left(\frac{\partial^2 v_x}{\partial x^2}+\frac{\partial^2 v_x}{\partial y^2}+\frac{\partial^2 v_x}{\partial z^2}\right)+\rho B_x\\&\quad=\rho\left(v_x\frac{\partial v_x}{\partial x}+v_y\frac{\partial v_x}{\partial y}+v_z\frac{\partial v_x}{\partial z}+\frac{\partial v_x}{\partial t}\right)\end{aligned} \tag{4.1.5a}$$

$$\begin{aligned}&-\frac{\partial p}{\partial y}+\frac{\mu}{3}\frac{\partial}{\partial y}\left(\frac{\partial v_x}{\partial x}+\frac{\partial v_y}{\partial y}+\frac{\partial v_z}{\partial z}\right)+\mu\left(\frac{\partial^2 v_y}{\partial x^2}+\frac{\partial^2 v_y}{\partial y^2}+\frac{\partial^2 v_y}{\partial z^2}\right)+\rho B_y\\&\quad=\rho\left(v_x\frac{\partial v_y}{\partial x}+v_y\frac{\partial v_y}{\partial y}+v_z\frac{\partial v_y}{\partial z}+\frac{\partial v_y}{\partial t}\right)\end{aligned} \tag{4.1.5b}$$

$$\begin{aligned}&-\frac{\partial p}{\partial z}+\frac{\mu}{3}\frac{\partial}{\partial z}\left(\frac{\partial v_x}{\partial x}+\frac{\partial v_y}{\partial y}+\frac{\partial v_z}{\partial z}\right)+\mu\left(\frac{\partial^2 v_z}{\partial x^2}+\frac{\partial^2 v_z}{\partial y^2}+\frac{\partial^2 v_z}{\partial z^2}\right)+\rho B_z\\&\quad=\rho\left(v_x\frac{\partial v_z}{\partial x}+v_y\frac{\partial v_z}{\partial y}+v_z\frac{\partial v_z}{\partial z}+\frac{\partial v_z}{\partial t}\right)\end{aligned} \tag{4.1.5c}$$

When the fluid is incompressible and no density change takes place, the continuity equation is simplified and the equation of state is unnecessary. Then, the incompressible fluid is governed by a simplified form of the Navier-Stokes equations. However, when we discuss acoustic waves, the two reminder Eqs. (4.1.3) and (4.1.4) must be fully incorporated since acoustic waves express the propagation of density changes.

4.2 Linearization

The nonlinearity stems from the acceleration terms in Navier-Stokes equations (4.1.5), the product (ρv_i) in the continuity equation (4.1.3) and the power of density in the equation of state (4.1.4). Thus, the governing equations for acoustic waves are fully nonlinear, coupled partial differential equations. In order to reduce the differential equations to tractable ones, we introduce the assumption that the acoustic wave is an infinitesimal small disturbance superimposed on a reference nonlinear fluid motion. This assumption states that the infinitesimal small disturbance is a deviation from the nonlinear reference state. We introduce a small parameter ε, and represent the disturbance with $(\bar{v}_i, \bar{p}, \bar{\rho})$. The five unknown functions, the velocity component v_i, the static hydro-pressure p, and the density ρ, are approximated as the power of the small parameter ε,

$$v_i = V_i + \varepsilon\bar{v}_i + O(\varepsilon^2), \quad p = p_0 + \varepsilon\bar{p} + O(\varepsilon^2), \quad \rho = \rho_0 + \varepsilon\bar{\rho} + O(\varepsilon^2) \tag{4.2.1}$$

We also assume that the wave source described by the body force is a small quantity of order ε^1,

$$B_i = \varepsilon\bar{B}_i + O(\varepsilon^2) \tag{4.2.2}$$

Substituting Eqs. (4.2.1–4.2.2) into Eq. (4.1.5a), and neglecting terms of higher order than $O(\varepsilon^1)$, the equation is reduced to two parts, the zeroth and first order terms. The zeroth order term is

$$\begin{aligned} &-\frac{\partial p_0}{\partial x} + \frac{\mu}{3}\frac{\partial}{\partial x}\left(\frac{\partial V_x}{\partial x} + \frac{\partial V_y}{\partial y} + \frac{\partial V_z}{\partial z}\right) + \mu\left(\frac{\partial^2 V_x}{\partial x^2} + \frac{\partial^2 V_x}{\partial y^2} + \frac{\partial^2 V_x}{\partial z^2}\right) \\ &\quad = +\rho_0\left(\frac{\partial V_x}{\partial t} + V_x\frac{\partial V_x}{\partial x} + V_y\frac{\partial V_x}{\partial y} + V_z\frac{\partial V_x}{\partial z}\right) \end{aligned} \tag{4.2.3}$$

and the first order term is

$$\begin{aligned} &\varepsilon\left[-\frac{\partial \bar{p}}{\partial x} + \frac{\mu}{3}\frac{\partial}{\partial x}\left(\frac{\partial \bar{v}_x}{\partial x} + \frac{\partial \bar{v}_y}{\partial y} + \frac{\partial \bar{v}_z}{\partial z}\right) + \mu\left(\frac{\partial^2 \bar{v}_x}{\partial x^2} + \frac{\partial^2 \bar{v}_x}{\partial y^2} + \frac{\partial^2 \bar{v}_x}{\partial z^2}\right) + \rho\bar{B}_x\right. \\ &= +\bar{\rho}\left(V_x\frac{\partial V_x}{\partial x} + V_y\frac{\partial V_x}{\partial y} + V_z\frac{\partial V_x}{\partial z}\right) \\ &\left. + \rho_0\frac{\partial \bar{v}_x}{\partial t} + \rho_0\left(V_x\frac{\partial \bar{v}_x}{\partial x} + \bar{v}_x\frac{\partial V_x}{\partial x}\right) + \rho_0\left(V_y\frac{\partial \bar{v}_x}{\partial y} + \bar{v}_y\frac{\partial V_x}{\partial y}\right) + \rho_0\left(V_z\frac{\partial \bar{v}_x}{\partial z} + \bar{v}_z\frac{\partial V_x}{\partial z}\right)\right] \end{aligned} \tag{4.2.4}$$

Similarly, we substitute Eqs. (4.2.1–4.2.2) into the reminder of Eq. (4.1.5). From Eq. (4.1.5b), we have the zeroth order term

$$\begin{aligned} & -\frac{\partial p_0}{\partial y} + \frac{\mu}{3}\frac{\partial}{\partial y}\left(\frac{\partial V_x}{\partial x} + \frac{\partial V_y}{\partial y} + \frac{\partial V_z}{\partial z}\right) + \mu\left(\frac{\partial^2 V_y}{\partial x^2} + \frac{\partial^2 V_y}{\partial y^2} + \frac{\partial^2 V_y}{\partial z^2}\right) \\ & \quad = +\rho_0\left(\frac{\partial V_y}{\partial t} + V_x\frac{\partial V_y}{\partial x} + V_y\frac{\partial V_y}{\partial y} + V_z\frac{\partial V_y}{\partial z}\right) \end{aligned} \tag{4.2.5}$$

and the first order term

$$\begin{aligned} \varepsilon\Bigg[& -\frac{\partial \bar{p}}{\partial y} + \frac{\mu}{3}\frac{\partial}{\partial y}\left(\frac{\partial \bar{v}_x}{\partial x} + \frac{\partial \bar{v}_y}{\partial y} + \frac{\partial \bar{v}_z}{\partial z}\right) + \mu\left(\frac{\partial^2 \bar{v}_y}{\partial x^2} + \frac{\partial^2 \bar{v}_y}{\partial y^2} + \frac{\partial^2 \bar{v}_y}{\partial z^2}\right) + \rho\bar{B}_y \\ & = +\bar{\rho}\left(V_x\frac{\partial V_y}{\partial x} + V_y\frac{\partial V_y}{\partial y} + V_z\frac{\partial V_y}{\partial z}\right) \\ & + \rho_0\frac{\partial \bar{v}_y}{\partial t} + \rho_0\left(V_x\frac{\partial \bar{v}_y}{\partial x} + \bar{v}_x\frac{\partial V_y}{\partial x}\right) + \rho_0\left(V_y\frac{\partial \bar{v}_y}{\partial y} + \bar{v}_y\frac{\partial V_y}{\partial y}\right) + \rho_0\left(V_z\frac{\partial \bar{v}_y}{\partial z} + \bar{v}_z\frac{\partial V_y}{\partial z}\right)\Bigg] \end{aligned} \tag{4.2.6}$$

From Eq. (4.1.5c), we also have for the zeroth order term

$$\begin{aligned} & -\frac{\partial p_0}{\partial z} + \frac{\mu}{3}\frac{\partial}{\partial z}\left(\frac{\partial V_x}{\partial x} + \frac{\partial V_y}{\partial y} + \frac{\partial V_z}{\partial z}\right) + \mu\left(\frac{\partial^2 V_z}{\partial x^2} + \frac{\partial^2 V_z}{\partial y^2} + \frac{\partial^2 V_z}{\partial z^2}\right) \\ & \quad = +\rho_0\left(\frac{\partial V_z}{\partial t} + V_x\frac{\partial V_z}{\partial x} + V_y\frac{\partial V_z}{\partial y} + V_z\frac{\partial V_z}{\partial z}\right) \end{aligned} \tag{4.2.7}$$

and for the first order term

$$\begin{aligned} \varepsilon\Bigg[& -\frac{\partial \bar{p}}{\partial z} + \frac{\mu}{3}\frac{\partial}{\partial z}\left(\frac{\partial \bar{v}_x}{\partial x} + \frac{\partial \bar{v}_y}{\partial y} + \frac{\partial \bar{v}_z}{\partial z}\right) + \mu\left(\frac{\partial^2 \bar{v}_z}{\partial x^2} + \frac{\partial^2 \bar{v}_z}{\partial y^2} + \frac{\partial^2 \bar{v}_z}{\partial z^2}\right) + \rho\bar{B}_z \\ & = +\bar{\rho}\left(V_x\frac{\partial V_z}{\partial x} + V_y\frac{\partial V_z}{\partial y} + V_z\frac{\partial V_z}{\partial z}\right) \\ & + \rho_0\frac{\partial \bar{v}_z}{\partial t} + \rho_0\left(V_x\frac{\partial \bar{v}_z}{\partial x} + \bar{v}_x\frac{\partial V_z}{\partial x}\right) + \rho_0\left(V_y\frac{\partial \bar{v}_z}{\partial y} + \bar{v}_y\frac{\partial V_z}{\partial y}\right) + \rho_0\left(V_z\frac{\partial \bar{v}_z}{\partial z} + \bar{v}_z\frac{\partial V_z}{\partial z}\right)\Bigg] \end{aligned} \tag{4.2.8}$$

Substituting Eq. (4.2.1) into Eq. (4.1.3), the continuity equation is decomposed into the two parts,

$$\begin{aligned} & \frac{\partial(\rho_0 + \varepsilon\bar{\rho})}{\partial t} + \frac{\partial(\rho_0 + \varepsilon\bar{\rho})(V_x + \varepsilon\bar{v}_x)}{\partial x} + \frac{\partial(\rho_0 + \varepsilon\bar{\rho})(V_y + \varepsilon\bar{v}_y)}{\partial y} + \frac{\partial(\rho_0 + \varepsilon\bar{\rho})(V_z + \varepsilon\bar{v}_z)}{\partial z} \\ & = \frac{\partial \rho_0}{\partial t} + \frac{\partial(\rho_0 V_x)}{\partial x} + \frac{\partial(\rho_0 V_y)}{\partial y} + \frac{\partial(\rho_0 V_z)}{\partial z} \\ & \quad + \varepsilon\left[\frac{\partial \bar{\rho}}{\partial t} + \frac{\partial(\rho_0\bar{v}_x)}{\partial x} + \frac{\partial(\rho_0\bar{v}_y)}{\partial y} + \frac{\partial(\rho_0\bar{v}_z)}{\partial z} + \frac{\partial(V_x\bar{\rho})}{\partial x} + \frac{\partial(V_y\bar{\rho})}{\partial y} + \frac{\partial(V_z\bar{\rho})}{\partial z}\right] \\ & \quad + O(\varepsilon^2) = 0 \end{aligned} \tag{4.2.9}$$

The equation of state for the adiabatic change in Eq. (4.1.4) is approximated as

$$\frac{p_0+\varepsilon\bar{p}}{p_0}=\left(\frac{\rho_0+\varepsilon\bar{\rho}}{\rho_0}\right)^{\kappa} \quad\Rightarrow\quad 1+\varepsilon\frac{\bar{p}}{p_0}=1+\varepsilon\frac{\kappa\bar{\rho}}{\rho_0}+O(\varepsilon^2) \quad\Rightarrow\quad \frac{\bar{p}}{p_0}=\frac{\kappa\bar{\rho}}{\rho_0} \tag{4.2.10}$$

Thus, we have the linear relation between the pressure and density deviations.

Consequently, we have the equations for the reference state as the $O(\varepsilon^0)$ part,

$$\begin{aligned} &-\frac{\partial p_0}{\partial x}+\frac{\mu}{3}\frac{\partial}{\partial x}\left(\frac{\partial V_x}{\partial x}+\frac{\partial V_y}{\partial y}+\frac{\partial V_z}{\partial z}\right)+\mu\left(\frac{\partial^2 V_x}{\partial x^2}+\frac{\partial^2 V_x}{\partial y^2}+\frac{\partial^2 V_x}{\partial z^2}\right)\\ &\quad=+\rho_0\left(\frac{\partial V_x}{\partial t}+V_x\frac{\partial V_x}{\partial x}+V_y\frac{\partial V_x}{\partial y}+V_z\frac{\partial V_x}{\partial z}\right) \end{aligned} \tag{4.2.11a}$$

$$\begin{aligned} &-\frac{\partial p_0}{\partial y}+\frac{\mu}{3}\frac{\partial}{\partial y}\left(\frac{\partial V_x}{\partial x}+\frac{\partial V_y}{\partial y}+\frac{\partial V_z}{\partial z}\right)+\mu\left(\frac{\partial^2 V_y}{\partial x^2}+\frac{\partial^2 V_y}{\partial y^2}+\frac{\partial^2 V_y}{\partial z^2}\right)\\ &\quad=+\rho_0\left(\frac{\partial V_y}{\partial t}+V_x\frac{\partial V_y}{\partial x}+V_y\frac{\partial V_y}{\partial y}+V_z\frac{\partial V_y}{\partial z}\right) \end{aligned} \tag{4.2.11b}$$

$$\begin{aligned} &-\frac{\partial p_0}{\partial z}+\frac{\mu}{3}\frac{\partial}{\partial z}\left(\frac{\partial V_x}{\partial x}+\frac{\partial V_y}{\partial y}+\frac{\partial V_z}{\partial z}\right)+\mu\left(\frac{\partial^2 V_z}{\partial x^2}+\frac{\partial^2 V_z}{\partial y^2}+\frac{\partial^2 V_z}{\partial z^2}\right)\\ &\quad=+\rho_0\left(\frac{\partial V_z}{\partial t}+V_x\frac{\partial V_z}{\partial x}+V_y\frac{\partial V_z}{\partial y}+V_z\frac{\partial V_z}{\partial z}\right) \end{aligned} \tag{4.2.11c}$$

$$\frac{\partial \rho_0}{\partial t}+\frac{\partial(\rho_0 V_x)}{\partial x}+\frac{\partial(\rho_0 V_y)}{\partial y}+\frac{\partial(\rho_0 V_z)}{\partial z}=0 \tag{4.2.12}$$

The disturbance of $O(\varepsilon^1)$ is governed by the equations

$$\begin{aligned} &-\frac{\partial \bar{p}}{\partial x}+\frac{\mu}{3}\frac{\partial}{\partial x}\left(\frac{\partial \bar{v}_x}{\partial x}+\frac{\partial \bar{v}_y}{\partial y}+\frac{\partial \bar{v}_z}{\partial z}\right)+\mu\left(\frac{\partial^2 \bar{v}_x}{\partial x^2}+\frac{\partial^2 \bar{v}_x}{\partial y^2}+\frac{\partial^2 \bar{v}_x}{\partial z^2}\right)+\rho_0\bar{B}_x\\ &\quad=+\bar{\rho}\left(V_x\frac{\partial V_x}{\partial x}+V_y\frac{\partial V_x}{\partial y}+V_z\frac{\partial V_x}{\partial z}\right)\\ &\qquad+\rho_0\frac{\partial \bar{v}_x}{\partial t}+\rho_0\left(V_x\frac{\partial \bar{v}_x}{\partial x}+\bar{v}_x\frac{\partial V_x}{\partial x}\right)+\rho_0\left(V_y\frac{\partial \bar{v}_x}{\partial y}+\bar{v}_y\frac{\partial V_x}{\partial y}\right)+\rho_0\left(V_z\frac{\partial \bar{v}_x}{\partial z}+\bar{v}_z\frac{\partial V_x}{\partial z}\right) \end{aligned} \tag{4.2.13a}$$

$$\begin{aligned} &-\frac{\partial \bar{p}}{\partial y}+\frac{\mu}{3}\frac{\partial}{\partial y}\left(\frac{\partial \bar{v}_x}{\partial x}+\frac{\partial \bar{v}_y}{\partial y}+\frac{\partial \bar{v}_z}{\partial z}\right)+\mu\left(\frac{\partial^2 \bar{v}_y}{\partial x^2}+\frac{\partial^2 \bar{v}_y}{\partial y^2}+\frac{\partial^2 \bar{v}_y}{\partial z^2}\right)+\rho_0\bar{B}_y\\ &\quad=+\bar{\rho}\left(V_x\frac{\partial V_y}{\partial x}+V_y\frac{\partial V_y}{\partial y}+V_z\frac{\partial V_y}{\partial z}\right)\\ &\qquad+\rho_0\frac{\partial \bar{v}_y}{\partial t}+\rho_0\left(V_x\frac{\partial \bar{v}_y}{\partial x}+\bar{v}_x\frac{\partial V_y}{\partial x}\right)+\rho_0\left(V_y\frac{\partial \bar{v}_y}{\partial y}+\bar{v}_y\frac{\partial V_y}{\partial y}\right)+\rho_0\left(V_z\frac{\partial \bar{v}_y}{\partial z}+\bar{v}_z\frac{\partial V_y}{\partial z}\right) \end{aligned} \tag{4.2.13b}$$

$$-\frac{\partial\bar{p}}{\partial z}+\frac{\mu}{3}\frac{\partial}{\partial z}\left(\frac{\partial\bar{v}_x}{\partial x}+\frac{\partial\bar{v}_y}{\partial y}+\frac{\partial\bar{v}_z}{\partial z}\right)+\mu\left(\frac{\partial^2\bar{v}_z}{\partial x^2}+\frac{\partial^2\bar{v}_z}{\partial y^2}+\frac{\partial^2\bar{v}_z}{\partial z^2}\right)+\rho_0\bar{B}_z$$
$$=+\bar{\rho}\left(V_x\frac{\partial V_z}{\partial x}+V_y\frac{\partial V_z}{\partial y}+V_z\frac{\partial V_z}{\partial z}\right)$$
$$+\rho_0\frac{\partial\bar{v}_z}{\partial t}+\rho_0\left(V_x\frac{\partial\bar{v}_z}{\partial x}+\bar{v}_x\frac{\partial V_z}{\partial x}\right)+\rho_0\left(V_y\frac{\partial\bar{v}_z}{\partial y}+\bar{v}_y\frac{\partial V_z}{\partial y}\right)+\rho_0\left(V_z\frac{\partial\bar{v}_z}{\partial z}+\bar{v}_z\frac{\partial V_z}{\partial z}\right)$$
(4.2.13c)

$$\frac{\partial\bar{\rho}}{\partial t}+\frac{\partial(\rho_0\bar{v}_x)}{\partial x}+\frac{\partial(\rho_0\bar{v}_y)}{\partial y}+\frac{\partial(\rho_0\bar{v}_z)}{\partial z}+\frac{\partial(V_x\bar{\rho})}{\partial x}+\frac{\partial(V_y\bar{\rho})}{\partial y}+\frac{\partial(V_z\bar{\rho})}{\partial z}=0 \quad (4.2.14)$$

$$\bar{\rho}=\frac{\rho_0}{\kappa p_0}\bar{p} \quad (4.2.15)$$

We should understand that the zeroth order equations, which govern the reference fluid flow, are already satisfied by the reference quantities (V_i, p_0, ρ_0). On the other hand, the first order equations, which include the change/deviation from the reference state, are the coupled differential equations for the 5 unknowns $(\bar{v}_i, \bar{p}, \bar{\rho})$.

4.3 Viscous Acoustic Fluid

When the reference state is a uniform flow with viscosity μ, we can assume $V_i, p_0, \rho_0 = const.$, and then the first order Eq. (4.2.13) are reduced to the simpler forms

$$-\frac{\partial\bar{p}}{\partial x}+\frac{\mu}{3}\frac{\partial}{\partial x}\left(\frac{\partial\bar{v}_x}{\partial x}+\frac{\partial\bar{v}_y}{\partial y}+\frac{\partial\bar{v}_z}{\partial z}\right)+\mu\left(\frac{\partial^2\bar{v}_x}{\partial x^2}+\frac{\partial^2\bar{v}_x}{\partial y^2}+\frac{\partial^2\bar{v}_x}{\partial z^2}\right)+\rho_0\bar{B}_x$$
$$=+\rho_0\frac{\partial\bar{v}_x}{\partial t}+\rho_0\left(V_x\frac{\partial\bar{v}_x}{\partial x}+V_y\frac{\partial\bar{v}_x}{\partial y}+V_z\frac{\partial\bar{v}_x}{\partial z}\right) \quad (4.3.1a)$$

$$-\frac{\partial\bar{p}}{\partial y}+\frac{\mu}{3}\frac{\partial}{\partial y}\left(\frac{\partial\bar{v}_x}{\partial x}+\frac{\partial\bar{v}_y}{\partial y}+\frac{\partial\bar{v}_z}{\partial z}\right)+\mu\left(\frac{\partial^2\bar{v}_y}{\partial x^2}+\frac{\partial^2\bar{v}_y}{\partial y^2}+\frac{\partial^2\bar{v}_y}{\partial z^2}\right)+\rho_0\bar{B}_y$$
$$=+\rho_0\frac{\partial\bar{v}_y}{\partial t}+\rho_0\left(V_x\frac{\partial\bar{v}_y}{\partial x}+V_y\frac{\partial\bar{v}_y}{\partial y}+V_z\frac{\partial\bar{v}_y}{\partial z}\right) \quad (4.3.1b)$$

$$-\frac{\partial\bar{p}}{\partial z}+\frac{\mu}{3}\frac{\partial}{\partial z}\left(\frac{\partial\bar{v}_x}{\partial x}+\frac{\partial\bar{v}_y}{\partial y}+\frac{\partial\bar{v}_z}{\partial z}\right)+\mu\left(\frac{\partial^2\bar{v}_z}{\partial x^2}+\frac{\partial^2\bar{v}_z}{\partial y^2}+\frac{\partial^2\bar{v}_z}{\partial z^2}\right)+\rho_0\bar{B}_z$$
$$=+\rho_0\frac{\partial\bar{v}_z}{\partial t}+\rho_0\left(V_x\frac{\partial\bar{v}_z}{\partial x}+V_y\frac{\partial\bar{v}_z}{\partial y}+V_z\frac{\partial\bar{v}_z}{\partial z}\right) \quad (4.3.1c)$$

With use of the linear relation between the pressure and density deviations, the continuity equation (4.2.14) can be rewritten in terms of the pressure and velocity gradients as

$$\frac{\partial \bar{p}}{\partial t} + V_x \frac{\partial \bar{p}}{\partial x} + V_y \frac{\partial \bar{p}}{\partial y} + V_z \frac{\partial \bar{p}}{\partial z} + \kappa p_0 \left(\frac{\partial \bar{v}_x}{\partial x} + \frac{\partial \bar{v}_y}{\partial y} + \frac{\partial \bar{v}_z}{\partial z} \right) = 0 \tag{4.3.2}$$

The above four Eqs. (4.3.1) and (4.3.2) are the governing equations for linearized acoustic waves in the uniformly flowing viscous fluid. The coupled differential equations, which have four unknowns $(\bar{v}_i, \bar{p})$, can be reduced to a single differential equation by the introduction of a velocity potential. We assume that the velocity components $(\bar{v}_i)$ can be derived from a single velocity potential $\phi(x, y, z, t)$ as

$$\bar{v}_x = \frac{\partial \phi}{\partial x}, \quad \bar{v}_y = \frac{\partial \phi}{\partial y}, \quad \bar{v}_z = \frac{\partial \phi}{\partial z} \tag{4.3.3}$$

As to the body force, we also assume that there exists a body force potential $B(x, y, z, t)$ and each body force component is derived as,

$$\bar{B}_x = \frac{\partial B}{\partial x}, \quad \bar{B}_y = \frac{\partial B}{\partial y}, \quad \bar{B}_z = \frac{\partial B}{\partial z} \tag{4.3.4}$$

Substituting Eqs. (4.3.3) and (4.3.4) into (4.3.1a), we obtain

$$\frac{\partial \bar{p}}{\partial x} + \rho_0 \frac{\partial}{\partial x} \left(\frac{\partial \phi}{\partial t} + V_x \frac{\partial \phi}{\partial x} + V_y \frac{\partial \phi}{\partial y} + V_z \frac{\partial \phi}{\partial z} \right) - \frac{4\mu}{3} \frac{\partial}{\partial x} \left(\frac{\partial^2 \phi}{\partial x^2} + \frac{\partial^2 \phi}{\partial y^2} + \frac{\partial^2 \phi}{\partial z^2} \right) - \rho_0 \frac{\partial B}{\partial x} = 0 \tag{4.3.5}$$

Since all terms in the above equation are derivatives with respect to the space variable x, we integrate it with respect to the variable x,

$$\bar{p} = -\rho_0 \left(\frac{\partial \phi}{\partial t} + V_x \frac{\partial \phi}{\partial x} + V_y \frac{\partial \phi}{\partial y} + V_z \frac{\partial \phi}{\partial z} \right) + \frac{4\mu}{3} \left(\frac{\partial^2 \phi}{\partial x^2} + \frac{\partial^2 \phi}{\partial y^2} + \frac{\partial^2 \phi}{\partial z^2} \right) + \rho_0 B \tag{4.3.6}$$

where the constant term is neglected since it is one of the reference quantities. Substitution into Eqs. (4.3.1b) and (4.3.1c) leads to the same equation and thus the equations of motion in all three directions are reduced to the single Eq. (4.3.6).

The velocity potential defined by Eq. (4.3.3) is also substituted into the continuity equation (4.3.2),

$$\frac{\partial \bar{p}}{\partial t} + V_x \frac{\partial \bar{p}}{\partial x} + V_y \frac{\partial \bar{p}}{\partial y} + V_z \frac{\partial \bar{p}}{\partial z} + \kappa p_0 \left(\frac{\partial^2 \phi}{\partial x^2} + \frac{\partial^2 \phi}{\partial y^2} + \frac{\partial^2 \phi}{\partial z^2} \right) = 0 \tag{4.3.7}$$

Then, we have the coupled differential Eqs. (4.3.6) and (4.3.7) in terms of only two unknowns, the velocity potential and the pressure deviation. Further, if we substitute Eq. (4.3.6) into Eq. (4.3.7), we have just the single differential equation for the velocity potential ϕ,

$$\nabla^2\phi + \frac{4\mu}{3\kappa p_0}\left(\frac{\partial}{\partial t} + V_x\frac{\partial}{\partial x} + V_y\frac{\partial}{\partial y} + V_z\frac{\partial}{\partial z}\right)\nabla^2\phi$$
$$= \frac{\rho_0}{\kappa p_0}\left(\frac{\partial}{\partial t} + V_x\frac{\partial}{\partial x} + V_y\frac{\partial}{\partial y} + V_z\frac{\partial}{\partial z}\right)^2\phi - \frac{\rho_0}{\kappa p_0}\left(\frac{\partial B}{\partial t} + V_x\frac{\partial B}{\partial x} + V_y\frac{\partial B}{\partial y} + V_z\frac{\partial B}{\partial z}\right) \tag{4.3.8}$$

where ∇^2 is the Laplacian operator defined by

$$\nabla^2 = \frac{\partial^2}{\partial x^2} + \frac{\partial^2}{\partial y^2} + \frac{\partial^2}{\partial z^2} \tag{4.3.9}$$

We can also derive the differential equation for the pressure deviation. We apply the Laplacian to both sides of Eq. (4.3.6),

$$\nabla^2\bar{p} = -\rho_0\left(\frac{\partial}{\partial t} + V_x\frac{\partial}{\partial x} + V_y\frac{\partial}{\partial y} + V_z\frac{\partial}{\partial z}\right)\nabla^2\phi + \frac{4\mu}{3}\nabla^2(\nabla^2\phi) + \rho_0\nabla^2 B \tag{4.3.10}$$

and derive the Laplacian of the velocity potential from Eq. (4.3.7) as

$$\nabla^2\phi = -\frac{1}{\kappa p_0}\left(\frac{\partial\bar{p}}{\partial t} + V_x\frac{\partial\bar{p}}{\partial x} + V_y\frac{\partial\bar{p}}{\partial y} + V_z\frac{\partial\bar{p}}{\partial z}\right) \tag{4.3.11}$$

The Laplacian of the velocity potential in Eq. (4.3.10) is replaced with the above pressure gradient. We therefore have the single differential equation for the pressure deviation as

$$\nabla^2\bar{p} + \frac{4\mu}{3\kappa p_0}\left(\frac{\partial}{\partial t} + V_x\frac{\partial}{\partial x} + V_y\frac{\partial}{\partial y} + V_z\frac{\partial}{\partial z}\right)\nabla^2\bar{p}$$
$$= \frac{\rho_0}{\kappa p_0}\left(\frac{\partial}{\partial t} + V_x\frac{\partial}{\partial x} + V_y\frac{\partial}{\partial y} + V_z\frac{\partial}{\partial z}\right)^2\bar{p} + \rho_0\left(\frac{\partial^2 B}{\partial x^2} + \frac{\partial^2 B}{\partial y^2} + \frac{\partial^2 B}{\partial z^2}\right) \tag{4.3.12}$$

Two differential Eqs. (4.3.8) and (4.3.12) for the potential and the pressure are mathematically the same, but the non-homogeneous terms of the body force are little bit different. We easily see that both equations are wave equations and their wave velocity is

$$c = \sqrt{(\kappa p_0)/\rho_0} \tag{4.3.13}$$

If we introduce a reference length l and define Reynolds number,

$$\mathrm{Re} = (\rho_0 c l)/\mu \tag{4.3.14}$$

and Mach numbers,

$$M_j = V_j/c, \quad j = x, y, z \tag{4.3.15}$$

the governing equation for the viscous acoustic media is simplified to

$$\begin{aligned}&\nabla^2\phi + \frac{4l}{3\mathrm{Re}}\left(\frac{1}{c}\frac{\partial}{\partial t} + M_x\frac{\partial}{\partial x} + M_y\frac{\partial}{\partial y} + M_z\frac{\partial}{\partial z}\right)\nabla^2\phi \\ &= \left(\frac{1}{c}\frac{\partial}{\partial t} + M_x\frac{\partial}{\partial x} + M_y\frac{\partial}{\partial y} + M_z\frac{\partial}{\partial z}\right)^2\phi - \frac{1}{c}\left(\frac{1}{c}\frac{\partial B}{\partial t} + M_x\frac{\partial B}{\partial x} + M_y\frac{\partial B}{\partial y} + M_z\frac{\partial B}{\partial z}\right)\end{aligned} \tag{4.3.16}$$

for the velocity potential, and to

$$\begin{aligned}&\nabla^2\bar{p} + \frac{4l}{3\mathrm{Re}}\left(\frac{1}{c}\frac{\partial}{\partial t} + M_x\frac{\partial}{\partial x} + M_y\frac{\partial}{\partial y} + M_z\frac{\partial}{\partial z}\right)\nabla^2\bar{p} \\ &= \left(\frac{1}{c}\frac{\partial}{\partial t} + M_x\frac{\partial}{\partial x} + M_y\frac{\partial}{\partial y} + M_z\frac{\partial}{\partial z}\right)^2\bar{p} + \rho_0\left(\frac{\partial^2 B}{\partial x^2} + \frac{\partial^2 B}{\partial y^2} + \frac{\partial^2 B}{\partial z^2}\right)\end{aligned} \tag{4.3.17}$$

for the pressure deviation. These two differential equations have only two parameters, Reynolds and Mach numbers. Thus, the nature of acoustic waves in the flowing viscous fluid is characterized by these two numbers.

4.4 Wave Radiation in a Uniform Flow

This section discusses a 2D wave propagation problem in a uniform flow. Take the (x, y) plane and assume that all quantities are independent of z, and assume that the non-viscous acoustic fluid ($\mu = 0$) is flowing along the x-axis with the uniform velocity V_x (see Fig. 4.1). Further, we also assume the Laplacian of the body force potential as the wave source Q,

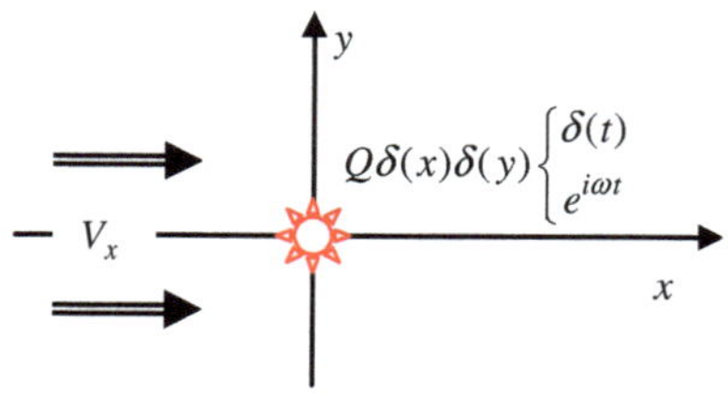

Fig. 4.1 A point source in a uniformly flowing fluid

$$\rho_0\left(\frac{\partial^2 B}{\partial x^2}+\frac{\partial^2 B}{\partial y^2}\right)=Q\delta(x)\delta(y)\delta(t) \tag{4.4.1}$$

Then, we substitute $\mathrm{Re}=\infty$, $M_y=M_z=0$ into Eq. (4.3.17). The governing equation for the pressure deviation is then reduced to the simple form

$$\nabla^2 p=\left(\frac{1}{c}\frac{\partial}{\partial t}+M_x\frac{\partial}{\partial x}\right)^2 p+Q\delta(x)\delta(y)\delta(t) \tag{4.4.2}$$

where c is the acoustic wave velocity defined by Eq. (4.3.13), and Mach number M_x is defined by Eq. (4.3.15). Furthermore, the pressure in this equation is the deviation from the reference state and the correct notation is $\bar{p}$ (over bar). However, we do not use the over bar for the pressure deviation since we apply the double Fourier transform and one of the transforms is classified by the over bar. So, in order to avoid the confusion, the over bar is dropped from the pressure deviation.

Since our acoustic field is of infinite extent, it is enough to obtain a particular solution of Eq. (4.4.2). Laplace transform with respect to the time t and double Fourier transform with respect to two space variables x and y are applied. The triple integral transform is defined as

$$\text{Laplace transform: } f^*(s)=\int_0^\infty f(t)\exp(-st)dt \tag{4.4.3}$$

$$\text{Fourier transform: } \bar{f}(\xi)=\int_0^\infty f(x)\exp(+i\xi x)dx,\quad f(x)=\frac{1}{2\pi}\int_0^\infty \bar{f}(\xi)\exp(-i\xi x)d\xi \tag{4.4.4}$$

$$\text{Fourier transform: } \tilde{f}(\eta)=\int_0^\infty f(y)\exp(+i\eta y)dy,\quad f(y)=\frac{1}{2\pi}\int_0^\infty \tilde{f}(\eta)\exp(-i\eta y)d\eta \tag{4.4.5}$$

Applying the triple transform to Eq. (4.4.2), the simple algebraic equation for the transformed pressure is obtained as

$$-(\xi^2+\eta^2)\tilde{\bar{p}}^* = (s/c - i\xi M_x)^2\tilde{\bar{p}}^* + Q \tag{4.4.6}$$

The pressure in the transformed domain is thus given by

$$\tilde{\bar{p}}^* = -\frac{Q}{\xi^2+\eta^2+(s/c-i\xi M_x)^2} = -\frac{Q}{\eta^2+q^2} \tag{4.4.7}$$

where q is written as

$$\begin{aligned} q^2 &= \xi^2 + (s/c - i\xi M_x)^2 \\ &= (1-M_x^2)\left\{\left(\xi - \frac{iM_x}{1-M_x^2}(s/c)\right)^2 + \left(\frac{s/c}{1-M_x^2}\right)^2\right\} \\ &= (1-M_x^2)\left\{(\xi - i\alpha M_x)^2 + \alpha^2\right\} \end{aligned} \tag{4.4.8}$$

The Laplace transform parameter s is included in the parameter α which is defined by

$$\alpha = \frac{s/c}{1-M_x^2} \tag{4.4.9}$$

As to the first inversion, we apply the Fourier inversion with respect to the parameter η, and reduce it to the semi-infinite integral,

$$\bar{p}^* = -\frac{1}{2\pi}\int_{-\infty}^{+\infty}\frac{Q}{\eta^2+q^2}\exp(-i\eta y)d\eta = -\frac{Q}{\pi}\int_0^{\infty}\frac{1}{\eta^2+q^2}\cos(\eta y)d\eta \tag{4.4.10}$$

The last integral is easily evaluated by the formula (2.1.22) in Chap. 2 and yields

$$\bar{p}^* = -\frac{Q}{2q}\exp(-q|y|) \tag{4.4.11}$$

The next step is to apply the Fourier inversion integral with respect to the parameter ξ,

$$p^* = -\frac{Q}{2}\frac{1}{2\pi}\int_{-\infty}^{+\infty}\frac{1}{q}\exp(-q|y|)\exp(-i\xi x)d\xi \tag{4.4.12}$$

Recalling q in Eq. (4.4.8), the integral in the above equation is rewritten in the explicit form,

$$p^* = -\frac{Q}{2\sqrt{1-M_x^2}}\frac{1}{2\pi}\int_{-\infty}^{+\infty}\frac{\exp\left[-|y|\sqrt{1-M_x^2}\sqrt{(\xi-i\alpha M_x)^2+\alpha^2}-i\xi x\right]}{\sqrt{(\xi-i\alpha M_x)^2+\alpha^2}}d\xi \tag{4.4.13}$$

Introducing the variable transform, $\xi \to \varsigma$, as

$$\varsigma = \xi - i\alpha M_x \tag{4.4.14}$$

we obtain the simpler integrand, but the integration path is shifted to the complex plane, not on the real axis, i.e.

$$p^* = -\frac{Q\exp(+\alpha M_x x)}{2}\frac{1}{\sqrt{1-M_x^2}}\frac{1}{2\pi}\int_{-\infty-i\alpha M_x}^{+\infty-i\alpha M_x}\frac{\exp\left(-|y|\sqrt{1-M_x^2}\sqrt{\varsigma^2+\alpha^2}-i\varsigma x\right)}{\sqrt{\varsigma^2+\alpha^2}}d\varsigma \tag{4.4.15}$$

The integration path for this integral is slightly shifted from the real axis in the complex ς-plane and is shown by the line CD in Fig. 4.2. In order to transform the integral to that along the real axis, we consider the complex integral Φ with the closed loop $L(\overrightarrow{\mathrm{CDBAC}})$ in the figure,

$$\Phi = -\frac{Q\exp(+\alpha M_x x)}{2}\frac{1}{\sqrt{1-M_x^2}}\frac{1}{2\pi}\int_L\frac{\exp\left(-|y|\sqrt{1-M_x^2}\sqrt{\varsigma^2+\alpha^2}-i\varsigma x\right)}{\sqrt{\varsigma^2+\alpha^2}}d\varsigma \tag{4.4.16}$$

The integrand has two branch points, at $\varsigma = \pm i\alpha$. Two branch cuts are introduced along the imaginary axis, as shown in the figure. Fortunately, if we assume subsonic

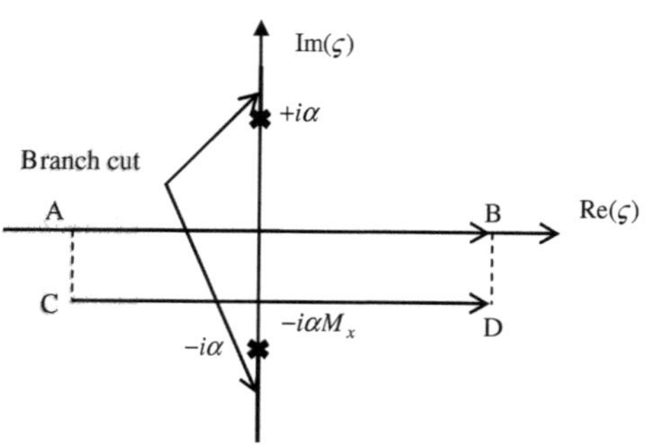

Fig. 4.2 Transform of integration path

motion of the flow, $M_x (= V_x/c) < 1$, these branch cuts do not cross the integration line CD and no singular point are then included in the closed loop L. In addition, the two integrals along the line AC and BD, with an infinite real part, vanish since $\mathrm{Re}\left(\sqrt{\varsigma^2 + \alpha^2}\right) > 0$. Then, applying Cauchy's integral theorem to the complex integral Φ in Eq. (4.4.16), the integral along the complex path CD is converted to that along the real axis AB in the ς-plane. That is

$$p^* = -\frac{Q}{2}\frac{\exp(+\alpha M_x x)}{\sqrt{1 - M_x^2}}\frac{1}{2\pi}\int_{-\infty}^{+\infty}\frac{\exp\left(-|y|\sqrt{1 - M_x^2}\sqrt{\varsigma^2 + \alpha^2} - i\varsigma x\right)}{\sqrt{\varsigma^2 + \alpha^2}}d\varsigma \quad (4.4.17)$$

The above integral can be further reduced to the semi-infinite integral,

$$p^* = -\frac{Q}{2}\frac{\exp(+\alpha M_x x)}{\sqrt{1 - M_x^2}}\frac{1}{\pi}\int_{0}^{\infty}\frac{\exp\left(-|y|\sqrt{1 - M_x^2}\sqrt{\varsigma^2 + \alpha^2}\right)}{\sqrt{\varsigma^2 + \alpha^2}}\cos(\varsigma x)d\varsigma \quad (4.4.18)$$

Applying the integration identity, which is the integral representation of the modified Bessel function (Erdélyi 1954, vol. I, pp. 17, 27; Watson 1966, p. 172),

$$\begin{aligned} K_0\left(\alpha\sqrt{x^2 + y^2}\right) &= \int_{0}^{\infty}\frac{\exp\left(-y\sqrt{\varsigma^2 + \alpha^2}\right)}{\sqrt{\varsigma^2 + \alpha^2}}\cos(\varsigma x)d\varsigma \\ &= \int_{1}^{\infty}\frac{\exp\left(-\alpha u\sqrt{x^2 + y^2}\right)}{\sqrt{u^2 - 1}}du \end{aligned} \quad (4.4.19)$$

to Eq. (4.4.18), we have the other integral form for the pressure,

$$p^* = -\frac{Q}{2\pi}\frac{\exp(+\alpha M_x x)}{\sqrt{1 - M_x^2}}\int_{1}^{\infty}\frac{1}{\sqrt{u^2 - 1}}\exp\left\{-\alpha u\sqrt{x^2 + (1 - M_x^2)y^2}\right\}du \quad (4.4.20)$$

Further, we recall the definition of α which includes the Laplace transform parameter s. The above Eq. (4.4.20) can be rewritten as

$$p^* = -\frac{Q}{2\pi}\frac{1}{\sqrt{1 - M_x^2}}\int_{1}^{\infty}\frac{\exp\left\{-\frac{s/c}{1 - M_x^2}\left(u\sqrt{x^2 + (1 - M_x^2)y^2} - M_x x\right)\right\}}{\sqrt{u^2 - 1}}du \quad (4.4.21)$$

Introducing the variable transform, $u \to t$, as defined by

$$t = \frac{u\sqrt{x^2 + (1 - M_x^2)y^2} - M_x x}{c(1 - M_x^2)}, \quad u = \frac{(1 - M_x^2)ct + M_x x}{\sqrt{x^2 + (1 - M_x^2)y^2}},$$
$$du = \frac{c(1 - M_x^2)}{\sqrt{x^2 + (1 - M_x^2)y^2}} dt \tag{4.4.22}$$

Equation (4.4.21) is converted to the form of the Laplace transform integral,

$$p^* = -\frac{cQ}{2\pi}\sqrt{1 - M_x^2} \int\limits_{\frac{\sqrt{x^2+(1-M_x^2)y^2}-M_x x}{c(1-M_x^2)}}^{\infty} \frac{\exp(-st)dt}{\sqrt{\{(1 - M_x^2)ct + M_x x\}^2 - \{x^2 + (1 - M_x^2)y^2\}}} \tag{4.4.23}$$

The above equation is the *Laplace transformed* pressure, but it is just in the form of the *Laplace transform integral*. Thus, the original pressure is its integrand with a shifted starting time since the lower limit is not zero. Inspecting Eq. (4.4.23), we can find the original pressure as

$$p = -\frac{cQ}{2\pi} \frac{\sqrt{1 - M_x^2}}{\sqrt{\{(1 - M_x^2)ct + M_x x\}^2 - \{x^2 + (1 - M_x^2)y^2\}}} \tag{4.4.24}$$

where its valid time range is

$$\frac{\sqrt{x^2 + (1 - M_x^2)y^2} - M_x x}{c(1 - M_x^2)} < t < \infty \tag{4.4.25}$$

In order to obtain a more compact expression, the argument in the radical and the time range are rewritten as

$$\sqrt{x^2 + (1 - M_x^2)y^2} < ct(1 - M_x^2) + M_x x \quad \Rightarrow (x - V_x t)^2 + y^2 < (ct)^2 \tag{4.4.26}$$

$$\{(1 - M_x^2)ct + M_x x\}^2 - \{x^2 + (1 - M_x^2)y^2\} = (1 - M_x^2)\left\{(ct)^2 - (x - V_x t)^2 - y^2\right\} \tag{4.4.27}$$

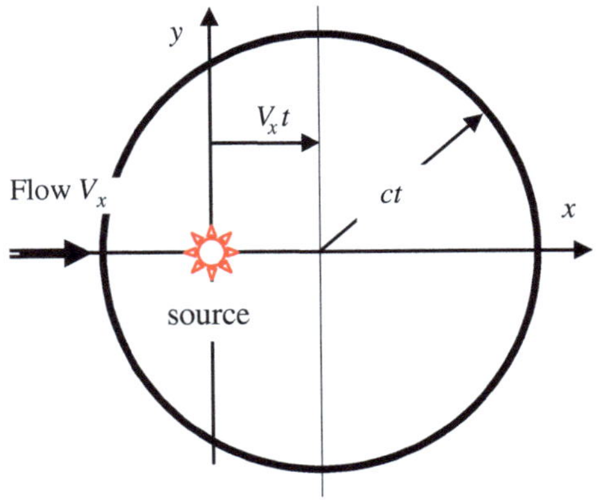

Fig. 4.3 Disturbed circular region in a uniform flow

Finally, the pressure deviation in the flowing fluid is expressed in the simple form,

$$p = -\frac{Q}{2\pi}\frac{1}{\sqrt{t^2 - \left\{(x - V_x t)^2 + y^2\right\}/c^2}} H\left(ct - \sqrt{(x - V_x t)^2 + y^2}\right) \tag{4.4.28}$$

We have just arrived at the final form of the solution that gives the pressure fluctuations in the flowing fluid with the stationary impulsive point source. The disturbed region is easily derived from the argument of the step function: it is a circle but its center is moving with the flow velocity,

$$(x - V_x t)^2 + y^2 = (ct)^2 \tag{4.4.29}$$

Figure 4.3 shows the typical wave front and the circular disturbed region in the flowing fluid.

4.5 Time-Harmonic Wave in a Uniform Flow

The steady-state acoustic response produced by a stationary time-harmonic source is governed by

$$\nabla^2 p - \left(\frac{1}{c}\frac{\partial}{\partial t} + M_x \frac{\partial}{\partial x}\right)^2 p = Q\delta(x)\delta(y)\exp(+i\omega t) \tag{4.5.1}$$

A solution to this equation can be obtained by the convolution integral of the impulsive response as

$$p = \lim_{t\to\infty} \int_0^t p^{(impulse)}(x, y, t') \exp\{i\omega(t - t')\}dt' \tag{4.5.2}$$

Substituting the impulsive response of Eq. (4.4.24), not the final one, we have

$$p = -\frac{cQ}{2\pi}\exp(i\omega t) \int_{\frac{\sqrt{x^2+(1-M_x^2)y^2}-M_x x}{c(1-M_x^2)}}^{\infty} \frac{\sqrt{1-M_x^2}\exp(-i\omega t')dt'}{\sqrt{\{(1-M_x^2)ct' + M_x x\}^2 - \{x^2 + (1-M_x^2)y^2\}}} \tag{4.5.3}$$

where the source magnitude Q is not the same as that of the impulsive source. For this integral, we make the change of variable, $t' \to u$, as

$$u = (1 - M_x^2)ct' + M_x x, \quad t' = \frac{u - M_x x}{(1 - M_x^2)c}, \quad dt' = \frac{du}{(1 - M_x^2)c} \tag{4.5.4}$$

Equation (4.5.3) is now rewritten as

$$p = -\frac{Q}{2\pi}\frac{1}{\sqrt{1-M_x^2}}\exp\left\{i\omega\left(t + \frac{M_x}{1-M_x^2}\frac{x}{c}\right)\right\} \int_{\sqrt{x^2+(1-M_x^2)y^2}}^{\infty} \frac{\exp\left\{-\frac{i\omega u}{(1-M_x^2)c}\right\}}{\sqrt{u^2 - \{x^2 + (1-M_x^2)y^2\}}}du \tag{4.5.5}$$

Since the integral in the above equation is just the definition integral of the Hankel function of the second kind (which is Eq. (3.2.5) in Sect. 3.2), we replace the integral with the Hankel function. The final form for the time-harmonic response is then given by

$$p = \frac{iQ}{4\sqrt{1-M_x^2}} H_0^{(2)}\left(\frac{\omega}{c}\frac{\sqrt{x^2 + (1-M_x^2)y^2}}{1-M_x^2}\right)\exp\left\{i\omega\left(t + \frac{M_x}{1-M_x^2}\frac{x}{c}\right)\right\} \tag{4.5.6}$$

Exercises

(4.1) When a wave source is moving with uniform velocity V along the x-axis, how do you change the nonhomogeneous wave source term in Eq. (4.4.1)?

(4.2) Using the asymptotic formula for the Hankel function (Watson 1966, p. 198),

$$H_{\nu}^{(2)}(z) \sim \sqrt{\frac{2}{\pi z}} \exp\{-i(z - \nu\pi/2 - \pi/4)\} \tag{a}$$

show that the time-harmonic response (4.5.6) is the out-going wave.

References

Erdélyi A (ed) (1954) Tables of integral transforms, vols I and II. McGraw-Hill, New York
Watson GN (1966) A treatise on the theory of bessel functions. Cambridge University Press, Cambridge

Chapter 5
Green's Functions for Beam and Plate

This chapter presents dynamic Green's functions for the elastic beam and plate. The Green's function is the deflection response produced by a point load. The deflection equation with the nonhomogeneous term of the applied load is discussed by applying the method of integral transform.

5.1 An Impulsive Load on a Beam

We shall obtain Green's function for an elastic beam. The deflection of the beam, $w(x, t)$, is governed by the well-known partial differential equation, so called beam/deflection equation,

$$EI\frac{\partial^4 w}{\partial x^4} + \rho A\frac{\partial^2 w}{\partial t^2} = p(x,t) \tag{5.1.1}$$

where EI is bending rigidity, ρA mass per unit length and $p(x, t)$ the load on the beam. An impulsive point loadwith magnitude P is assumed by

$$p(x,t) = P\delta(x)\delta(t) \tag{5.1.2}$$

where $\delta(.)$ is Dirac's delta function. The nonhomogeneous deflection equation (5.1.1) is rewritten as

$$\alpha^4\frac{\partial^4 w}{\partial x^4} + \frac{\partial^2 w}{\partial t^2} = Q\delta(x)\delta(t) \tag{5.1.3}$$

where

$$\alpha^4 = \frac{EI}{\rho A}, \quad Q = \frac{P}{\rho A} \tag{5.1.4}$$

K. Watanabe, *Integral Transform Techniques for Green's Function*,
Lecture Notes in Applied and Computational Mechanics 76,
DOI 10.1007/978-3-319-17455-6_5

For solving the deflection equation, we employ the quiescent condition at an initial time,

$$w|_{t=0} = \left.\frac{\partial w}{\partial x}\right|_{t=0} = 0 \tag{5.1.5}$$

and the convergence condition at infinity,

$$w|_{|x|\to\infty} = \left.\frac{\partial w}{\partial x}\right|_{|x|\to\infty} = \left.\frac{\partial^2 w}{\partial x^2}\right|_{|x|\to\infty} = \left.\frac{\partial^3 w}{\partial x^3}\right|_{|x|\to\infty} = 0 \tag{5.1.6}$$

The higher derivatives in the above equation mean the vanishing of the moment and shear force at infinity.

In order to solve the nonhomogeneous deflection equation, we employ the method of integral transform. Since the deflection depends on two variables, we apply the following double transform: Laplace transform with respect to the time variable,

$$f^*(s) = \int_0^\infty f(t)\exp(-st)dt \tag{5.1.7}$$

and Fourier transform with respect to the space variable,

$$\bar{f}(\xi) = \int_{-\infty}^{+\infty} f(x)\exp(+i\xi x)dx, \qquad f(x) = \frac{1}{2\pi}\int_{-\infty}^{+\infty} \bar{f}(\xi)\exp(-i\xi x)d\xi \tag{5.1.8}$$

Applying the double transform to the deflection equation (5.1.3), we have the algebraic equation for the transformed deflection and then the deflection in the transformed domain is given by

$$\bar{w}^* = \frac{Q}{(\alpha\xi)^4 + s^2} \tag{5.1.9}$$

The Laplace inversion is applied firstly. The Laplace inversion formula (Erdélyi 1954, vol. I, pp. 150, 1),

$$L^{-1}\left(\frac{1}{s^2 + a^2}\right) = \frac{1}{a}\sin(at) \tag{5.1.10}$$

is applied to Eq. (5.1.9) and thus we have for the Fourier transformed deflection,

$$\bar{w} = \frac{Q}{(\alpha\xi)^2}\sin(\alpha^2 t\xi^2) \tag{5.1.11}$$

The Fourier inversion integral of the above equation is reduced to the semi-infinite integral, as

$$w = \frac{1}{2\pi}\int_{-\infty}^{+\infty}\frac{Q}{(\alpha\xi)^2}\sin(\alpha^2 t\xi^2)\exp(-i\xi x)d\xi = \frac{Q}{\pi\alpha^2}\int_0^{\infty}\frac{1}{\xi^2}\sin(\alpha^2 t\xi^2)\cos(x\xi)d\xi \tag{5.1.12}$$

Fortunately, we have a convenient formula for the inversion integral. That is the formula (Erdélyi 1954, vol. I, pp. 23, 3)

$$\int_0^{\infty}\frac{1}{x^2}\sin(ax^2)\cos(xy)dx = \frac{\pi}{2}y\left\{S\left(\frac{y}{\sqrt{2\pi a}}\right) - C\left(\frac{y}{\sqrt{2\pi a}}\right)\right\} + \sqrt{\pi a}\sin\left(\frac{y^2}{4a} + \frac{\pi}{4}\right) \tag{5.1.13}$$

where $C(x)$ and $S(x)$ are Fresnel integrals/functions (Erdélyi 1954, vol. II, p. 431) defined by

$$\left.\begin{matrix} C(x) \\ S(x) \end{matrix}\right\} = \frac{1}{\sqrt{2\pi}}\int_0^x \frac{1}{\sqrt{u}}\left\{\begin{matrix} \sin(u) \\ \cos(u) \end{matrix}\right\}du \tag{5.1.14}$$

Applying this formula to the last integral in Eq. (5.1.12) and rewriting the expression, we obtain the Green's function for the dynamic deflection of the beam as

$$w(x,t) = \frac{Q}{\alpha}\sqrt{\frac{t}{\pi}}\left[\frac{\pi z}{\sqrt{2}}\{S(z) - C(z)\} + \sin\left\{\frac{\pi}{4}(2z^2+1)\right\}\right] \tag{5.1.15}$$

where the dimensionless variable z is defined by

$$z = \frac{x}{\alpha\sqrt{2\pi t}} \tag{5.1.16}$$

Note that the Green's function given by Eq. (5.1.15) is the solution for the impulsive response; however, it does not show any wave nature since the deflection wave is dispersive and its wave velocity depends on the frequency. The impulsive source includes an infinite frequency, i.e. the initial disturbance spreads all over the beam at once without showing any wave nature.

5.2 A Moving Time-Harmonic Load on a Beam

When a time-harmonic load with frequency ω is moving with the uniform velocity V, the deflection equation for the beam is given by,

$$EI\frac{\partial^4 w}{\partial x^4} + \rho A\frac{\partial^2 w}{\partial t^2} = P\delta(x - Vt)\exp(+i\omega t) \tag{5.2.1}$$

where the load location is expressed by the delta function and is at $x = Vt$. The convergence condition given by Eq. (5.1.6) is also assumed.

Since the time variable t is included not only in the argument of the exponential function, but also in that of delta function, it is not good to assume a simple time-harmonic vibration such as $w(x,t) = w^{\#}(x)\exp(+i\omega t)$. We directly apply the Fourier transform defined by Eq. (5.1.8) to the deflection equation (5.2.1),

$$\frac{d^2\bar{w}}{dt^2} + (\alpha\xi)^4\bar{w} = Q\exp\{+i(V\xi + \omega)t\} \tag{5.2.2}$$

where α and Q are defined by Eq. (5.1.4). The particular solution corresponding to the nonhomogeneous loading term is easily obtained as

$$\bar{w} = \frac{Q}{(\alpha\xi)^4 - (\xi V + \omega)^2}\exp\{+i(\xi V + \omega)t\} \tag{5.2.3}$$

Factorizing the denominator, we get

$$(\alpha\xi)^4 - (\xi V + \omega)^2 = \alpha^4(\xi - \xi_1^+)(\xi - \xi_1^-)(\xi - \xi_2^+)(\xi - \xi_2^-) \tag{5.2.4}$$

where the eigenvalues are given by

$$\xi_1^{\pm} = \frac{1}{\alpha}\left\{\pm\sqrt{\omega + \left(\frac{V}{2\alpha}\right)^2} + \frac{V}{2\alpha}\right\}, \quad \xi_2^{\pm} = \frac{1}{\alpha}\left\{\pm i\sqrt{\omega - \left(\frac{V}{2\alpha}\right)^2} - \frac{V}{2\alpha}\right\} \tag{5.2.5}$$

the Fourier inversion integral with respect to the parameter ξ is expressed as

$$w = \frac{Q}{2\pi\alpha^4}\exp(+i\omega t)\int_{-\infty}^{+\infty}\frac{\exp\{-i\xi(x - Vt)\}}{(\xi - \xi_1^+)(\xi - \xi_1^-)(\xi - \xi_2^+)(\xi - \xi_2^-)}d\xi \tag{5.2.6}$$

In order to evaluate the integral in Eq. (5.2.6), we apply the theory of complex integrals (the residue theorem). Following the discussion in Sect. 2.2, the complex frequency with a small negative imaginary part is assumed. Due to this assumption, all singular points do not lie on the real axis. Two singular points, ξ_1^- and ξ_2^+, are in the upper complex ξ-plane and the other two, ξ_1^+ and ξ_2^-, are in the lower plane. Then, the complex integral

$$\Phi = \frac{1}{2\pi}\int\limits_{C^{\pm}} \frac{\exp\{-i\xi(x - Vt)\}}{(\xi - \xi_1^+)(\xi - \xi_1^-)(\xi - \xi_2^+)(\xi - \xi_2^-)} d\xi \tag{5.2.7}$$

is to be discussed. Two closed loops $C^{\pm}$ are shown in Fig. 5.1. We employ the circuit C^+ in the case of $x - Vt < 0$ and C^- in that of $x - Vt > 0$ in order to guarantee the convergence on the large semi-circle. Jordan's lemma is applied to the complex integral Φ and after somewhat lengthy calculation we arrive at the expressions,

$$\begin{aligned}
&\frac{1}{2\pi}\int\limits_{-\infty}^{+\infty} \frac{\exp\{-i\xi(x - Vt)\}}{(\xi - \xi_1^+)(\xi - \xi_1^-)(\xi - \xi_2^+)(\xi - \xi_2^-)} d\xi \\
&= -\frac{i\alpha^3}{4}\frac{\exp\left[-(i/\alpha)\left\{\sqrt{\omega + (V/2\alpha)^2} + (V/2\alpha)\right\}(x - Vt)\right]}{\sqrt{\omega + (V/2\alpha)^2}\left\{\sqrt{\omega + (V/2\alpha)^2} + (V/2\alpha)\right\}^2} \\
&\quad - \frac{\alpha^3}{4}\frac{\exp\left[(1/\alpha)\left\{-\sqrt{\omega - (V/2\alpha)^2} + i(V/2\alpha)\right\}(x - Vt)\right]}{\sqrt{\omega - (V/2\alpha)^2}\left\{\sqrt{\omega - (V/2\alpha)^2} - i(V/2\alpha)\right\}^2}; \quad x - Vt > 0 \\
&= -\frac{i\alpha^3}{4}\frac{\exp\left[(i/\alpha)\left\{\sqrt{\omega + (V/2\alpha)^2} - (V/2\alpha)\right\}(x - Vt)\right]}{\sqrt{\omega + (V/2\alpha)^2}\left\{\sqrt{\omega + (V/2\alpha)^2} - (V/2\alpha)\right\}^2} \\
&\quad - \frac{\alpha^3}{4}\frac{\exp\left[(1/\alpha)\left\{\sqrt{\omega - (V/2\alpha)^2} + i(V/2\alpha)\right\}(x - Vt)\right]}{\sqrt{\omega - (V/2\alpha)^2}\left\{\sqrt{\omega - (V/2\alpha)^2} + i(V/2\alpha)\right\}^2}; \quad x - Vt < 0
\end{aligned} \tag{5.2.8}$$

Then, the deflection produced by the moving time-harmonic load is given by

$$\begin{aligned}
w(x,t) = &-\frac{iQ}{4\alpha}\frac{\exp\left\{-(i/\alpha)\left(\sqrt{\omega + \Omega^2} + \Omega\right)(x - Vt) + i\omega t\right\}}{\sqrt{\omega + \Omega^2}\left(\sqrt{\omega - \Omega^2} + \Omega\right)^2} \\
&- \frac{Q}{4\alpha}\frac{\exp\left\{(1/\alpha)\left(-\sqrt{\omega - \Omega^2} + i\Omega\right)(x - Vt) + i\omega t\right\}}{\sqrt{\omega - \Omega^2}\left(\sqrt{\omega - \Omega^2} - i\Omega\right)^2}; \quad x - Vt > 0
\end{aligned} \tag{5.2.9a}$$

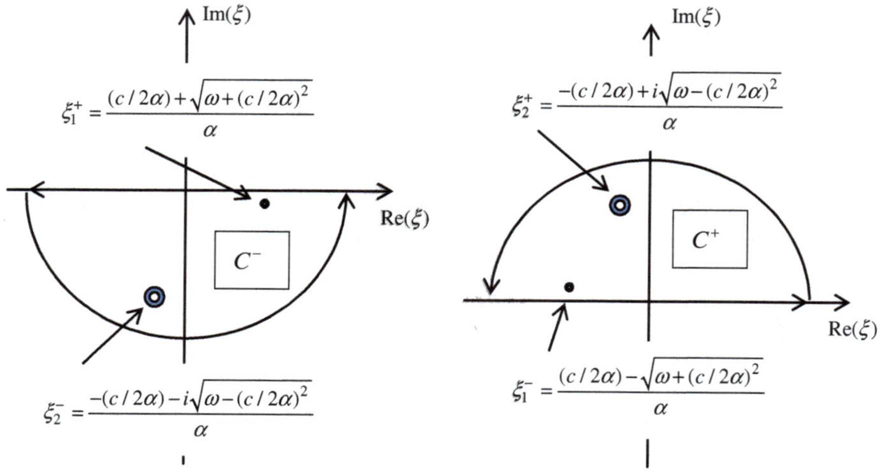

Fig. 5.1 Closed loops $C^{\pm}$ for the complex integral Φ

$$
\begin{aligned}
w(x,t) = & -\frac{iQ}{4\alpha}\frac{\exp\left\{(i/\alpha)\left(\sqrt{\omega+\Omega^2}-\Omega\right)(x-Vt)+i\omega t\right\}}{\sqrt{\omega+\Omega^2}\left(\sqrt{\omega-\Omega^2}-\Omega\right)^2} \\
& +\frac{Q}{4\alpha}\frac{\exp\left\{(1/\alpha)\left(\sqrt{\omega-\Omega^2}+i\Omega\right)(x-Vt)+i\omega t\right\}}{\sqrt{\omega-\Omega^2}\left(\sqrt{\omega-\Omega^2}+i\Omega\right)^2}; \quad x-Vt<0
\end{aligned}
\tag{5.2.9b}
$$

where

$$
\Omega = \frac{V}{2\alpha} \tag{5.2.10}
$$

Inspecting the above equations, the radical $\sqrt{\omega-\Omega^2}$ in the denominator gives a critical velocity

$$
V_{cr} = 2\alpha\sqrt{\omega} \tag{5.2.11}
$$

at which the deflection divergence takes place.

When the load is stationary at the coordinate origin, the simple time-harmonic response is obtained by taking the limit $\Omega \to 0$ $(V \to 0)$ in Eq. (5.2.10). It yields

$$
w(x,t) = -\frac{Q}{4\alpha\sqrt{\omega}^3}\left[i\exp\left\{+i\left(\omega t-\frac{\sqrt{\omega}}{\alpha}|x|\right)\right\}+\exp\left(-\frac{\sqrt{\omega}}{\alpha}|x|+i\omega t\right)\right] \tag{5.2.12}
$$

This is the particular solution of the deflection equation with the time-harmonic stationary load,

$$\alpha^4 \frac{\partial^4 w}{\partial x^4} + \frac{\partial^2 w}{\partial t^2} = Q\delta(x)\exp(+i\omega t) \tag{5.2.13}$$

5.3 An Impulsive Load on a Plate

Let us consider an impulsive response of an infinite elastic plate. Taking the (x, y) coordinate on the neutral plane of the plate, the deflection equation for the plate with an impulsive point loadis given by

$$D\left(\frac{\partial^4 w}{\partial x^4} + 2\frac{\partial^4 w}{\partial x^2 \partial y^2} + \frac{\partial^4 w}{\partial y^4}\right) + \rho h \frac{\partial^2 w}{\partial t^2} = P\delta(x)\delta(y)\delta(t) \tag{5.3.1}$$

where D is bending rigidity, ρh density per unit area and P magnitude of the load. The initial condition,

$$w|_{t=0} = \left.\frac{\partial w}{\partial t}\right|_{t=0} = 0 \tag{5.3.2}$$

and the convergence condition,

$$\begin{aligned} w|_{\sqrt{x^2+y^2}\to\infty} = \left.\frac{\partial w}{\partial x}\right|_{\sqrt{x^2+y^2}\to\infty} = \left.\frac{\partial^2 w}{\partial x^2}\right|_{\sqrt{x^2+y^2}\to\infty} = \left.\frac{\partial^3 w}{\partial x^3}\right|_{\sqrt{x^2+y^2}\to\infty} = 0 \\ \left.\frac{\partial w}{\partial y}\right|_{\sqrt{x^2+y^2}\to\infty} = \left.\frac{\partial^2 w}{\partial y^2}\right|_{\sqrt{x^2+y^2}\to\infty} = \left.\frac{\partial^3 w}{\partial y^3}\right|_{\sqrt{x^2+y^2}\to\infty} = \left.\frac{\partial^2 w}{\partial x \partial y}\right|_{\sqrt{x^2+y^2}\to\infty} = 0 \end{aligned} \tag{5.3.3}$$

are applied to the deflection equation.

In order to obtain the particular solution corresponding to the non-homogeneous loading term, we apply the triple integral transform to Eq. (5.3.1). The Laplace transform with respect to the time t,

$$f^*(s) = \int_0^\infty f(t)\exp(-st)dt \tag{5.3.4}$$

and the double Fourier transform with respect to the space variables (x, y),

$$\bar{f}(\xi) = \int_{-\infty}^{+\infty} f(x)\exp(+i\xi x)dx, \qquad f(x) = \frac{1}{2\pi}\int_{-\infty}^{+\infty} \bar{f}(\xi)\exp(-i\xi x)d\xi \tag{5.3.5}$$

$$\tilde{f}(\eta) = \int_{-\infty}^{+\infty} f(y)\exp(+i\eta y)dy, \qquad f(y) = \frac{1}{2\pi}\int_{-\infty}^{+\infty} \tilde{f}(\eta)\exp(-i\eta y)d\eta \tag{5.3.6}$$

are applied to Eq. (5.3.1). The exact solution in the transformed domain is easily obtained as

$$\tilde{\bar{w}}^* = \frac{Q}{s^2 + \{\beta^2(\xi^2+\eta^2)\}^2} \tag{5.3.7}$$

where

$$\beta^4 = \frac{D}{\rho h}, \quad Q = \frac{P}{\rho h} \tag{5.3.8}$$

Now, let us consider the inversion. The Laplace inversion of Eq. (5.3.7) is easily performed with use of the formula (5.1.10). It follows that

$$\tilde{\bar{w}} = \frac{Q}{\beta^2(\xi^2+\eta^2)}\sin\{\beta^2(\xi^2+\eta^2)t\}, \quad t > 0 \tag{5.3.9}$$

The double Fourier inversion is expressed by the double integral as

$$w = \frac{1}{(2\pi)^2}\int_{-\infty}^{+\infty}\int_{-\infty}^{+\infty} \frac{Q}{\beta^2(\xi^2+\eta^2)}\sin\{\beta^2(\xi^2+\eta^2)t\}\exp\{-i(\xi x+\eta y)\}d\xi d\eta \tag{5.3.10}$$

If we introduce the polar coordinates $(r,\ \theta)$ and the polar integration variables $(u,\ \varphi)$ as

$$x = r\cos\theta, \quad y = r\sin\theta \tag{5.3.11}$$

$$\xi = u\sin\varphi, \quad \eta = u\cos\varphi, \quad d\xi d\eta = ud\varphi du \tag{5.3.12}$$

the double integral can be rewritten as

$$w = \frac{Q}{(2\pi)^2\beta^2}\int_0^{\infty}\int_0^{2\pi} \frac{1}{u}\sin(\beta^2 tu^2)\exp\{-iur\sin(\varphi+\theta)\}d\varphi du \tag{5.3.13}$$

The exponential function in the integrand is the generating function of Bessel function (Watson 1966, p. 22),

$$\exp\{-iur\sin(\varphi+\theta)\} = \sum_{-\infty}^{+\infty} J_n(ur)\exp\{-in(\varphi+\theta)\} \tag{5.3.14}$$

where $J_n(.)$ is the n-th order Bessel function of the first kind. We substitute the above into Eq. (5.3.13) and exchange the order of the summation and integration as

$$\begin{aligned} w &= \frac{Q}{(2\pi)^2\beta^2}\int_0^\infty\int_0^{2\pi}\frac{1}{u}\sin(\beta^2tu^2)\sum_{-\infty}^{+\infty}J_n(ur)\exp\{-in(\varphi+\theta)\}d\varphi du \\ &= \frac{Q}{(2\pi)^2\beta^2}\sum_{-\infty}^{+\infty}\int_0^\infty\frac{1}{u}\sin(\beta^2tu^2)J_n(ur)du\int_0^{2\pi}\exp\{-in(\varphi+\theta)\}d\varphi \end{aligned} \tag{5.3.15}$$

The inner integral with respect to the angle variable φ is easily evaluated as

$$\int_0^{2\pi}\exp\{-in(\varphi+\theta)\}d\varphi = \begin{cases} 2\pi; & n=0 \\ 0; & n\neq 0 \end{cases} \tag{5.3.16}$$

Then, we have the single integral for the deflection

$$w = \frac{Q}{2\pi\beta^2}\int_0^\infty\frac{1}{u}\sin(\beta^2tu^2)J_0(ur)du \tag{5.3.17}$$

Lastly, the integration formula (Erdélyi 1954, vol. II, pp. 11, 39),

$$\int_0^\infty\frac{1}{x}\sin(ax^2)J_0(bx)dx = \frac{1}{2}\mathrm{si}\left(\frac{b^2}{4a}\right) \tag{5.3.18}$$

where si(x) is one of sine integrals (Erdélyi 1954, vol. II, p. 430) defined by

$$\mathrm{si}(x) = -\int_x^\infty\frac{\sin(u)}{u}du \tag{5.3.19}$$

is applied to Eq. (5.3.17). The Green's function for the dynamic plate deflection is thus given by

$$w = \frac{Q}{4\pi\beta^2}\mathrm{si}\left(\frac{r^2}{4\beta^2 t}\right), \quad t > 0; \quad r = \sqrt{x^2 + y^2} \tag{5.3.20}$$

5.4 A Time-Harmonic Load on a Plate

Let us consider an elastic plate on which a time-harmonic point load is applied. The dynamic deflectionof the plate is governed by the deflection equation with the nonhomogeneous loading term,

$$D\left(\frac{\partial^4 w}{\partial x^4} + 2\frac{\partial^4 w}{\partial x^2 \partial y^2} + \frac{\partial^4 w}{\partial y^4}\right) + \rho h \frac{\partial^2 w}{\partial t^2} = P\delta(x)\delta(y)\exp(+i\omega t) \tag{5.4.1}$$

where P is the magnitude of the load (but is not of the same dimension as that of the impulsive load in the previous section). The convergence condition at infinity,

$$\begin{aligned} w|_{\sqrt{x^2+y^2}\to\infty} &= \left.\frac{\partial w}{\partial x}\right|_{\sqrt{x^2+y^2}\to\infty} = \left.\frac{\partial^2 w}{\partial x^2}\right|_{\sqrt{x^2+y^2}\to\infty} = \left.\frac{\partial^3 w}{\partial x^3}\right|_{\sqrt{x^2+y^2}\to\infty} = 0 \\ \left.\frac{\partial w}{\partial y}\right|_{\sqrt{x^2+y^2}\to\infty} &= \left.\frac{\partial^2 w}{\partial y^2}\right|_{\sqrt{x^2+y^2}\to\infty} = \left.\frac{\partial^3 w}{\partial y^3}\right|_{\sqrt{x^2+y^2}\to\infty} = \left.\frac{\partial^2 w}{\partial x \partial y}\right|_{\sqrt{x^2+y^2}\to\infty} = 0 \end{aligned} \tag{5.4.2}$$

is also assumed. The double Fourier transform defined by Eqs. (5.3.5) and (5.3.6) is applied to the deflection equation (5.4.1). We then have the simple differential equation,

$$\frac{d^2\tilde{\tilde{w}}}{dt^2} + \beta^4(\xi^2 + \eta^2)^2\tilde{\tilde{w}} = Q\exp(+i\omega t) \tag{5.4.3}$$

and its particular solution is given by

$$\tilde{\tilde{w}} = \frac{Q}{\beta^4(\xi^2 + \eta^2)^2 - \omega^2}\exp(+i\omega t) \tag{5.4.4}$$

where β and Q are defined by Eq. (5.3.8).

The formal Fourier inversion is given by the double integral

$$w = \frac{Q}{(2\pi)^2}\exp(+i\omega t)\int_{-\infty}^{+\infty}\int_{-\infty}^{+\infty}\frac{\exp(-i\xi x - i\eta y)}{\beta^4(\xi^2 + \eta^2)^2 - \omega^2}d\xi d\eta \tag{5.4.5}$$

Introducing the polar coordinates defined by Eqs. (5.3.11) and (5.3.12) in the previous section, the double Fourier inversion integral can be rewritten as

$$w = \frac{Q}{(2\pi)^2}\exp(+i\omega t)\int_0^{\infty}\int_0^{2\pi}\frac{u}{(\beta u)^4 - \omega^2}\exp\{-iur\sin(\varphi+\theta)\}du d\varphi \quad (5.4.6)$$

The inner integral with respect to the angular variable φ is evaluated with the aid of the formulas (5.3.14) and (5.3.16), and the above Eq. (5.4.6) is reduced to the single integral,

$$w = \frac{Q}{2\pi\beta^4}\exp(+i\omega t)\int_0^{\infty}\frac{u}{u^4 - \left(\sqrt{\omega}/\beta\right)^4}J_0(ur)du \quad (5.4.7)$$

where the radial distance r is $\sqrt{x^2+y^2}$.

The last integral with respect to the variable u is evaluated by applying the complex integral theory. As was done in the former time-harmonic problem in Sect. 5.2 and earlier too, we introduce a small negative imaginary part to the frequency, and then consider the complex integral

$$\Phi = \int_C \frac{z}{z^4 - \left(\sqrt{\omega}/\beta\right)^4}H_0^{(1)}(zr)dz \quad (5.4.8)$$

where $H_0^{(1)}(.)$ is Hankel function of the first kind. The closed loop C is shown in Fig. 5.2. Since the Hankel function has a logarithmic singularity and its branch point is at the origin, one branch cut along the negative real axis is introduced. The integrand has four poles, whereof two are in the upper z-plane due to the introduction of the imaginary part to the frequency. They are marked in the figure and these two poles are specified with argument

$$z = -\frac{\sqrt{\omega}}{\beta} = \frac{\sqrt{\omega}}{\beta}\exp(+\pi i), \quad z = +i\frac{\sqrt{\omega}}{\beta} = \frac{\sqrt{\omega}}{\beta}\exp(+\pi i/2) \quad (5.4.9)$$

It should be understood that the argument of z along the positive real axis is zero, but that along the negative real axis is $+\pi$. Then, we apply Cauchy's theorem to the complex integral Φ in Eq. (5.4.8). The integral along the large semi-circle vanishes as its radius tends to infinity, and that along the small semi-circle also vanishes as its radius tends to zero. Thus, the integral along the real axis is converted to the sum of two residues at the poles. That is

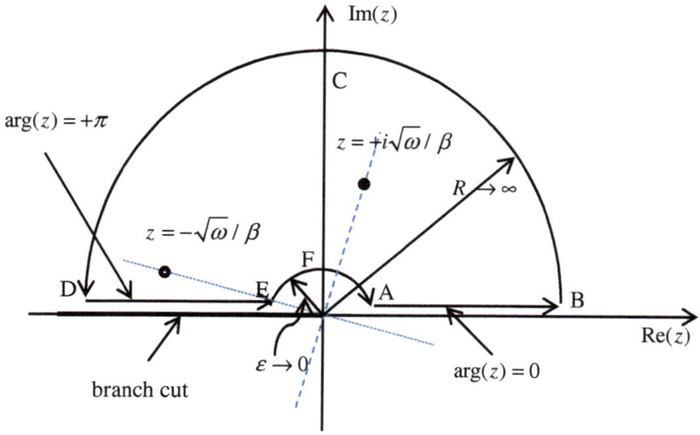

Fig. 5.2 A closed loop C for the complex integral Φ

$$
\begin{aligned}
&\int_{\varepsilon\to 0}^{R\to\infty} \frac{u}{u^4-(\sqrt{\omega}/\beta)^4}\left\{H_0^{(1)}(ur)-H_0^{(1)}(ure^{\pi i})\right\}du \\
&= 2\pi i\left\{-\frac{\beta^2}{4\omega}H_0^{(1)}\left(\frac{r\sqrt{\omega}}{\beta}e^{\pi i/2}\right)+\frac{\beta^2}{4\omega}H_0^{(1)}\left(\frac{r\sqrt{\omega}}{\beta}e^{\pi i}\right)\right\}
\end{aligned}
\tag{5.4.10}
$$

The formulas for the Hankel function (Watson 1966, pp. 75 and 78),

$$
H_0^{(1)}(xe^{\pi i}) = -H_0^{(2)}(x), \quad H_0^{(1)}(xe^{\pi i/2}) = -\frac{2i}{\pi}K_0(x) \tag{5.4.11}
$$

are very useful for arranging Eq. (5.4.10). Then, we have the integration formula for our use,

$$
\int_0^\infty \frac{u}{u^4-(\sqrt{\omega}/\beta)^4}J_0(ru)du = -\frac{\beta^2}{2\omega}\left\{\frac{\pi i}{2}H_0^{(2)}\left(\frac{r\sqrt{\omega}}{\beta}\right)+K_0\left(\frac{r\sqrt{\omega}}{\beta}\right)\right\} \tag{5.4.12}
$$

Applying this formula to Eq. (5.4.7), the time-harmonic response of the plate, i.e. the time-harmonic Green's function, is given by

$$
w(x,y,t) = -\frac{Q}{4\pi\beta^2\omega}\left\{\frac{\pi i}{2}H_0^{(2)}\left(\frac{r\sqrt{\omega}}{\beta}\right)+K_0\left(\frac{r\sqrt{\omega}}{\beta}\right)\right\}\exp(+i\omega t) \tag{5.4.13}
$$

where r is the radial distance from the load.

Table 5.1 Green's functions for beam and plate

	Deflection equation	Source $p(x, t)$	Green's function
1D beam	$EI\frac{\partial^4 w}{\partial x^4} + \rho A\frac{\partial^2 w}{\partial t^2} = p(x,t)$	$P\delta(x)\delta(t)$	$w(x,t) = \frac{Q}{\alpha}\sqrt{\frac{t}{\pi}}\Big[\frac{\pi z}{\sqrt{2}}\{S(z) - C(z)\} + \sin\{\frac{\pi}{4}(2z^2+1)\}\Big]$; $z = \frac{x}{\alpha\sqrt{2\pi t}}$
	$\alpha = \left(\frac{EI}{\rho A}\right)^{1/4}, \quad Q = \frac{P}{\rho A}$	$P\delta(x)\exp(+i\omega t)$	$w(x,t) = -\frac{Q}{4\alpha\sqrt{\omega}^3}\left[i\exp\left\{+i\left(\omega t - \frac{\sqrt{\omega}}{\alpha}\lvert x\rvert\right)\right\} + \exp\left(-\frac{\sqrt{\omega}}{\alpha}\lvert x\rvert + i\omega t\right)\right]$
2D plate	$D\left(\frac{\partial^4 w}{\partial x^4} + 2\frac{\partial^4 w}{\partial x^2 \partial y^2} + \frac{\partial^4 w}{\partial y^4}\right) + \rho h\frac{\partial^2 w}{\partial t^2} = p(x,y,t)$	$P\delta(x)\delta(y)\delta(t)$	$w(x,y,t) = \frac{Q}{4\pi\beta^2}\mathrm{si}\left(\frac{r^2}{4\beta^2 t}\right); \quad r = \sqrt{x^2+y^2}$
	$\beta = \left(\frac{D}{\rho h}\right)^{1/4}, \quad Q = \frac{P}{\rho h}$	$P\delta(x)\delta(y)\exp(+i\omega t)$	$w(x,y,t) = -\frac{Q}{4\pi\beta^2\omega}\left\{\frac{\pi i}{2}H_0^{(2)}\left(\frac{r\sqrt{\omega}}{\beta}\right) + K_0\left(\frac{r\sqrt{\omega}}{\beta}\right)\right\}\exp(+i\omega t)$; $r = \sqrt{x^2+y^2}$

The reader will find that the deflection has two components: One is a time-harmonic wave which is given by the product $H_0^{(2)}(r\sqrt{\omega}/\beta)\exp(+i\omega t)$. The other is a simple time-harmonic vibration $K_0(r\sqrt{\omega}/\beta)\exp(+i\omega t)$ whose amplitude decays exponentially with the distance. The same nature can be found in the 1D beam response in Eq. (5.2.12). Table 5.1 shows the summary for the deflection Green's functions.

Exercises

(5.1) Derive the wave velocity from the time-harmonic responses (5.2.9), (5.2.12) and (5.4.13) and show that all velocities depend on the frequency.

(5.2) Using the generating function for Bessel function (5.3.14), derive the integral representation for the Bessel function as

$$J_n(z) = \frac{1}{2\pi}\int_{-\pi}^{+\pi} \exp(-iz\sin\varphi + in\varphi)d\varphi = \frac{1}{\pi}\int_{0}^{+\pi} \cos(n\varphi - z\sin\varphi)d\varphi \quad \text{(a)}$$

(5.3) The asymptotic formula for the Hankel function is given by

$$H_0^{(1)}(z) \sim \sqrt{\frac{2}{\pi z}}\exp\{+i(z-\pi/4)\}; \quad z \to \infty \quad \text{(b)}$$

Explain why the integral along the semi-circle BCD in Fig. 5.2 vanishes for the complex integral (5.4.8).

Appendix

See Table 5.1.

References

Erdélyi A (ed) (1954) Tables of integral transforms, vol I and II. McGraw-Hill, New-York

Watson GN (1966) A treatise on the theory of bessel functions. Cambridge University Press, Cambridge

Chapter 6
Cagniard-de Hoop Technique

The success of the integral transform method hinges on the evaluation of inversion integrals. It is not always easy to find a suitable integration formula. If we cannot find any suitable formula, the inversion is left in its integral form and some numerical techniques must be applied for the evaluation. However, in the case of double inversion, such as Laplace and Fourier inversions, if we could convert the first Fourier inversion integral to a definition integral of Laplace transform, the next Laplace inversion can be carried out by inspection without evaluating its Laplace inversion integral. For example, when we have the double transformed function $F(\xi, s)$ and its Fourier inversion integral

$$f^*(x,s) = \int_{-\infty}^{+\infty} F(\xi,s)\exp(-i\xi x)d\xi \tag{6.1}$$

if we could apply some mathematical techniques to the integral and convert the integral to the definition form of Laplace transform,

$$f^*(x,s) = \int_{a}^{\infty} G(x,t)\exp(-st)dt \tag{6.2}$$

i.e. the Laplace transform integral, the Laplace inversion is easily carried out by inspection and its inversion is given by

$$f(x,t) = \begin{cases} G(x,t); & t > a \\ 0; & t < a \end{cases} = H(t-a)G(x,t) \tag{6.3}$$

The most important and substantial part of this technique is to convert the Fourier inversion integral to the form of the Laplace transform,

$$\int_{-\infty}^{+\infty} F(\xi,s)\exp(-i\xi x)d\xi = \int_{a}^{\infty} G(x,t)\exp(-st)dt \tag{6.4}$$

K. Watanabe, *Integral Transform Techniques for Green's Function*,
Lecture Notes in Applied and Computational Mechanics 76,
DOI 10.1007/978-3-319-17455-6_6

This conversion is carried out with aids of the theory of complex integrals and is called "Cagniard-de Hoop technique." Cagniard (1962) is the author of this technique, but his original technique is to map the inversion integral onto the integral in another complex t-plane. De Hoop (1961) has modified the technique to use the same complex plane without mapping. The present chapter explains the technique developed by de Hoop, but uses the name "Cagniard-de Hoop technique." The essential idea is to convert the first inversion integral to the definition integral of the second integral transform. This technique is applicable to other combinations of two integral transforms, not limited to Laplace-Fourier transforms. In addition, it will be worth to cite textbooks (Achenbach 1973; Fung 1970; Graff 1975; Miklowitz 1978), which treat elastic waves and explain the Cagniard-de Hoop technique in some depths.

The first section in the present chapter treats a very simple problem in order to demonstrate the Cagniard-de Hoop technique in details. In the following sections, 2D and 3D Green's functions (dyadic) for the coupled elastodynamic equations are discussed and some additional explanations for the Cagniard-de-Hoop technique are described.

6.1 2D Anti-plane Deformation

As the first example of the Cagniard-de Hoop technique, the simplest elastodynamic problem is discussed. We consider an elastic half space and take the Cartesian coordinate (x, y, z) as shown in Fig. 6.1 where the z-axis is normal to the paper plane. The surface of the solid is denoted by $y = 0$ and its interior $y > 0$. An impulsive anti-plane shear load directed to the negative z-direction is applied on the surface and is expressed by

$$\sigma_{yz}\big|_{y=0} = P_z\delta(x)\delta(t) \tag{6.1.1}$$

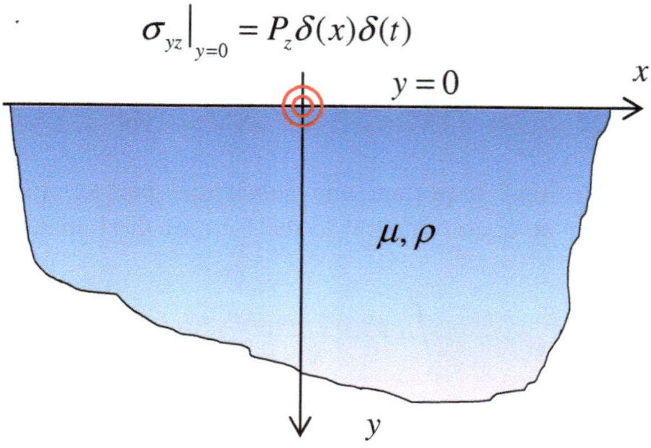

Fig. 6.1 An impulsive anti-plane shear load on the surface of a semi-infinite elastic solid

where P_z is the load magnitude. Since the load of infinite length is directed to the anti-plane z-direction, an anti-plane deformation is induced. The equation of motion is given by

$$\frac{\partial \sigma_{xz}}{\partial x} + \frac{\partial \sigma_{yz}}{\partial y} = \rho \frac{\partial^2 u_z}{\partial t^2} \tag{6.1.2}$$

where ρ is the density. Hooke's law for two shear stresses is given by

$$\sigma_{xz} = \mu \frac{\partial u_z}{\partial x}, \quad \sigma_{yz} = \mu \frac{\partial u_z}{\partial y} \tag{6.1.3}$$

where μ is the shear rigidity. We also employ the quiescent condition at an initial time,

$$u_z|_{t=0} = \left.\frac{\partial u_z}{\partial t}\right|_{t=0} = 0 \tag{6.1.4}$$

and the convergence condition at infinity

$$u_z|_{\sqrt{x^2+y^2}\to\infty} = \left.\frac{\partial u_z}{\partial x}\right|_{\sqrt{x^2+y^2}\to\infty} = \left.\frac{\partial u_z}{\partial y}\right|_{\sqrt{x^2+y^2}\to\infty} = 0 \tag{6.1.5}$$

Equations (6.1.1)–(6.1.5) constitute the present elastodynamic problem. Substituting Hooke's law of Eq. (6.1.3) into the equation of motion (6.1.2), we obtain the typical wave equation for the anti-plane displacement u_z,

$$\frac{\partial^2 u_z}{\partial x^2} + \frac{\partial^2 u_z}{\partial y^2} = \frac{1}{c_s^2} \frac{\partial^2 u_z}{\partial t^2} \tag{6.1.6}$$

where the shear wave velocity c_s is defined by

$$c_s = \sqrt{\mu/\rho} \tag{6.1.7}$$

In order to solve our elastodynamic problem, Laplace and Fourier transforms are employed; Laplace transform with respect to the time t,

$$f^*(s) = \int_0^\infty f(t) \exp(-st) dt \tag{6.1.8}$$

and Fourier transform with respect to the space variable x,

$$\bar{f}(\xi) = \int_{-\infty}^{+\infty} f(x) \exp(+i\xi x) dx, \quad f(x) = \frac{1}{2\pi} \int_{-\infty}^{+\infty} \bar{f}(\xi) \exp(-i\xi x) d\xi \tag{6.1.9}$$

With the aid of the quiescent and convergence conditions, the displacement equation (6.1.6) and Hooke's law (6.1.3) are transformed to

$$\frac{d^2 \bar{u}_z^*}{dy^2} - \left\{ \xi^2 + (s/c_s)^2 \right\} \bar{u}_z^* = 0 \tag{6.1.10}$$

$$\bar{\sigma}_{xz}^* = -i\xi\mu\bar{u}_z^*, \quad \bar{\sigma}_{yz}^* = \mu \frac{d\bar{u}_z^*}{dy} \tag{6.1.11}$$

After solving the transformed displacement equation (6.1.10), the displacement and stress, which satisfy the convergence condition at infinity, are given by

$$\bar{u}_z^* = A(\xi, s) \exp(-\alpha_s y), \quad \bar{\sigma}_{yz}^* = -\mu\alpha_s A(\xi, s) \exp(-\alpha_s y) \tag{6.1.12}$$

Here $A(\xi, s)$ is an unknown coefficient to be determined by the loading condition, and the radical is defined by

$$\alpha_s = \sqrt{\xi^2 + (s/c_s)^2}; \quad \mathrm{Re}(\alpha_s) > 0 \tag{6.1.13}$$

The boundary condition is also transformed, to

$$\left. \bar{\sigma}_{yz}^* \right|_{y=0} = P_z \tag{6.1.14}$$

Substituting the second equation in Eq. (6.1.12) into the above transformed boundary condition, the coefficient is determined as

$$A(\xi, s) = -\frac{P_z}{\mu\alpha_s} \tag{6.1.15}$$

and thus, the displacement in the transformed domain is explicitly determined as

$$\bar{u}_z^* = -\frac{P_z}{\mu\alpha_s} \exp(-\alpha_s y) \tag{6.1.16}$$

Now, we shall consider the inversion. The formal Fourier inversion for the displacement is given by the infinite integral

$$u_z^* = -\frac{P_z}{2\pi\mu}\int_{-\infty}^{+\infty}\frac{1}{\alpha_s}\exp(-\alpha_s y - i\xi x)d\xi \tag{6.1.17}$$

The above Eq. (6.1.17) is the Laplace transformed displacement, but in the form of Fourier inversion integral. If we could convert the integral into the definition form of Laplace transform integral, its Laplace inversion can be performed by inspection (i.e. the Cagniard-de Hoop technique).

Let us start to apply the technique. First of all, in order to eliminate the Laplace transform parameter s, leaving it only in the argument of the exponential function in the integrand, a variable transform from ξ to ς is introduced as

$$\xi = s\varsigma \tag{6.1.18}$$

Here we assume that the Laplace transform parameter s is a positive real constant. The integral in Eq. (6.1.17) is rewritten as

$$u_z^* = -\frac{P_z}{2\pi\mu}\int_{-\infty}^{+\infty}\frac{1}{\sqrt{\varsigma^2 + 1/c_s^2}}\exp\left\{-s\left(y\sqrt{\varsigma^2 + 1/c_s^2} + i\varsigma x\right)\right\}d\varsigma \tag{6.1.19}$$

where the radical must satisfy the radiation condition,

$$\mathrm{Re}\left(\sqrt{\varsigma^2 + 1/c_s^2}\right) > 0 \tag{6.1.20}$$

Examining the integral in Eq. (6.1.19), the Laplace transform parameter is included only in the argument of the exponential function as a simple multiplier. If we could change the argument to a simple product, such as st where t is a new variable, the integral will have the form of the definition integral of the Laplace transform. So, the variable transform from ς to the new variable t is introduced as

$$t = y\sqrt{\varsigma^2 + 1/c_s^2} + i\varsigma x \tag{6.1.21}$$

However, its inverse gives multiple values for ς

$$\varsigma_s^{(\pm)} = \frac{-itx \pm y\sqrt{t^2 - (x^2 + y^2)/c_s^2}}{x^2 + y^2} \tag{6.1.22}$$

Due to the multi-valuedness, we are puzzled as to which one is the suitable inversion for ς. In order to solve this puzzle, we examine the inverse as a function of the real parameter t. We separate the real and imaginary parts of the inverse as $\varsigma_s^{(\pm)} = \pm X - iY$, where

$$X = \frac{y\sqrt{t^2 - (x^2 + y^2)/c_s^2}}{x^2 + y^2}, \quad Y = \frac{tx}{x^2 + y^2} \tag{6.1.23}$$

and eliminate the parameter t. An equation for the hyperbolic curve in (X, Y) plane is derived as

$$\left(\frac{X}{y}\right)^2 - \left(\frac{Y}{x}\right)^2 = -\frac{1}{c_s^2} \tag{6.1.24}$$

Thus, the inverse $\varsigma_s^{(\pm)}$ has the form of two semi-hyperbolic curves in the complex ς-plane. When $x > 0$, $\varsigma_s^{(-)}$ is the left half and $\varsigma_s^{(+)}$ is the right half of the hyperbola in the lower ς-plane, as shown in Fig. 6.2. On the other hand, when $x<0$, the hyperbolic curves $\varsigma_s^{(\pm)}$ are in the upper ς-plane. The connected two semi-hyperbolic curves CAB shown in the figure is called Cagniard's path.

If we could connect the real line with the edges of two semi-hyperbolas in the complex plane, a closed loop for the complex integral is formed; and then Cauchy's integral theorem can be applied. Thus, we consider the complex integral Φ, whose integrand is the same as that of the Fourier inversion integral in Eq. (6.1.19),

$$\Phi = \frac{1}{2\pi}\oint_C \frac{1}{\sqrt{\varsigma^2 + 1/c_s^2}} \exp\left\{-s\left(y\sqrt{\varsigma^2 + 1/c_s^2} + i\varsigma x\right)\right\} d\varsigma \tag{6.1.25}$$

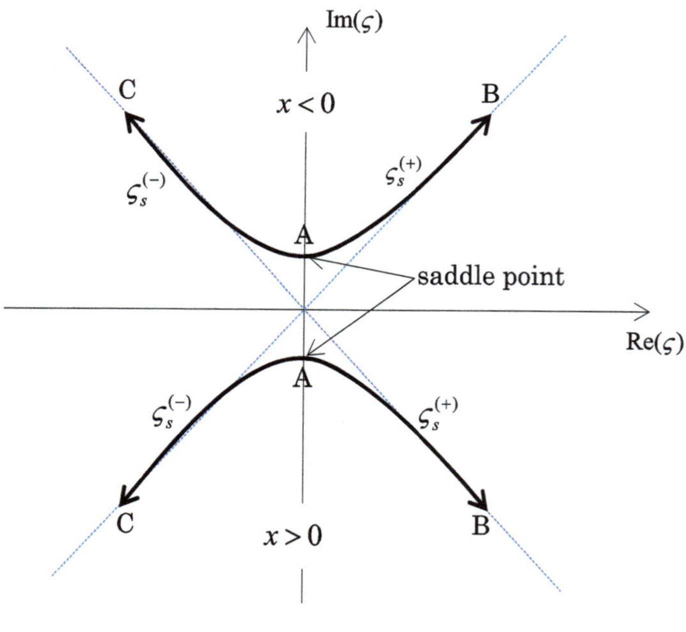

Fig. 6.2 Cagniard's path

The integrand has two branch points at

$$\varsigma = \pm i/c_s \tag{6.1.26}$$

In order to guarantee the radiation condition of Eq. (6.1.20), two branch cuts, which are discussed in Sect. 1.3.2 in Chap. 1, are introduced along the imaginary axis in the complex ς-plane as shown in Fig. 6.3. When the real parameter t varies from $\sqrt{x^2+y^2}/c_s$ to $+\infty$, the inverse $\varsigma_s^{(+)}$ moves on the semi-hyperbola from A to B, and the other inverse $\varsigma_s^{(-)}$ moves on the other semi-hyperbola from A to C. These two semi-hyperbolas are connected at the saddle point A,

$$\varsigma_{saddle} = -\frac{ix}{c_s\sqrt{x^2+y^2}} \text{ at } t = \sqrt{x^2+y^2}/c_s \tag{6.1.27}$$

This saddle point is always smaller in magnitude than the branch point $\varsigma = \pm i/c_s$. The other edges at the infinite t must be connected with the real line. The edge B is connected to the line at the positive infinity F with the large arc. The edge C is also connected to the line at the negative infinity D with the large arc, as shown in Fig. 6.3. Then, the closed loop for the complex integral is formed by the loop $\overline{\text{CABFEDC}}$. The closed loop $C^{(-)}$ in the lower ς-plane is employed in the case of positive x and the loop $C^{(+)}$ in the upper plane is in the case of negative x. In either

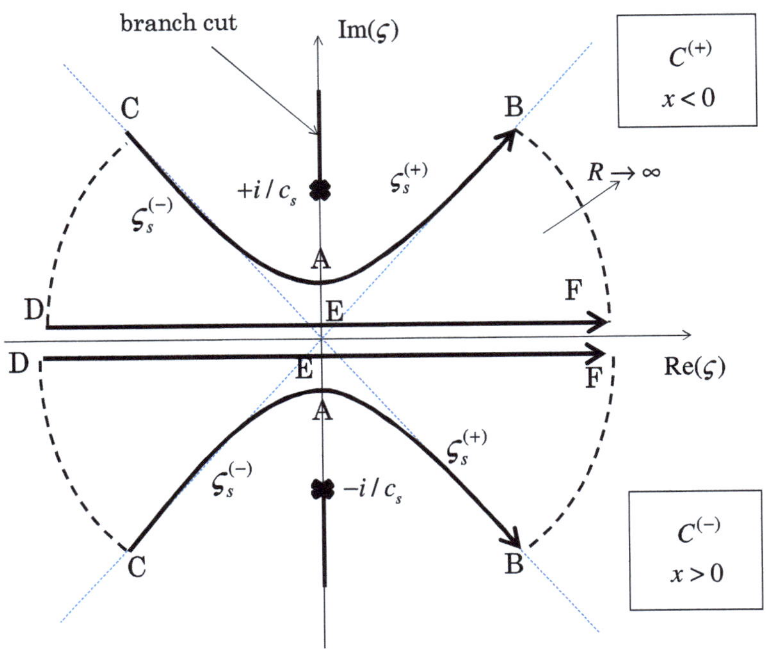

Fig. 6.3 Closed loop $C^{(\pm)}$ for the complex integral Φ

case, no singular point is included in the closed loop. The integral along the large arc vanishes as its radius tends to infinity, since the radiation condition of Eq. (6.1.20) is kept. Thus, the integral along the real axis is converted to that along the two semi-hyperbolas, CA and AB, due to Cauchy's integral theorem.

The integral along the hyperbola is one of the parametric integrals and its parameter is the variable t. The identity between two integrals along the real axis and along the hyperbolas is thus given by

$$\begin{aligned}&\frac{1}{2\pi}\int_{-\infty}^{+\infty}\frac{1}{\sqrt{\varsigma^2+1/c_s^2}}\exp\left\{-s\left(y\sqrt{\varsigma^2+1/c_s^2}+i\varsigma x\right)\right\}d\varsigma\\&=\frac{1}{2\pi}\int_{\sqrt{x^2+y^2}/c_s}^{\infty}\left\{\frac{1}{\sqrt{\varsigma^2+1/c_s^2}}\frac{d\varsigma}{dt}\bigg|_{\varsigma=\varsigma_s^{(+)}}-\frac{1}{\sqrt{\varsigma^2+1/c_s^2}}\frac{d\varsigma}{dt}\bigg|_{\varsigma=\varsigma_s^{(-)}}\right\}\exp(-st)dt\end{aligned} \tag{6.1.28}$$

The integrand in the right hand side of Eq. (6.1.28) can be simplified. Explicit expressions for the radical and the gradient are derived from the definition of Eq. (6.1.21) and its inverse Eq. (6.1.22). They are

$$\begin{aligned}\frac{1}{\sqrt{\left(\varsigma_s^{(\pm)}\right)^2+1/c_s^2}}&=\frac{x^2+y^2}{yt\pm ix\sqrt{t^2-(x^2+y^2)/c_s^2}}\\\frac{d\varsigma_s^{(\pm)}}{dt}&=\pm\frac{yt\pm ix\sqrt{t^2-(x^2+y^2)/c_s^2}}{(x^2+y^2)\sqrt{t^2-(x^2+y^2)/c_s^2}}\end{aligned} \tag{6.1.29}$$

Substituting these into the integral in the right hand side of Eq. (6.1.28), the parametric integral along the hyperbola is simplified as

$$\begin{aligned}&\frac{1}{2\pi}\int_{\sqrt{x^2+y^2}/c_s}^{\infty}\left\{\frac{1}{\sqrt{\varsigma^2+1/c_s^2}}\frac{d\varsigma}{dt}\bigg|_{\varsigma=\varsigma_s^{(+)}}-\frac{1}{\sqrt{\varsigma^2+1/c_s^2}}\frac{d\varsigma}{dt}\bigg|_{\varsigma=\varsigma_s^{(-)}}\right\}\exp(-st)dt\\&=\frac{1}{\pi}\int_{\sqrt{x^2+y^2}/c_s}^{\infty}\frac{1}{\sqrt{t^2-(x^2+y^2)/c_s^2}}\exp(-st)dt\end{aligned} \tag{6.1.30}$$

Thus, we have converted the Fourier inversion integral to the integral form of the Laplace transform,

$$\begin{aligned}
&\frac{1}{2\pi}\int_{-\infty}^{+\infty}\frac{1}{\sqrt{\varsigma^2+1/c_s^2}}\exp\left\{-s\left(y\sqrt{\varsigma^2+1/c_s^2}+i\varsigma x\right)\right\}d\varsigma \\
&=\frac{1}{\pi}\int_{\sqrt{x^2+y^2}/c_s}^{\infty}\frac{1}{\sqrt{t^2-(x^2+y^2)/c_s^2}}\exp(-st)dt
\end{aligned} \tag{6.1.31}$$

Substituting the right hand side of the above equation into the Laplace transformed displacement in Eq. (6.1.19), we have the Laplace transformed displacement in the form of Laplace transform integral,

$$u_z^*(x,y,s)=-\frac{P_z}{\pi\mu}\int_{\sqrt{x^2+y^2}/c_s}^{\infty}\frac{1}{\sqrt{t^2-(x^2+y^2)/c_s^2}}\exp(-st)dt \tag{6.1.32}$$

This integral is just the form of Laplace transform, but its lower limit is not zero. The integral states that the original function is vanishing before $t=\sqrt{x^2+y^2}/c_s$ and after this time the function has the form of the integrand. Then, we can easily anticipate the original displacement function before Laplace transform, i.e. Laplace inversion of Eq. (6.1.32),

$$\begin{aligned}
u_z(x,y,t)&=L^{-1}\left\{u_z^*(x,y,s)\right\} \\
&=-\frac{P_z}{\pi\mu}\begin{cases}\frac{1}{\sqrt{t^2-(x^2+y^2)/c_s^2}}; & t>\sqrt{x^2+y^2}/c_s \\ 0; & t<\sqrt{x^2+y^2}/c_s\end{cases}
\end{aligned} \tag{6.1.33}$$

Utilizing Heaviside's unit step function, the displacement is expressed in the compact form,

$$u_z(x,y,t)=-\frac{P_z}{\pi\mu}\frac{1}{\sqrt{t^2-(x^2+y^2)/c_s^2}}H\left(c_st-\sqrt{x^2+y^2}\right) \tag{6.1.34}$$

This is our final result for the double inversion. The reader should notice that we did not use any integration formula for Laplace and Fourier inversions but we did the inversion. This is the Cagniard-de Hoop technique!

The conditional for the step function gives the circular region disturbed by the transient SH-wave,

$$(c_st)^2>x^2+y^2 \tag{6.1.35}$$

and its edge is a circular (cylindrical in 3D) wave front with the center at the source point,

$$(c_s t)^2 = x^2 + y^2 \tag{6.1.36}$$

6.2 2D In-plane Deformation

We consider a 2D transient response of an semi-infinite elastic solid. Take the coordinate system (x, y) so that the surface of the solid is at $y = 0$ and its interior in $y > 0$ as shown in Fig. 6.4. An impulsive point load is applied on the surface and is expressed by

$$\sigma_{yx}\big|_{y=0} = P_x\delta(x)\delta(t), \quad \sigma_{yy}\big|_{y=0} = P_y\delta(x)\delta(t) \tag{6.2.1}$$

where P_j, $j = x$, y are the components of the load. The in-plane deformation of the elastic solid is governed by the equations of motion,

$$\begin{aligned} \frac{\partial\sigma_{xx}}{\partial x} + \frac{\partial\sigma_{yx}}{\partial y} &= \rho\frac{\partial^2 u_x}{\partial t^2} \\ \frac{\partial\sigma_{xy}}{\partial x} + \frac{\partial\sigma_{yy}}{\partial y} &= \rho\frac{\partial^2 u_y}{\partial t^2} \end{aligned} \tag{6.2.2}$$

where ρ is density. Hooke's law for the plane strain is given by

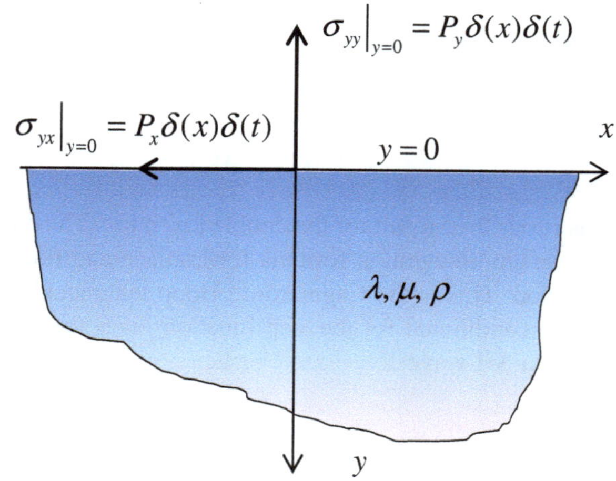

Fig. 6.4 An impulsive load on the surface of a semi-infinite elastic solid

$$\begin{aligned}
\sigma_{xx} &= \lambda\left(\frac{\partial u_x}{\partial x} + \frac{\partial u_y}{\partial y}\right) + 2\mu\frac{\partial u_x}{\partial x} \\
\sigma_{yy} &= \lambda\left(\frac{\partial u_x}{\partial x} + \frac{\partial u_y}{\partial y}\right) + 2\mu\frac{\partial u_y}{\partial y} \\
\sigma_{xy} &= \sigma_{yx} = \mu\left(\frac{\partial u_x}{\partial y} + \frac{\partial u_y}{\partial x}\right)
\end{aligned} \tag{6.2.3}$$

where λ and μ are Lame's constants. We employ the quiescent condition at an initial time,

$$u_i|_{t=0} = \left.\frac{\partial u_i}{\partial t}\right|_{t=0} = 0; \quad i = x,\, y \tag{6.2.4}$$

and the convergence (radiation) condition at infinity

$$u_i|_{\sqrt{x^2+y^2}\to\infty} = \left.\frac{\partial u_i}{\partial x}\right|_{\sqrt{x^2+y^2}\to\infty} = \left.\frac{\partial u_i}{\partial y}\right|_{\sqrt{x^2+y^2}\to\infty} = 0; \quad i = x, y \tag{6.2.5}$$

Equations from (6.2.1) to (6.2.5) constitute the impulsive Lamb's problem (Fung 1970) for the 2D semi-infinite elastic solid. Substituting Hooke's law into the equations of motion, the displacement equations are obtained as

$$\begin{aligned}
\gamma^2\frac{\partial^2 u_x}{\partial x^2} + \frac{\partial^2 u_x}{\partial y^2} + (\gamma^2 - 1)\frac{\partial^2 u_y}{\partial x \partial y} &= \frac{1}{c_s^2}\frac{\partial^2 u_x}{\partial t^2} \\
(\gamma^2 - 1)\frac{\partial^2 u_x}{\partial x \partial y} + \frac{\partial^2 u_y}{\partial x^2} + \gamma^2\frac{\partial^2 u_y}{\partial y^2} &= \frac{1}{c_s^2}\frac{\partial^2 u_y}{\partial t^2}
\end{aligned} \tag{6.2.6}$$

where γ is the velocity ratio defined by Eq. (3.8) in Chap. 3. We apply the double integral transform: Laplace transform with respect to the time t,

$$f^*(s) = \int_0^\infty f(t)\exp(-st)dt \tag{6.2.7}$$

and Fourier transform with respect to the space variable x,

$$\bar{f}(\xi) = \int_{-\infty}^{+\infty} f(x)\exp(+i\xi x)dx, \quad f(x) = \frac{1}{2\pi}\int_{-\infty}^{+\infty} \bar{f}(\xi)\exp(-i\xi x)d\xi \tag{6.2.8}$$

With use of the quiescent and radiation conditions, the displacement equations are transformed to the coupled ordinary differential equations with constant coefficients,

$$\begin{aligned}
&\frac{d^2\bar{u}_x^*}{dy^2} - \{\gamma^2\xi^2 + (s/c_s)^2\}\bar{u}_x^* - i\xi(\gamma^2 - 1)\frac{d\bar{u}_y^*}{dy} = 0 \\
&-i\xi(\gamma^2 - 1)\frac{d\bar{u}_x^*}{dy} + \gamma^2\frac{d^2\bar{u}_y^*}{dy^2} - \{\xi^2 + (s/c_s)^2\}\bar{u}_y^* = 0
\end{aligned} \tag{6.2.9}$$

Hooke's law and the boundary condition are also transformed to

$$\begin{aligned}
\frac{1}{\mu}\bar{\sigma}_{xy}^* &= \frac{1}{\mu}\bar{\sigma}_{yx}^* = \frac{d\bar{u}_x^*}{dy} - i\xi\bar{u}_y^* \\
\frac{1}{\mu}\bar{\sigma}_{yy}^* &= -i\xi(\gamma^2 - 2)\bar{u}_x^* + \gamma^2\frac{d\bar{u}_y^*}{dy}
\end{aligned} \tag{6.2.10}$$

and

$$\bar{\sigma}_{yx}^*\Big|_{y=0} = P_x, \quad \bar{\sigma}_{yy}^*\Big|_{y=0} = P_y \tag{6.2.11}$$

The solution for the displacement equation (6.2.9) is given by

$$\begin{aligned}
\bar{u}_x^* &= \frac{i\xi}{\alpha_d}A(\xi, s)\exp(-\alpha_d y) + B(\xi, s)\exp(-\alpha_s y) \\
\bar{u}_y^* &= A(\xi, s)\exp(-\alpha_d y) - \frac{i\xi}{\alpha_s}B(\xi, s)\exp(-\alpha_s y)
\end{aligned} \tag{6.2.12}$$

where $A(\xi, s)$ and $B(\xi, s)$ are unknown coefficients, and the two radicals are defined by

$$\alpha_j = \sqrt{\xi^2 + (s/c_j)^2}; \quad j = d, s \tag{6.2.13}$$

with the radiation condition $\mathrm{Re}(\alpha_j) > 0$. Substituting Eq. (6.2.12) into Eq. (6.2.10), the stress components are obtained as

$$\begin{aligned}
\frac{1}{\mu}\bar{\sigma}_{yx}^* &= -2i\xi A(\xi, s)\exp(-\alpha_d y) - \frac{\alpha_s^2 + \xi^2}{\alpha_s}B(\xi, s)\exp(-\alpha_s y) \\
\frac{1}{\mu}\bar{\sigma}_{yy}^* &= -\frac{\alpha_s^2 + \xi^2}{\alpha_d}A(\xi, s)\exp(-\alpha_d y) + 2i\xi B(\xi, s)\exp(-\alpha_s y)
\end{aligned} \tag{6.2.14}$$

Applying the boundary condition of Eq. (6.2.11) to the above stresses, we obtain the following simple algebraic equations for the unknown coefficients,

$$
\begin{aligned}
-2i\xi A(\xi,s) - \frac{\alpha_s^2+\xi^2}{\alpha_s}B(\xi,s) &= P_x/\mu \\
-\frac{\alpha_s^2+\xi^2}{\alpha_d}A(\xi,s) + 2i\xi B(\xi,s) &= P_y/\mu
\end{aligned}
\tag{6.2.15}
$$

Their solutions are given by

$$
\begin{aligned}
A(\xi,s) &= -\frac{2i\xi\alpha_d\alpha_s(P_x/\mu) + \alpha_d(\alpha_s^2+\xi^2)(P_y/\mu)}{R(\xi,s)} \\
B(\xi,s) &= -\frac{\alpha_s(\alpha_s^2+\xi^2)(P_x/\mu) - 2i\xi\alpha_d\alpha_s(P_y/\mu)}{R(\xi,s)}
\end{aligned}
\tag{6.2.16}
$$

where $R(\xi,s)$ is Rayleigh equation, defined by

$$
R(\xi,s) = (\alpha_s^2+\xi^2)^2 - 4\alpha_d\alpha_s\xi^2 \tag{6.2.17}
$$

Substitution of Eq. (6.2.16) into Eq. (6.2.12) gives the displacement in the double transformed domain. The transformed displacement can also be expressed as the sum of the dilatational and shear wave contributions,

$$
\begin{aligned}
\bar{u}_x^* &= \frac{P_x}{\mu}\left\{\bar{I}_{xx}^{(d)*}(\xi,y,s) + \bar{I}_{xx}^{(s)*}(\xi,y,s)\right\} + \frac{P_y}{\mu}\left\{\bar{I}_{xy}^{(d)*}(\xi,y,s) + \bar{I}_{xy}^{(s)*}(\xi,y,s)\right\} \\
\bar{u}_y^* &= \frac{P_x}{\mu}\left\{\bar{I}_{yx}^{(d)*}(\xi,y,s) + \bar{I}_{yx}^{(s)*}(\xi,y,s)\right\} + \frac{P_y}{\mu}\left\{\bar{I}_{yy}^{(d)*}(\xi,y,s) + \bar{I}_{yy}^{(s)*}(\xi,y,s)\right\}
\end{aligned}
\tag{6.2.18}
$$

where the superscripts "*d*" and "*s*" for *k* in $\bar{I}_{ij}^{(k)*}$ stand for the dilatational and shear wave contributions, respectively. The first subscript *i* stands for the direction of the displacement component and the second *j* for the component of the load. More explicit expressions for each of the contributions are

$$
\begin{aligned}
\bar{I}_{xx}^{(d)*}(\xi,y,s) &= \frac{2\xi^2\alpha_s}{R(\xi,s)}\exp(-\alpha_d y) \\
\bar{I}_{xx}^{(s)*}(\xi,y,s) &= -\frac{\alpha_s(\alpha_s^2+\xi^2)}{R(\xi,s)}\exp(-\alpha_s y) \\
\bar{I}_{xy}^{(d)*}(\xi,y,s) &= -\frac{i\xi(\alpha_s^2+\xi^2)}{R(\xi,s)}\exp(-\alpha_d y) \\
\bar{I}_{xy}^{(s)*}(\xi,y,s) &= \frac{2i\xi\alpha_d\alpha_s}{R(\xi,s)}\exp(-\alpha_s y)
\end{aligned}
\tag{6.2.19}
$$

for $\bar{u}_x^*$, and

$$
\begin{aligned}
\bar{I}_{yx}^{(d)*}(\xi, y, s) &= -\frac{2i\xi\alpha_d\alpha_s}{R(\xi, s)}\exp(-\alpha_d y) \\
\bar{I}_{yx}^{(s)*}(\xi, y, s) &= \frac{i\xi(\alpha_s^2 + \xi^2)}{R(\xi, s)}\exp(-\alpha_s y) \\
\bar{I}_{yy}^{(d)*}(\xi, y, s) &= -\frac{\alpha_d(\alpha_s^2 + \xi^2)}{R(\xi, s)}\exp(-\alpha_d y) \\
\bar{I}_{yy}^{(s)*}(\xi, y, s) &= \frac{2\xi^2\alpha_d}{R(\xi, s)}\exp(-\alpha_s y)
\end{aligned}
\tag{6.2.20}
$$

for $\bar{u}_y^*$.

We have obtained explicit expressions for the transformed displacement components. Our next task is to invert the displacement into the real domain (x, y, t). However, it might be impossible to invert the displacement as a whole. The inversion must be carried out for each wave contribution one by one. As all dilatational wave contributions given by $\bar{I}_{ij}^{(d)*}$ have the same exponential function $\exp(-\alpha_d y)$ and the shear wave contributions $\bar{I}_{ij}^{(s)*}$ does the function $\exp(-\alpha_s y)$, we invert these two wave contributions separately.

(1) Inversion of the dilatational wave contribution

As a representative for the dilatational wave contribution, we consider the inversion of $\bar{I}_{xx}^{(d)*}$. The formal Fourier inversion integral with respect to the parameter ξ is given by

$$
I_{xx}^{(d)*}(x, y, s) = \frac{1}{2\pi}\int_{-\infty}^{+\infty} \frac{2\xi^2\alpha_s}{R(\xi, s)}\exp(-\alpha_d y - i\xi x)d\xi \tag{6.2.21}
$$

In order to eliminate the Laplace transform parameter s in the integrand except the exponential function, the variable transform from ξ to the new variable ς is introduced as

$$
\xi = (s/c_d)\varsigma \tag{6.2.22}
$$

The inversion integral is rewritten as

$$
I_{xx}^{(d)*}(x, y, s) = \frac{1}{2\pi}\int_{-\infty}^{+\infty} F_{xx}^{(d)}(\varsigma)\exp\left\{-(s/c_d)\left(y\sqrt{\varsigma^2 + 1} + ix\varsigma\right)\right\}d\varsigma \tag{6.2.23}
$$

where

$$F_{xx}^{(d)}(\varsigma) = \frac{2\varsigma^2\sqrt{\varsigma^2+\gamma^2}}{R(\varsigma)} \tag{6.2.24}$$

and its denominator, which is called the Rayleigh equation/function, is redefined by

$$R(\varsigma) = (2\varsigma^2+\gamma^2)^2 - 4\varsigma^2\sqrt{\varsigma^2+1}\sqrt{\varsigma^2+\gamma^2} \tag{6.2.25}$$

Note that the Rayleigh equation $R(\varsigma)$ differs from $R(\xi, s)$ in Eq. (6.2.17).

Subsequently, we consider the complex integral Φ whose integrand is the same as that in Eq. (6.2.23),

$$\Phi = \frac{1}{2\pi}\oint_{C^{(\pm)}} F_{xx}^{(d)}(\varsigma)\exp\left\{-(s/c_d)\left(y\sqrt{\varsigma^2+1}+ix\varsigma\right)\right\}d\varsigma \tag{6.2.26}$$

The closed loops $C^{\pm}$ are discussed here. The two radicals $\sqrt{\varsigma^2+1}$ and $\sqrt{\varsigma^2+\gamma^2}$ in the integrand have each two branch points at $\varsigma = \pm i, \quad \pm i\gamma$, respectively. Thus, four branch cuts are introduced along the imaginary axis in the complex ς-plane as shown in Fig. 6.5. Only the lower half plane is shown since the path and singular point are symmetric about the real axis. The Rayleigh equation has two pure imaginary roots, $\varsigma = \pm i\gamma_R$, where γ_R is the velocity ratio of the dilatational wave c_d to the Rayleigh wave c_R,

$$\gamma_R = c_d/c_R \tag{6.2.27}$$

This velocity ration γ_R is always greater than γ, since the following inequality holds for the isotropic media:

$$c_R < c_s < c_d \tag{6.2.28}$$

As to the argument of the exponential function, we hope to transform it to the product of the new (time) variable t and the transform parameter s. So, we introduce the new variable t as

$$t = \left(y\sqrt{\varsigma^2+1}+ix\varsigma\right)/c_d \tag{6.2.29}$$

Solving for the variable ς, we have two solutions, the so-called "Cagniard's path",

$$\varsigma_d^{(\pm)} = \frac{-ix(c_dt) \pm y\sqrt{(c_dt)^2-(x^2+y^2)}}{x^2+y^2} \tag{6.2.30}$$

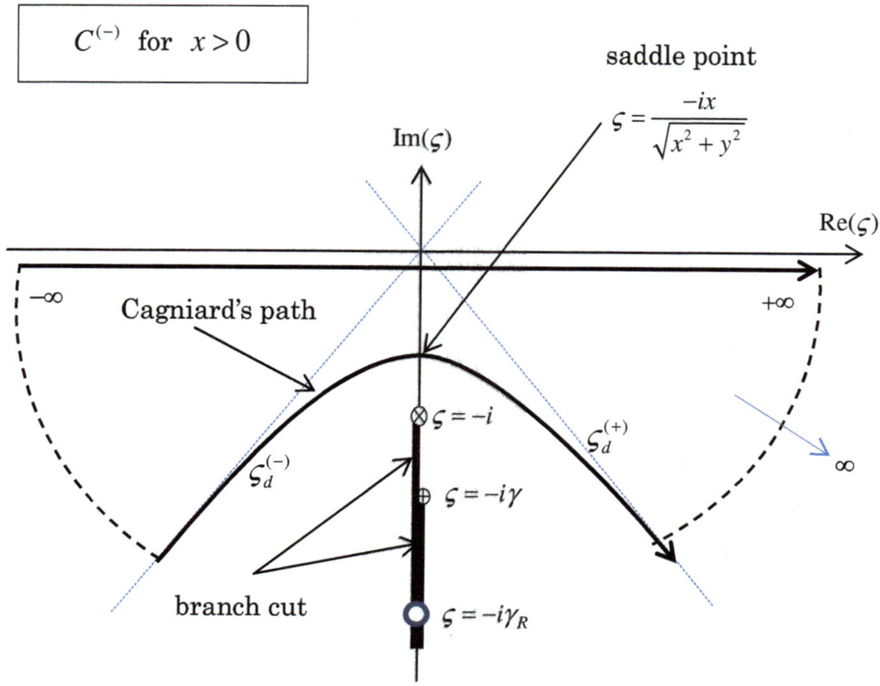

Fig. 6.5 Closed loop $C^{(-)}$ and Cagniard's path for the dilatational wave contribution

These solutions $\varsigma_d^{(\pm)}$ take the form of two symmetrical semi-hyperbolas in the complex ς-plane as the real parameter t varies from $\sqrt{x^2+y^2}/c_d$ to infinity. The two semi-hyperbolas are connected to each other at the saddle point

$$\varsigma_{saddle}^{(d)} = -\frac{ix}{\sqrt{x^2+y^2}} \text{ at } t = \sqrt{x^2+y^2}/c_d \tag{6.2.31}$$

This connected hyperbola is called Cagniard's path as shown in Fig. 6.5. Fortunately, the saddle point is not on any branch cut. Then, we can determine the closed loop in the complex ς-plane.

When the Cagniard's path is in the lower plane, the closed loop is composed of an infinite line along the real axis, Caginiard's path and two large arcs which connect the line to the Cagniard's path. When the Caginiard's path is in the upper plane, the closed loop is composed of the similar ones in the lower plane and is symmetric with the lower loop about the real axis. The lower loop $C^{(-)}$, and the branch cuts and points are shown in Fig. 6.5. The closed loop $C^{(-)}$ is employed when the space variable x is positive, and the loop $C^{(+)}$ when it is negative, due to the convergence at the large arc. In any case, positive or negative x, i.e. the upper or lower circuit in the complex plane, the closed loop does not include any singular point. Then we can apply Cauchy's integral theorem to the complex integral Φ. Since the integral along the large arc vanishes as its radius tends to infinity, we can convert the integral along the real axis to that along the Cagniard's path. That is

$$\begin{aligned} I_{xx}^{(d)*}(x,y,s) &= \frac{1}{2\pi}\int_{-\infty}^{+\infty} F_{xx}^{(d)}(\varsigma)\exp\left\{-(s/c_d)\left(y\sqrt{\varsigma^2+1}+ix\varsigma\right)\right\}d\varsigma \\ &= \frac{1}{2\pi}\int_{\sqrt{x^2+y^2}/c_d}^{+\infty}\left\{F_{xx}^{(d)}(\varsigma_d^{(+)})\frac{d\varsigma_d^{(+)}}{dt} - F_{xx}^{(d)}(\varsigma_d^{(-)})\frac{d\varsigma_d^{(-)}}{dt}\right\}\exp(-st)dt \end{aligned} \tag{6.2.32}$$

If we understand that the integration variable t as the real time variable, the integral has just the form of Laplace transform integral, but with the shifted starting time $\sqrt{x^2+y^2}/c_d$. As the Laplace inversion of $I_{xx}^{(d)*}(x,y,s)$ is the original function before Laplace transform, the inversion is carried out by inspection on the last line of Eq. (6.2.32). Utilizing the step function, the original function is given by

$$L^{-1}\left[I_{xx}^{(d)*}(x,y,s)\right] = \frac{1}{2\pi}\left\{F_{xx}^{(d)}(\varsigma_d^{(+)})\frac{d\varsigma_d^{(+)}}{dt} - F_{xx}^{(d)}(\varsigma_d^{(-)})\frac{d\varsigma_d^{(-)}}{dt}\right\}H\left(t-\sqrt{x^2+y^2}/c_d\right) \tag{6.2.33}$$

Consequently, the double inversion for $\bar{I}_{xx}^{(d)*}(\xi,y,s)$ is given by

$$I_{xx}^{(d)}(x,y,t) = \frac{1}{2\pi}\left\{F_{xx}^{(d)}(\varsigma_d^{(+)})\frac{d\varsigma_d^{(+)}}{dt} - F_{xx}^{(d)}(\varsigma_d^{(-)})\frac{d\varsigma_d^{(-)}}{dt}\right\}H\left(t-\sqrt{x^2+y^2}/c_d\right) \tag{6.2.34}$$

The same Cagniard-de Hoop technique can be applied to the other dilatational wave contributions $\bar{I}_{ij}^{(d)*}$. Thus, we have the unified expression for the dilatational wave contributions as

$$I_{ij}^{(d)}(x,y,t) = \frac{1}{2\pi}\left\{F_{ij}^{(d)}(\varsigma_d^{(+)})\frac{d\varsigma_d^{(+)}}{dt} - F_{ij}^{(d)}(\varsigma_d^{(-)})\frac{d\varsigma_d^{(-)}}{dt}\right\}H\left(t-\sqrt{x^2+y^2}/c_d\right) \tag{6.2.35}$$

where $F_{ij}^{(d)}(\varsigma)$ are given by

$$\begin{aligned} F_{xx}^{(d)}(\varsigma) &= \frac{2\varsigma^2\sqrt{\varsigma^2+\gamma^2}}{R(\varsigma)}, \quad F_{xy}^{(d)}(\varsigma) = -\frac{i\varsigma(2\varsigma^2+\gamma^2)}{R(\varsigma)} \\ F_{yx}^{(d)}(\varsigma) &= -\frac{2i\varsigma\sqrt{\varsigma^2+1}\sqrt{\varsigma^2+\gamma^2}}{R(\varsigma)}, \quad F_{yy}^{(d)}(\varsigma) = -\frac{(2\varsigma^2+\gamma^2)\sqrt{\varsigma^2+1}}{R(\varsigma)} \end{aligned} \tag{6.2.36}$$

(2) Inversion of the shear wave contribution

We consider the double inversion for the shear wave contribution, which has the exponential function $\exp(-\alpha_s y)$. As a representative, we consider the inversion of

$$\bar{I}_{xx}^{(s)*}(\xi, y, s) = -\frac{\alpha_s(\alpha_s^2+\xi^2)}{R(\xi, s)}\exp(-\alpha_s y) \tag{6.2.37}$$

Its formal Fourier inversion is given by

$$I_{xx}^{(s)*}(x, y, s) = \frac{1}{2\pi}\int_{-\infty}^{+\infty}\left\{-\frac{\alpha_s(\alpha_s^2+\xi^2)}{R(\xi, s)}\right\}\exp(-\alpha_s y - i\xi x)d\xi \tag{6.2.38}$$

Introducing the variable transform defined by Eq. (6.2.22), the inversion integral is rewritten as

$$I_{xx}^{(s)*}(x, y, s) = \frac{1}{2\pi}\int_{-\infty}^{+\infty} F_{xx}^{(s)}(\varsigma)\exp\left\{-(s/c_d)\left(y\sqrt{\varsigma^2+\gamma^2}+i\varsigma x\right)\right\}d\varsigma \tag{6.2.39}$$

where

$$F_{xx}^{(s)}(\varsigma) = -\frac{(2\varsigma^2+\gamma^2)\sqrt{\varsigma^2+\gamma^2}}{R(\varsigma)} \tag{6.2.40}$$

and the Rayleigh equation $R(\varsigma)$ is defined by Eq. (6.2.25).

We shall consider the complex integral Φ whose integrand is the same as that in Eq. (6.2.39),

$$\Phi = \frac{1}{2\pi}\oint_{C^{(\pm)}} F_{xx}^{(s)}(\varsigma)\exp\left\{-(s/c_d)\left(y\sqrt{\varsigma^2+\gamma^2}+i\varsigma x\right)\right\}d\varsigma \tag{6.2.41}$$

The closed loop $C^{(\pm)}$ is discussed here. Four branch points at $\varsigma = \pm i, \pm i\gamma$ are found from the two radicals, and corresponding branch cuts are introduced along the imaginary axis in the ς-plane. The denominator, i.e. the Rayleigh equation, has two symmetric zeros at $\varsigma = \pm i\gamma_R$. These zeros give the simple poles for the integrand and the poles are on the imaginary axis (on the branch cut), but is greater in magnitude than any of the branch points. The branch points, cuts and poles are shown in Fig. 6.6a, b.

As to the Cagniard's path, we introduce a new parameter t, defined by

$$t = \left(y\sqrt{\varsigma^2 + \gamma^2} + i\varsigma x\right)/c_d \tag{6.2.42}$$

Its inversion gives the Cagniard's path in the complex ς-plane,

$$\varsigma_s^{(\pm)} = \frac{-ix(c_d t) \pm y\sqrt{(c_d t)^2 - \gamma^2(x^2 + y^2)}}{x^2 + y^2} \tag{6.2.43}$$

The saddle point at $t = \sqrt{x^2 + y^2}/c_s$ is

$$\varsigma_{saddle}^{(s)} = -\frac{i\gamma x}{\sqrt{x^2 + y^2}} \tag{6.2.44}$$

Comparing the magnitude of the saddle point with those of two branch points, we find that the saddle point is always smaller than the branch points $\varsigma = \pm i\gamma$ of the shear wave, but greater than the branch points $\varsigma = \pm i$ of the dilatational wave, if the inequality

$$\frac{\gamma|x|}{\sqrt{x^2 + y^2}} > 1 \quad \Rightarrow \quad y < |x|\sqrt{\gamma^2 - 1} \tag{6.2.45}$$

holds. Thus, we have two different closed loops for the complex integral.

When the saddle point is smaller in magnitude than the branch points $\varsigma = \pm i$, the Cagniard's path does not cross any branch cut; the closed loop is then composed of the line along the real axis, the Cagniard's path, and two large arcs which connect the line with the Cagniard's path. This closed loop is similar to that in the case of the dilatational wave contribution and the Cagniard's path is denoted by the path I in Fig. 6.6a.

On the other hand, when the saddle point is greater in magnitude than the branch point of the dilatational wave, i.e. the inequality of Eq. (6.2.45) holds, the Cagniard's path crosses the branch cut. The Cagniard's path has to be deformed along the imaginary axis, so that the closed loop does not include any branch points and cuts. This deformed Cagniard's path is shown as the path II in Fig. 6.6b, where the regular Cagniard's path is deformed by two short lines along the dilatational branch cut and a small circle around the dilatational branch point. Thus, the closed loop for this case is composed of the line along the real axis, the deformed Cagniard's path II, and two large arcs. Then, two closed loops are considered for the complex integral Φ and its choice depends on the inequality in Eq. (6.2.45).

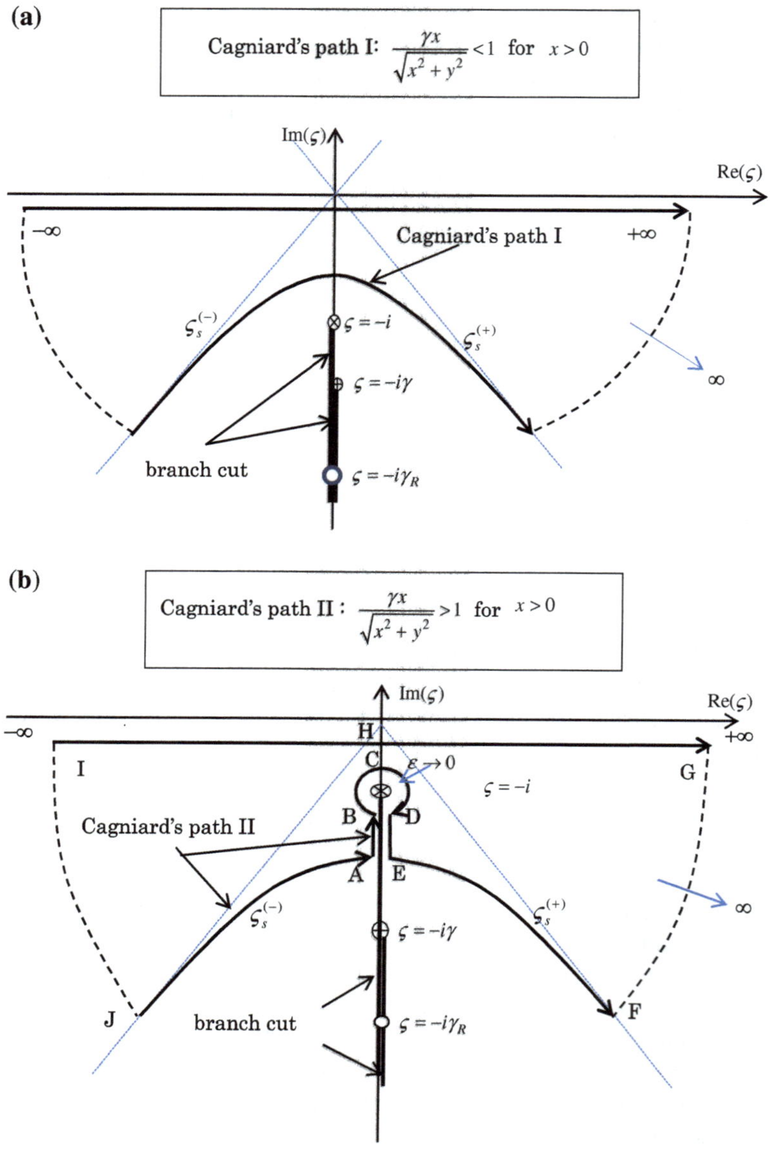

Fig. 6.6 Two closed loops for the 2D Cagniard's technique. **a** Loop with Cagniard's path I. **b** Loop with deformed Cagniard's path II

When the inequality of Eq. (6.2.45) does not hold, we employ the closed loop with the Cagniard's path I, and the Cauchy's integral theorem is applied to the complex integral Φ in Eq. (6.2.41). The integral along the real axis is converted to that along the Cagniard's path I, since no singular point is included in the loop and the integral along the large arc vanishes. That is,

$$I_{xx}^{(s)*}(x,y,s) = \frac{1}{2\pi}\int_{-\infty}^{+\infty} F_{xx}^{(s)}(\varsigma)\exp\left\{-(s/c_d)\left(y\sqrt{\varsigma^2+\gamma^2}+i\varsigma x\right)\right\}d\varsigma$$
$$= \frac{1}{2\pi}\int_{\sqrt{x^2+y^2}/c_s}^{\infty}\left\{F_{xx}^{(s)}(\varsigma_s^{(+)})\frac{d\varsigma_s^{(+)}}{dt} - F_{xx}^{(s)}(\varsigma_s^{(-)})\frac{d\varsigma_s^{(-)}}{dt}\right\}\exp(-st)dt \tag{6.2.46}$$

Then, the Laplace inversion is carried out by inspection and the double inversion for $\bar{I}_{xx}^{(s)*}$ is given by

$$I_{xx}^{(s)}(x,y,t)$$
$$= \frac{1}{2\pi}\left\{F_{xx}^{(s)}(\varsigma_s^{(+)})\frac{d\varsigma_s^{(+)}}{dt} - F_{xx}^{(s)}(\varsigma_s^{(-)})\frac{d\varsigma_s^{(-)}}{dt}\right\}H\left(t-\sqrt{x^2+y^2}/c_s\right)H\left(1-\frac{\gamma|x|}{\sqrt{x^2+y^2}}\right) \tag{6.2.47}$$

where the last step function is the conditional for the inequality.

When the inequality of Eq. (6.2.45) holds, we have to employ the loop with the Cagniard's path II. In addition to the regular hyperbolic path, the deformed Cagniard's path II has one small circle around the branch point $\varsigma = \pm i$ and two short lines along the branch cut. We perform the complex integral along the closed loop shown in Fig. 6.6b. The integral around the small circle vanishes as its radius tends to zero, but the line integral along the branch cut does not vanish and must be discussed. As was discussed in Sect. 1.3.2 in Chap. 1, the radical for the dilatational wave, $\sqrt{\varsigma^2+1}$, has different arguments, depending on the side of the cut. However, the radical $\sqrt{\varsigma^2+\gamma^2}$ for the shear wave does not change its argument since the saddle point is smaller in magnitude than the branch points $\pm i\gamma$. The sign and argument of the two radicals on the deformed Cagniard's path II are summarized in Table 6.1, where the new radicals are introduced as

$$\beta_d = \sqrt{\eta^2-1},\quad \beta_s = \sqrt{\gamma^2-\eta^2} \tag{6.2.48}$$

Using the new notations, the complex integral Φ in Eq. (6.2.45) is carried out along the closed loop with the deformed Cagniard's path II. Since no singular point is included in the loop and the integral along the large arc vanishes, the line integral

Table 6.1 Argument and magnitude of the radical and Rayleigh functions

Radicals	Path AB: $\varsigma = -i\eta$	Path DE: $\varsigma = -i\eta$
$\sqrt{\varsigma^2+1}$	$+i\sqrt{\eta^2-1} = +i\beta_d$	$-i\sqrt{\eta^2-1} = -i\beta_d$
$\sqrt{\varsigma^2+\gamma^2}$	$+\sqrt{\gamma^2-\eta^2} = +\beta_s$	$+\sqrt{\gamma^2-\eta^2} = +\beta_s$
$R(\varsigma)$	$(\gamma^2-2\eta^2)^2+4i\eta^2\beta_d\beta_s$	$(\gamma^2-2\eta^2)^2-4i\eta^2\beta_d\beta_s$
$\beta_d = \sqrt{\eta^2-1},\quad \beta_s = \sqrt{\gamma^2-\eta^2}$		

along the real axis is converted to that along the Cagniard's path II. The deformed Cagniard path is composed of two lines along the imaginary axis with different argument for the radicals and the regular Cagniard's path. Since the integral along the small circle $\overset{\frown}{BCD}$ vanishes as its radius tends to zero, the Fourier inversion integral, i.e. the integral along the real axis, is converted to the three integrals

$$\begin{aligned}
I_{xx}^{(s)*}(x,y,s) &= \frac{1}{2\pi}\int_{-\infty}^{+\infty} F_{xx}^{(s)}(\varsigma)\exp\left\{-(s/c_d)\left(y\sqrt{\varsigma^2+\gamma^2}+i\varsigma x\right)\right\}d\varsigma \\
&= \frac{1}{2\pi}\int_{\sqrt{x^2+y^2}/c_s}^{+\infty}\left\{F_{xx}^{(s)}(\varsigma_s^{(+)})\frac{d\varsigma_s^{(+)}}{dt} - F_{xx}^{(s)}(\varsigma_s^{(-)})\frac{d\varsigma_s^{(-)}}{dt}\right\}\exp(-st)dt \\
&\quad + \frac{1}{2\pi}\int_{AB} F_{xx}^{(s)}(\varsigma)\exp\left\{-(s/c_d)\left(y\sqrt{\varsigma^2+\gamma^2}+i\varsigma x\right)\right\}d\varsigma \\
&\quad + \frac{1}{2\pi}\int_{DE} F_{xx}^{(s)}(\varsigma)\exp\left\{-(s/c_d)\left(y\sqrt{\varsigma^2+\gamma^2}+i\varsigma x\right)\right\}d\varsigma
\end{aligned} \tag{6.2.49}$$

The last two integrals along the branch line are rearranged further. With the use of Table 6.1, the two branch line integrals are unified as

$$\begin{aligned}
&\frac{1}{2\pi}\int_{AB} F_{xx}^{(s)}(\varsigma)\exp\left\{-(s/c_d)\left(y\sqrt{\varsigma^2+\gamma^2}+i\varsigma x\right)\right\}d\varsigma \\
&\quad + \frac{1}{2\pi}\int_{DE} F_{xx}^{(s)}(\varsigma)\exp\left\{-(s/c_d)\left(y\sqrt{\varsigma^2+\gamma^2}+i\varsigma x\right)\right\}d\varsigma \\
&= \frac{1}{2\pi}\int_{1}^{\frac{\gamma x}{\sqrt{x^2+y^2}}} f_{xx}^{(s)}(\eta)\exp\{-(s/c_d)(\beta_s y + x\eta)\}d\eta
\end{aligned} \tag{6.2.50}$$

where the new integrand is

$$f_{xx}^{(s)}(\eta) = -\frac{8\eta^2(\gamma^2-2\eta^2)\beta_d\beta_s^2}{(\gamma^2-2\eta^2)^4+16\eta^4\beta_d^2\beta_s^2} \tag{6.2.51}$$

In order to transform the last integral in Eq. (6.2.50) to the form of Laplace transform integral, we introduce the variable transform from η to the new variable t as

$$t = (\beta_s y + x\eta)/c_d \tag{6.2.52}$$

Solving for η, we have

$$\eta_H = \frac{x(c_d t) - y\sqrt{\gamma^2(x^2+y^2) - (c_d t)^2}}{x^2+y^2} \tag{6.2.53}$$

Then, the last integral in Eq. (6.2.50) is transformed to that of the Laplace transform,

$$\begin{aligned} &\frac{1}{2\pi}\int_1^{\frac{\gamma x}{\sqrt{x^2+y^2}}} f_{xx}^{(s)}(\eta)\exp\{-(s/c_d)(\beta_s y + x\eta)\}d\eta \\ &= \frac{1}{2\pi}\int_{\left(x+y\sqrt{\gamma^2-1}\right)/c_d}^{\sqrt{x^2+y^2}/c_s} f_{xx}^{(s)}(\eta_H)\frac{d\eta_H}{dt}\exp(-st)dt \end{aligned} \tag{6.2.54}$$

Finally, we have converted the Fourier inversion integral to the form of Laplace transform integral as

$$\begin{aligned} I_{xx}^{(s)*}(x,y,s) &= \frac{1}{2\pi}\int_{-\infty}^{+\infty} F_{xx}^{(s)}(\varsigma)\exp\left\{-(s/c_d)\left(y\sqrt{\varsigma^2+\gamma^2} + i\varsigma x\right)\right\}d\varsigma \\ &= \frac{1}{2\pi}\int_{\sqrt{x^2+y^2}/c_s}^{+\infty}\left\{F_{xx}^{(s)}(\varsigma_s^{(+)})\frac{d\varsigma_s^{(+)}}{dt} - F_{xx}^{(s)}(\varsigma_s^{(-)})\frac{d\varsigma_s^{(-)}}{dt}\right\}\exp(-st)dt \\ &\quad + \frac{1}{2\pi}\int_{\left(x+y\sqrt{\gamma^2-1}\right)/c_d}^{\sqrt{x^2+y^2}/c_s} f_{xx}^{(s)}(\eta_H)\frac{d\eta_H}{dt}\exp(-st)dt \end{aligned} \tag{6.2.55}$$

The Laplace inversion is carried out by inspection. The branch line integral of Eq. (6.2.54) appears only when the conditional of Eq. (6.2.45) holds. Using the step function for this conditional, the Laplace inversion is finally expressed by

$$\begin{aligned} &I_{xx}^{(s)}(x,y,t) \\ &= H\left(\frac{\gamma x}{\sqrt{x^2+y^2}} - 1\right)\left[\frac{1}{2\pi}\left\{F_{xx}^{(s)}(\varsigma_s^{(+)})\frac{d\varsigma_s^{(+)}}{dt} - F_{xx}^{(s)}(\varsigma_s^{(-)})\frac{d\varsigma_s^{(-)}}{dt}\right\}H\left(t - \sqrt{x^2+y^2}/c_s\right)\right. \\ &\qquad \left. + \frac{1}{2\pi}f_{xx}^{(s)}(\eta_H)\frac{d\eta_H}{dt}H\left(\sqrt{x^2+y^2}/c_s - t\right)H\left(t - \frac{x+y\sqrt{\gamma^2-1}}{c_d}\right)\right] \end{aligned} \tag{6.2.56}$$

Comparing this expression with that of Eq. (6.2.47), we see that the above equation includes the full conditional. Thus, the final result for the double inversions of the shear wave contribution is given by

$$\begin{aligned}
&I_{xx}^{(s)}(x,y,t)\\
&=\frac{1}{2\pi}\left\{F_{xx}^{(s)}\left(\varsigma_s^{(+)}\right)\frac{d\varsigma_s^{(+)}}{dt}-F_{xx}^{(s)}\left(\varsigma_s^{(-)}\right)\frac{d\varsigma_s^{(-)}}{dt}\right\}H\left(t-\sqrt{x^2+y^2}/c_s\right)\\
&\quad+\frac{1}{2\pi}f_{xx}^{(s)}(\eta_H)\frac{d\eta_H}{dt}H\left(\frac{\gamma x}{\sqrt{x^2+y^2}}-1\right)H\left(\sqrt{x^2+y^2}/c_s-t\right)H\left(t-\frac{x+y\sqrt{\gamma^2-1}}{c_d}\right)
\end{aligned}\tag{6.2.57}$$

The same inversion technique is applied to another double inversion for the shear wave contribution and we have the unified expression for the shear wave contribution as

$$\begin{aligned}
&I_{ij}^{(s)}(x,y,t)\\
&=\frac{1}{2\pi}\left\{F_{ij}^{(s)}\left(\varsigma_s^{(+)}\right)\frac{d\varsigma_s^{(+)}}{dt}-F_{ij}^{(s)}\left(\varsigma_s^{(-)}\right)\frac{d\varsigma_s^{(-)}}{dt}\right\}H\left(c_st-\sqrt{x^2+y^2}\right)\\
&\quad+\frac{1}{2\pi}f_{ij}^{(s)}(\eta_H)\frac{d\eta_H}{dt}H\left(\frac{\gamma|x|}{\sqrt{x^2+y^2}}-1\right)H\left(\sqrt{x^2+y^2}-c_st\right)H\left(c_dt-|x|-y\sqrt{\gamma^2-1}\right)
\end{aligned}\tag{6.2.58}$$

where the integrands are given by

$$\begin{aligned}
F_{xx}^{(s)}(\varsigma)&=-\frac{(2\varsigma^2+\gamma^2)\sqrt{\varsigma^2+\gamma^2}}{R(\varsigma)},\quad F_{xy}^{(s)}(\varsigma)=\frac{2i\varsigma\sqrt{\varsigma^2+1}\sqrt{\varsigma^2+\gamma^2}}{R(\varsigma)}\\
F_{yx}^{(s)}(\varsigma)&=\frac{i\varsigma(2\varsigma^2+\gamma^2)}{R(\varsigma)},\quad F_{yy}^{(s)}(\varsigma)=\frac{2\varsigma^2\sqrt{\varsigma^2+1}}{R(\varsigma)}
\end{aligned}\tag{6.2.59}$$

$$\begin{aligned}
f_{xx}^{(s)}(\eta)&=-\frac{8\eta^2(\gamma^2-2\eta^2)\beta_d\beta_s^2}{R_H(\eta)},\quad f_{xy}^{(s)}(\eta)=-\frac{4\eta(\gamma^2-2\eta^2)^2\beta_d\beta_s}{R_H(\eta)}\\
f_{yx}^{(s)}(\eta)&=\frac{8\eta^3(\gamma^2-2\eta^2)\beta_d\beta_s}{R_H(\eta)},\quad f_{yy}^{(s)}(\eta)=\frac{4\eta^2(\gamma^2-2\eta^2)^2\beta_d}{R_H(\eta)}
\end{aligned}\tag{6.2.60}$$

and

$$R_H(\eta)=(\gamma^2-2\eta^2)^4+16\eta^4\beta_d^2\beta_s^2\tag{6.2.61}$$

(3) Green's dyadic

We have just inverted the two wave contributions and thus the displacement response is given by

$$\begin{aligned} u_x &= \frac{P_x}{\mu}\left\{I_{xx}^{(d)}(x,y,t) + I_{xx}^{(s)}(x,y,t)\right\} + \frac{P_y}{\mu}\left\{I_{xy}^{(d)}(x,y,t) + I_{xy}^{(s)}(x,y,t)\right\} \\ u_y &= \frac{P_x}{\mu}\left\{I_{yx}^{(d)}(x,y,t) + I_{yx}^{(s)}(x,y,t)\right\} + \frac{P_y}{\mu}\left\{I_{yy}^{(d)}(x,y,t) + I_{yy}^{(s)}(x,y,t)\right\} \end{aligned} \tag{6.2.62}$$

When we express the displacement in terms of Green's dyadic $G_{ij}(x,y,t)$,

$$\begin{aligned} u_x &= P_x G_{xx}(x,y,t) + P_y G_{xx}(x,y,t) \\ u_y &= P_x G_{yx}(x,y,t) + P_y G_{yy}(x,y,t) \end{aligned} \tag{6.2.63}$$

the dyadic is given by

$$G_{ij}(x,y,t) = \frac{1}{\mu}\left\{I_{ij}^{(d)}(x,y,t) + I_{ij}^{(s)}(x,y,t)\right\} \tag{6.2.64}$$

A more detailed expression is

$$\begin{aligned} &G_{ij}(x,y,t) \\ &= \frac{1}{2\pi\mu}\left[\left\{F_{xx}^{(d)}\left(\varsigma_d^{(+)}\right)\frac{d\varsigma_d^{(+)}}{dt} - F_{xx}^{(d)}\left(\varsigma_d^{(-)}\right)\frac{d\varsigma_d^{(-)}}{dt}\right\}H(c_d t - r)\right. \\ &\quad + \left\{F_{ij}^{(s)}\left(\varsigma_s^{(+)}\right)\frac{d\varsigma_s^{(+)}}{dt} - F_{ij}^{(s)}\left(\varsigma_s^{(-)}\right)\frac{d\varsigma_s^{(-)}}{dt}\right\}H(c_s t - r) \\ &\quad \left. + f_{ij}^{(s)}(\eta_H)\frac{d\eta_H}{dt}H(\gamma|x| - r)H(r - c_s t)H\left(c_d t - |x| - y\sqrt{\gamma^2 - 1}\right)\right] \end{aligned} \tag{6.2.65}$$

where the radial distance $r = \sqrt{x^2 + y^2}$ is introduced.

It should be noticed that the step function $H(c_d t - r)$ shows the disturbed circular region produced by the dilatational wave and $H(c_s t - r)$ does the region disturbed by the shear wave. These two circular waves are emanating from the source point. The triple product of the step functions

$$H(\gamma|x| - r)H(r - c_s t)H\left(c_d t - |x| - y\sqrt{\gamma^2 - 1}\right)$$

gives the region disturbed by the shear wave which is induced by the precursor dilatational wave. We call this region "von Schmidt wave". Its wave region is shown in Fig. 6.7.

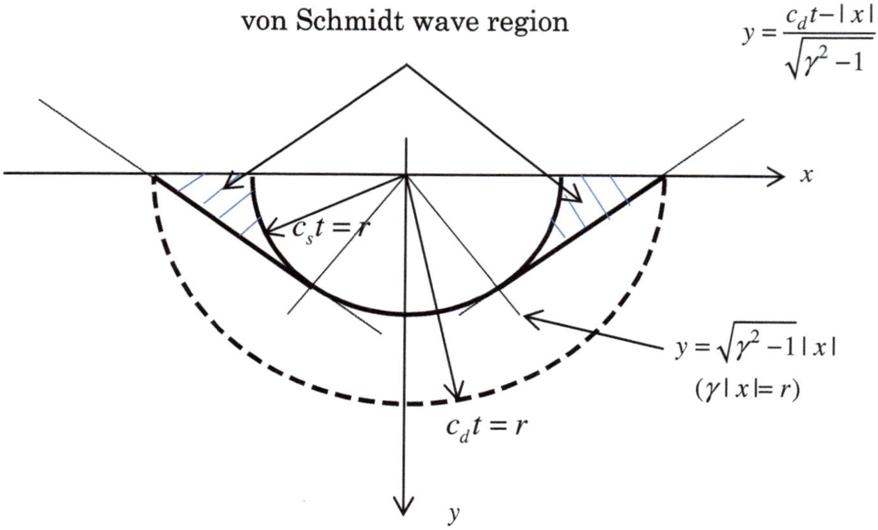

Fig. 6.7 von Schmidt wave region

6.3 3D Dynamic Lamb's Problem

The transient response of a fully 3D elastic half space is discussed here. We take the 3D coordinate system (x, y, z). The surface is denoted by $z = 0$ and the interior by $z > 0$ as shown in Fig. 6.8. The 3D deformation of the isotropic elastic solid is governed by equations of motion,

$$
\begin{aligned}
\frac{\partial \sigma_{xx}}{\partial x} + \frac{\partial \sigma_{yx}}{\partial y} + \frac{\partial \sigma_{zx}}{\partial z} &= \rho \frac{\partial^2 u_x}{\partial t^2} \\
\frac{\partial \sigma_{xy}}{\partial x} + \frac{\partial \sigma_{yy}}{\partial y} + \frac{\partial \sigma_{zy}}{\partial z} &= \rho \frac{\partial^2 u_y}{\partial t^2} \\
\frac{\partial \sigma_{xz}}{\partial x} + \frac{\partial \sigma_{yz}}{\partial y} + \frac{\partial \sigma_{zz}}{\partial z} &= \rho \frac{\partial^2 u_z}{\partial t^2}
\end{aligned}
\tag{6.3.1}
$$

and Hooke's law for the isotropic elastic solid,

$$
\begin{aligned}
\sigma_{xx} &= \lambda\left(\frac{\partial u_x}{\partial x} + \frac{\partial u_y}{\partial y} + \frac{\partial u_z}{\partial z}\right) + 2\mu \frac{\partial u_x}{\partial x} \\
\sigma_{yy} &= \lambda\left(\frac{\partial u_x}{\partial x} + \frac{\partial u_y}{\partial y} + \frac{\partial u_z}{\partial z}\right) + 2\mu \frac{\partial u_y}{\partial y} \\
\sigma_{zz} &= \lambda\left(\frac{\partial u_x}{\partial x} + \frac{\partial u_y}{\partial y} + \frac{\partial u_z}{\partial z}\right) + 2\mu \frac{\partial u_z}{\partial z}
\end{aligned}
$$

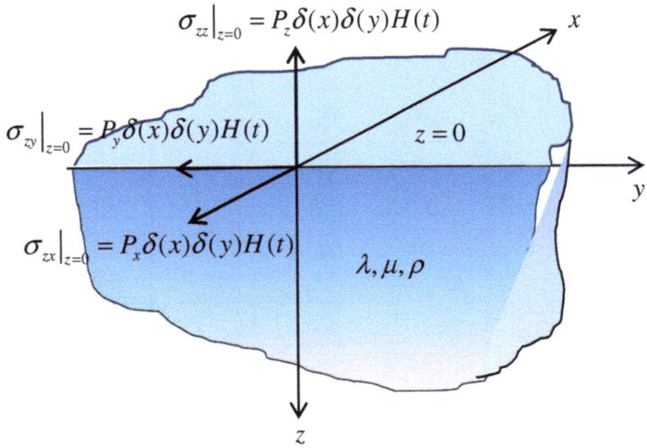

Fig. 6.8 A suddenly applied load P_i on the surface of a semi-infinite elastic solid

$$\begin{aligned}
\sigma_{xy} &= \sigma_{yx} = \mu\left(\frac{\partial u_x}{\partial y} + \frac{\partial u_y}{\partial x}\right) \\
\sigma_{yz} &= \sigma_{zy} = \mu\left(\frac{\partial u_z}{\partial y} + \frac{\partial u_y}{\partial z}\right) \\
\sigma_{zx} &= \sigma_{xz} = \mu\left(\frac{\partial u_x}{\partial z} + \frac{\partial u_z}{\partial x}\right)
\end{aligned} \tag{6.3.2}$$

where λ and μ are Lame's constants.

We assume that a point load is suddenly applied at the coordinate origin on the surface $z = 0$, i.e.

$$\begin{aligned}
\sigma_{zx}|_{z=0} &= P_x\delta(x)\delta(y)H(t) \\
\sigma_{zy}|_{z=0} &= P_y\delta(x)\delta(y)H(t) \\
\sigma_{zz}|_{z=0} &= P_z\delta(x)\delta(y)H(t)
\end{aligned} \tag{6.3.3}$$

where P_j, $j = x, y, z$ are components of the load. Note that the loading time function is Heaviside's unit step function, not Dirac's delta function. This is because that we get the simpler solution for the dynamic response. The quiescent condition at an initial time

$$u_i|_{t=0} = \left.\frac{\partial u_i}{\partial t}\right|_{t=0} = 0; \quad i = x, y, z \tag{6.3.4}$$

and the convergence condition at infinity

$$u_i|_{\sqrt{x^2+y^2+z^2}\to\infty} = \left.\frac{\partial u_i}{\partial x}\right|_{\sqrt{x^2+y^2+z^2}\to\infty} = \left.\frac{\partial u_i}{\partial y}\right|_{\sqrt{x^2+y^2+z^2}\to\infty} = \left.\frac{\partial u_i}{\partial z}\right|_{\sqrt{x^2+y^2+z^2}\to\infty} = 0; \quad i = x, y, z \tag{6.3.5}$$

are employed.

Substituting Hooke's law of Eq. (6.3.2) into the equation of motion (6.3.1), the displacement equations are given by

$$
\begin{aligned}
&\gamma^2 \frac{\partial^2 u_x}{\partial x^2} + \frac{\partial^2 u_x}{\partial y^2} + \frac{\partial^2 u_x}{\partial z^2} + (\gamma^2 - 1)\left(\frac{\partial^2 u_y}{\partial x \partial y} + \frac{\partial^2 u_z}{\partial x \partial z}\right) = \frac{1}{c_s^2}\frac{\partial^2 u_x}{\partial t^2} \\
&\frac{\partial^2 u_y}{\partial x^2} + \gamma^2 \frac{\partial^2 u_y}{\partial y^2} + \frac{\partial^2 u_y}{\partial z^2} + (\gamma^2 - 1)\left(\frac{\partial^2 u_x}{\partial x \partial y} + \frac{\partial^2 u_z}{\partial y \partial z}\right) = \frac{1}{c_s^2}\frac{\partial^2 u_y}{\partial t^2} \\
&\frac{\partial^2 u_z}{\partial x^2} + \frac{\partial^2 u_z}{\partial y^2} + \gamma^2 \frac{\partial^2 u_z}{\partial z^2} + (\gamma^2 - 1)\left(\frac{\partial^2 u_x}{\partial x \partial z} + \frac{\partial^2 u_y}{\partial y \partial z}\right) = \frac{1}{c_s^2}\frac{\partial^2 u_z}{\partial t^2}
\end{aligned}
\tag{6.3.6}
$$

The present 3D elastodynamic problem is also discussed by the integral transform method. With use of the quiescent and convergence conditions, we apply Laplace transform with respect to the time, defined by

$$
f^*(s) = \int_0^\infty f(t)\exp(-st)dt \tag{6.3.7}
$$

and the double Fourier transform with respect to two space variables x and y, defined by

$$
\begin{aligned}
\bar{\bar{f}}(\xi, \eta) &= \int_{-\infty}^{+\infty}\int_{-\infty}^{+\infty} f(x, y)\exp(+i\xi x + i\eta y)dxdy \\
f(x, y) &= \frac{1}{(2\pi)^2}\int_{-\infty}^{+\infty}\int_{-\infty}^{+\infty} \bar{\bar{f}}(\xi, \eta)\exp(-i\xi x - i\eta y)d\xi d\eta
\end{aligned}
\tag{6.3.8}
$$

to the displacement equations (6.3.6) and Hooke's law of Eq. (6.3.2). The displacement equations are transformed to the coupled ordinary differential equations,

$$
\begin{aligned}
&\frac{d^2\bar{\bar{u}}_x^*}{dz^2} - \{\gamma^2\xi^2 + \eta^2 + (s/c_s)^2\}\bar{\bar{u}}_x^* - \xi\eta(\gamma^2 - 1)\bar{\bar{u}}_y^* - i\xi(\gamma^2 - 1)\frac{d\bar{\bar{u}}_z^*}{dz} = 0 \\
&\frac{d^2\bar{\bar{u}}_y^*}{dz^2} - \{\xi^2 + \gamma^2\eta^2 + (s/c_s)^2\}\bar{\bar{u}}_y^* - \xi\eta(\gamma^2 - 1)\bar{\bar{u}}_x^* - i\eta(\gamma^2 - 1)\frac{d\bar{\bar{u}}_z^*}{dz} = 0 \\
&\gamma^2\frac{d^2\bar{\bar{u}}_z^*}{dz^2} - \{\xi^2 + \eta^2 + (s/c_s)^2\}\bar{\bar{u}}_z^* - i\xi(\gamma^2 - 1)\frac{d\bar{\bar{u}}_x^*}{dz} - i\eta(\gamma^2 - 1)\frac{d\bar{\bar{u}}_y^*}{dz} = 0
\end{aligned}
\tag{6.3.9}
$$

The stress components to be used for the boundary condition are transformed as follows:

$$\begin{aligned}\frac{1}{\mu}\tilde{\bar{\sigma}}_{zx}^{*} &= \frac{d\tilde{\bar{u}}_{x}^{*}}{dz} - i\xi\tilde{\bar{u}}_{z}^{*}\\ \frac{1}{\mu}\tilde{\bar{\sigma}}_{zy}^{*} &= \frac{d\tilde{\bar{u}}_{y}^{*}}{dz} - i\eta\tilde{\bar{u}}_{z}^{*}\\ \frac{1}{\mu}\tilde{\bar{\sigma}}_{zz}^{*} &= \gamma^{2}\frac{d\tilde{\bar{u}}_{z}^{*}}{dz} - i\xi(\gamma^{2}-2)\tilde{\bar{u}}_{x}^{*} - i\eta(\gamma^{2}-2)\tilde{\bar{u}}_{y}^{*}\end{aligned} \tag{6.3.10}$$

The boundary condition on the surface is also transformed to

$$\frac{1}{\mu}\tilde{\bar{\sigma}}_{zj}^{*}\Big|_{z=0} = \frac{P_j}{\mu s}; \quad j = x,\, y,\, z \tag{6.3.11}$$

The general solution for the displacement equations (6.3.9) is obtained as

$$\begin{aligned}\tilde{\bar{u}}_{x}^{*} &= \frac{i\xi}{\alpha_d}A(\xi,\eta,s)\exp(-\alpha_d z) + B(\xi,\eta,s)\exp(-\alpha_s z)\\ \tilde{\bar{u}}_{y}^{*} &= \frac{i\eta}{\alpha_d}A(\xi,\eta,s)\exp(-\alpha_d z) + C(\xi,\eta,s)\exp(-\alpha_s z)\\ \tilde{\bar{u}}_{z}^{*} &= A(\xi,\eta,s)\exp(-\alpha_d z) - \frac{i}{\alpha_s}\{\xi B(\xi,\eta,s) + \eta C(\xi,\eta,s)\}\exp(-\alpha_s z)\end{aligned} \tag{6.3.12}$$

The stress components are

$$\begin{aligned}\frac{1}{\mu}\tilde{\bar{\sigma}}_{zx}^{*} &= -2i\xi A(\xi,\eta,s)\exp(-\alpha_d z) - \frac{1}{\alpha_s}\{(\alpha_s^2+\xi^2)B(\xi,\eta,s) + \xi\eta C(\xi,\eta,s)\}\exp(-\alpha_s z)\\ \frac{1}{\mu}\tilde{\bar{\sigma}}_{zy}^{*} &= -2i\eta A(\xi,\eta,s)\exp(-\alpha_d z) - \frac{1}{\alpha_s}\{\xi\eta B(\xi,\eta,s) + (\alpha_s^2+\xi^2)C(\xi,\eta,s)\}\exp(-\alpha_s z)\\ \frac{1}{\mu}\tilde{\bar{\sigma}}_{zz}^{*} &= -\frac{\alpha_s^2+\xi^2+\eta^2}{\alpha_d}A(\xi,\eta,s)\exp(-\alpha_d z) + 2i\{\xi B(\xi,\eta,s) + \eta C(\xi,\eta,s)\}\exp(-\alpha_s z)\end{aligned} \tag{6.3.13}$$

where $A(\xi,\eta,s)$, $B(\xi,\eta,s)$ and $C(\xi,\eta,s)$ are unknown coefficients to be determined by the boundary condition. The radicals are defined by

$$\alpha_j = \sqrt{\xi^2+\eta^2+(s/c_j)^2}, \quad \mathrm{Re}(\alpha_j) > 0; \quad j = d, s \tag{6.3.14}$$

where the conditional $\mathrm{Re}(\alpha_j) > 0$ guarantees the convergence condition at infinity. Applying the boundary condition (6.3.11) to the stresses of Eq. (6.3.13), we obtain the following simple algebraic equations for the coefficients, $A(\xi,\eta,s)$, $B(\xi,\eta,s)$ and $C(\xi,\eta,s)$,

$$\begin{aligned}
&-2i\xi A(\xi,\eta,s)-\frac{1}{\alpha_s}\left\{(\alpha_s^2+\xi^2)B(\xi,\eta,s)+\xi\eta C(\xi,\eta,s)\right\}=\frac{P_x}{\mu s}\\
&-2i\eta A(\xi,\eta,s)-\frac{1}{\alpha_s}\left\{\xi\eta B(\xi,\eta,s)+(\alpha_s^2+\xi^2)C(\xi,\eta,s)\right\}=\frac{P_y}{\mu s}\\
&-\frac{\alpha_s^2+\xi^2+\eta^2}{\alpha_d}A(\xi,\eta,s)+2i\{\xi B(\xi,\eta,s)+\eta C(\xi,\eta,s)\}=\frac{P_z}{\mu s}
\end{aligned} \tag{6.3.15}$$

The coefficients are determined as

$$A(\xi,\eta,s)=-\frac{2i\xi\alpha_d\alpha_s}{sR(\xi,\eta,s)}\frac{P_x}{\mu}-\frac{2i\eta\alpha_d\alpha_s}{sR(\xi,\eta,s)}\frac{P_y}{\mu}-\frac{\alpha_d(\alpha_s^2+\xi^2+\eta^2)}{sR(\xi,\eta,s)}\frac{P_z}{\mu} \tag{6.3.16}$$

$$\begin{aligned}
B(\xi,\eta,s)=&\left\{\frac{\xi^2(\alpha_s^2+\xi^2+\eta^2-4\alpha_d\alpha_s)}{s\alpha_sR(\xi,\eta,s)}-\frac{1}{s\alpha_s}\right\}\frac{P_x}{\mu}\\
&+\frac{\xi\eta(\alpha_s^2+\xi^2+\eta^2-4\alpha_d\alpha_s)}{s\alpha_sR(\xi,\eta,s)}\frac{P_y}{\mu}+\frac{2i\xi\alpha_d\alpha_s}{sR(\xi,\eta,s)}\frac{P_z}{\mu}
\end{aligned} \tag{6.3.17}$$

$$\begin{aligned}
C(\xi,\eta,s)=&\frac{\xi\eta(\alpha_s^2+\xi^2+\eta^2-4\alpha_d\alpha_s)}{s\alpha_sR(\xi,\eta,s)}\frac{P_x}{\mu}\\
&+\left\{\frac{\eta^2(\alpha_s^2+\xi^2+\eta^2-4\alpha_d\alpha_s)}{s\alpha_sR(\xi,\eta,s)}-\frac{1}{s\alpha_s}\right\}\frac{P_y}{\mu}+\frac{2i\eta\alpha_d\alpha_s}{sR(\xi,\eta,s)}\frac{P_z}{\mu}
\end{aligned} \tag{6.3.18}$$

where Rayleigh equation is defined by

$$R(\xi,\eta,s)=(\xi^2+\eta^2+\alpha_s^2)^2-4\alpha_d\alpha_s(\xi^2+\eta^2) \tag{6.3.19}$$

Then, the transformed displacement components are given by

$$\begin{aligned}
\tilde{\bar{u}}_x^*=&\frac{P_x}{\mu}\left\{\tilde{\bar{I}}_{xx}^{*(d)}(\xi,\eta,z,s)+\tilde{\bar{I}}_{xx}^{*(s)}(\xi,\eta,z,s)+\tilde{\bar{I}}_{xx}^{*(SH)}(\xi,\eta,z,s)\right\}\\
&+\frac{P_y}{\mu}\left\{\tilde{\bar{I}}_{xy}^{*(d)}(\xi,\eta,z,s)+\tilde{\bar{I}}_{xy}^{*(s)}(\xi,\eta,z,s)\right\}\\
&+\frac{P_z}{\mu}\left\{\tilde{\bar{I}}_{xz}^{*(d)}(\xi,\eta,z,s)+\tilde{\bar{I}}_{xz}^{*(s)}(\xi,\eta,z,s)\right\}
\end{aligned} \tag{6.3.20a}$$

$$\begin{aligned}
\tilde{\bar{u}}_y^*=&\frac{P_x}{\mu}\left\{\tilde{\bar{I}}_{yx}^{*(d)}(\xi,\eta,z,s)+\tilde{\bar{I}}_{yx}^{*(s)}(\xi,\eta,z,s)\right\}\\
&+\frac{P_y}{\mu}\left\{\tilde{\bar{I}}_{yy}^{*(d)}(\xi,\eta,z,s)+\tilde{\bar{I}}_{yy}^{*(s)}(\xi,\eta,z,s)+\tilde{\bar{I}}_{yy}^{*(SH)}(\xi,\eta,z,s)\right\}\\
&+\frac{P_z}{\mu}\left\{\tilde{\bar{I}}_{yz}^{*(d)}(\xi,\eta,z,s)+\tilde{\bar{I}}_{yz}^{*(s)}(\xi,\eta,z,s)\right\}
\end{aligned} \tag{6.3.20b}$$

$$\begin{aligned}\tilde{\bar{u}}_z^* &= \frac{P_x}{\mu}\left\{\tilde{\bar{I}}_{zx}^{*(d)}(\xi,\eta,z,s) + \tilde{\bar{I}}_{zx}^{*(s)}(\xi,\eta,z,s)\right\} \\ &+ \frac{P_y}{\mu}\left\{\tilde{\bar{I}}_{zy}^{*(d)}(\xi,\eta,z,s) + \tilde{\bar{I}}_{zy}^{*(s)}(\xi,\eta,z,s)\right\} \\ &+ \frac{P_z}{\mu}\left\{\tilde{\bar{I}}_{zz}^{*(d)}(\xi,\eta,z,s) + \tilde{\bar{I}}_{zz}^{*(s)}(\xi,\eta,z,s)\right\}\end{aligned} \tag{6.3.20c}$$

where

$$\begin{aligned}\tilde{\bar{I}}_{xx}^{*(d)}(\xi,\eta,z,s) &= \frac{2\xi^2\alpha_s}{sR(\xi,\eta,s)}\exp(-\alpha_d z) \\ \tilde{\bar{I}}_{xx}^{*(s)}(\xi,\eta,z,s) &= \frac{\xi^2(\alpha_s^2+\xi^2+\eta^2-4\alpha_d\alpha_s)}{s\alpha_s R(\xi,\eta,s)}\exp(-\alpha_s z) \\ \tilde{\bar{I}}_{xx}^{*(SH)}(\xi,\eta,z,s) &= -\frac{1}{s\alpha_s}\exp(-\alpha_s z) \\ \tilde{\bar{I}}_{xy}^{*(d)}(\xi,\eta,z,s) &= \frac{2\xi\eta\alpha_s}{sR(\xi,\eta,s)}\exp(-\alpha_d z) \\ \tilde{\bar{I}}_{xy}^{*(s)}(\xi,\eta,z,s) &= \frac{\xi\eta(\alpha_s^2+\xi^2+\eta^2-4\alpha_d\alpha_s)}{s\alpha_s R(\xi,\eta,s)}\exp(-\alpha_s z) \\ \tilde{\bar{I}}_{xz}^{*(d)}(\xi,\eta,z,s) &= -\frac{i\xi(\alpha_s^2+\xi^2+\eta^2)}{sR(\xi,\eta,s)}\exp(-\alpha_d z) \\ \tilde{\bar{I}}_{xz}^{*(s)}(\xi,\eta,z,s) &= \frac{2i\xi\alpha_d\alpha_s}{sR(\xi,\eta,s)}\exp(-\alpha_s z)\end{aligned} \tag{6.3.21a}$$

$$\begin{aligned}\tilde{\bar{I}}_{yx}^{*(d)}(\xi,\eta,z,s) &= \frac{2\xi\eta\alpha_s}{sR(\xi,\eta,s)}\exp(-\alpha_d z) \\ \tilde{\bar{I}}_{yx}^{*(s)}(\xi,\eta,z,s) &= \frac{\xi\eta(\alpha_s^2+\xi^2+\eta^2-4\alpha_d\alpha_s)}{s\alpha_s R(\xi,\eta,s)}\exp(-\alpha_s z) \\ \tilde{\bar{I}}_{yy}^{*(d)}(\xi,\eta,z,s) &= \frac{2\eta^2\alpha_s}{sR(\xi,\eta,s)}\exp(-\alpha_d z) \\ \tilde{\bar{I}}_{yy}^{*(s)}(\xi,\eta,z,s) &= \frac{\eta^2(\alpha_s^2+\xi^2+\eta^2-4\alpha_d\alpha_s)}{s\alpha_s R(\xi,\eta,s)}\exp(-\alpha_s z) \\ \tilde{\bar{I}}_{yy}^{*(SH)}(\xi,\eta,z,s) &= -\frac{1}{s\alpha_s}\exp(-\alpha_s z) \\ \tilde{\bar{I}}_{yz}^{*(d)}(\xi,\eta,z,s) &= -\frac{i\eta(\alpha_s^2+\xi^2+\eta^2)}{sR(\xi,\eta,s)}\exp(-\alpha_d z) \\ \tilde{\bar{I}}_{yz}^{*(s)}(\xi,\eta,z,s) &= \frac{2i\eta\alpha_d\alpha_s}{sR(\xi,\eta,s)}\exp(-\alpha_s z)\end{aligned} \tag{6.3.21b}$$

$$
\begin{aligned}
\tilde{\bar{I}}_{zx}^{*(d)}(\xi,\eta,z,s) &= -\frac{2i\xi\alpha_d\alpha_s}{sR(\xi,\eta,s)}\exp(-\alpha_d z) \\
\tilde{\bar{I}}_{zx}^{*(s)}(\xi,\eta,z,s) &= \frac{i\xi(\alpha_s^2+\xi^2+\eta^2)}{sR(\xi,\eta,s)}\exp(-\alpha_s z) \\
\tilde{\bar{I}}_{zy}^{*(d)}(\xi,\eta,z,s) &= -\frac{2i\eta\alpha_d\alpha_s}{sR(\xi,\eta,s)}\exp(-\alpha_d z) \\
\tilde{\bar{I}}_{zy}^{*(s)}(\xi,\eta,z,s) &= \frac{i\eta(\alpha_s^2+\xi^2+\eta^2)}{sR(\xi,\eta,s)}\exp(-\alpha_s z) \\
\tilde{\bar{I}}_{zz}^{*(d)}(\xi,\eta,z,s) &= -\frac{\alpha_d(\alpha_s^2+\xi^2+\eta^2)}{sR(\xi,\eta,s)}\exp(-\alpha_d z) \\
\tilde{\bar{I}}_{zz}^{*(s)}(\xi,\eta,z,s) &= \frac{2\alpha_d(\xi^2+\eta^2)}{sR(\xi,\eta,s)}\exp(-\alpha_s z)
\end{aligned}
\tag{6.3.21c}
$$

In the above equations, the superscript k in $\tilde{\bar{I}}_{ij}^{*(k)}$ indicates the wave type. The superscript "*d*" stands for the dilatational wave (P-wave), "*s*" for the vertically polarized shear wave (SV-wave), and "*SH*" for the horizontally polarized shear wave (SH-wave). The meaning of the subscripts *i* and *j* are the same as those in the 2D problem. Note that the SH-wave contribution appears only in the horizontal components and that no SH-wave contribution is included in the vertical component which is normal to the surface.

Now, we shall invert each wave component separately.

(1) Inversion of P wave contribution: $\tilde{\bar{I}}_{ij}^{*(d)}(\xi,\eta,z,s)$

As a representative, we consider the inversion of $\tilde{\bar{I}}_{xx}^{*(d)}(\xi,\eta,z,s)$. Its formal Fourier double inversion with respect to the parameters ξ and η is given by

$$
\bar{I}_{xx}^{*(d)}(x,y,z,s) = \frac{1}{(2\pi)^2}\int_{-\infty}^{+\infty}\int_{-\infty}^{+\infty}\frac{2\xi^2\alpha_s}{sR(\xi,\eta,s)}\exp(-\alpha_d z - i\xi x - i\eta y)d\xi d\eta \tag{6.3.22}
$$

We introduce the variable transform from (ξ,η) to the new variables $(p,\ q)$ defined by

$$
\xi = (s/r)(xp - yq), \quad \eta = (s/r)(yp + xq) \tag{6.3.23}
$$

where the horizontal distance defined by $r = \sqrt{x^2+y^2}$ is introduced. This variable transform corresponds to the rotation of the axes (ξ,η). The equivalent area element is given by $d\xi d\eta = s^2 dpdq$. The double integral in Eq. (6.3.22) is converted to another double integral,

$$I_{xx}^{*(d)}(x,y,z,s) = \frac{1}{(2\pi r)^2}\int_{-\infty}^{+\infty}\int_{-\infty}^{+\infty}\frac{2(x^2p^2+y^2q^2-2xypq)\beta_s}{R(p,q)}\exp\{-s(\beta_d z+irp)\}dpdq \tag{6.3.24}$$

where

$$R(p,q) = (\beta_s^2+p^2+q^2)^2 - 4\beta_d\beta_s(p^2+q^2) \tag{6.3.25}$$

$$\beta_j = \sqrt{p^2+q^2+1/c_j^2}\,,\quad \mathrm{Re}(\beta_j) > 0;\quad j = d,s \tag{6.3.26}$$

Note that the Rayleigh equation $R(p,\ q)$ is different from the former Rayleigh equation $R(\xi,\eta,s)$.

The integrand in Eq. (6.3.24) includes both odd and even functions of the new variable q. As the integration with respect to q is carried out over the whole range from $-\infty$ to $+\infty$, the integral whose integrand has an odd power (or the anti-symmetric function of q) vanishes. Equation (6.3.24) is then slightly simplified, to

$$I_{xx}^{*(d)}(x,y,z,s) = \int_0^{\infty}dq\int_{-\infty}^{+\infty}F_{xx}^{(d)}(p,q)\exp\{-s(\beta_d z+irp)\}dp \tag{6.3.27}$$

where the new notation for the integrand is introduced as

$$F_{xx}^{(d)}(p,q) = \frac{(x^2p^2+y^2q^2)\beta_s}{(\pi r)^2R(p,q)} \tag{6.3.28}$$

The inner integral in Eq. (6.3.27) is in a convenient form for the application of Cagniard-de Hoop technique; thus we apply the technique to the inner integral with respect to the variable p.

Now, we shall consider the complex integral Φ whose integrand is the same as that of the inner integral in Eq. (6.3.27),

$$\Phi = \int_C F_{xx}^{(d)}(p,q)\exp\{-s(\beta_d z+irp)\}dp \tag{6.3.29}$$

In order to transform the argument of the exponential function to the simple product of the new variable t and the Laplace transform parameter s, the new variable t is introduced as

$$t = \beta_d z + irp \tag{6.3.30}$$

Solving for p, the Cagniard's path is given by

$$p_d^{(\pm)} = \frac{-irt \pm z\sqrt{t^2 - (q^2 + 1/c_d^2)R^2}}{R^2} \tag{6.3.31}$$

and its gradient is

$$\frac{dp_d^{(\pm)}}{dt} = \pm \frac{zt \mp ir\sqrt{t^2 - (q^2 + 1/c_d^2)R^2}}{R^2\sqrt{t^2 - (q^2 + 1/c_d^2)R^2}} \tag{6.3.32}$$

where the 3D radial distance R is defined by

$$R = \sqrt{x^2 + y^2 + z^2} \tag{6.3.33}$$

The integrand has four branch points at

$$p = \pm i\sqrt{q^2 + 1/c_d^2}, \quad \pm i\sqrt{q^2 + 1/c_s^2} \tag{6.3.34}$$

From these branch points, four branch cuts are introduced along the imaginary axis in the complex p-plane as was discussed in Sect. 1.3.2 in Chap. 1. Since the saddle point of the Cagniard's path at $t = R\sqrt{q^2 + 1/c_d^2}$,

$$p_{saddle}^{(d)} = -\frac{ir}{R}\sqrt{q^2 + 1/c_d^2} \tag{6.3.35}$$

is smaller in magnitude than any of the branch points, the Cagniard's path does not cross any cut. In addition, the Rayleigh poles $p = \pm i\sqrt{q^2 + 1/c_R^2}$, which are derived from $R(p, q) = 0$, are on the imaginary axis and is greater in magnitude than the saddle point and all branch points. Then, the closed loop C is composed of the infinite line along the real p-axis, the Cagniard's path, and two large arcs which connect the Cagniard's path to the line on the real axis.

The closed loop C is shown in Fig. 6.9. The complex integral Φ in Eq. (6.3.29) is carried out along the closed loop. Since no singular point is included in the loop and the integrals along the large arc vanish, the integral along the real axis is converted to that along the Cagniard's path, i.e.

$$\begin{aligned}
&\int_{-\infty}^{+\infty} F_{xx}^{(d)}(p, q)\exp\{-s(\beta_d z + irp)\}dp \\
&= \int_{R\sqrt{q^2+1/c_d^2}}^{\infty} \left\{F_{xx}^{(d)}(p_d^{(+)}, q)\frac{dp_d^{(+)}}{dt} - F_{xx}^{(d)}(p_d^{(-)}, q)\frac{dp_d^{(-)}}{dt}\right\}\exp(-st)dt
\end{aligned} \tag{6.3.36}$$

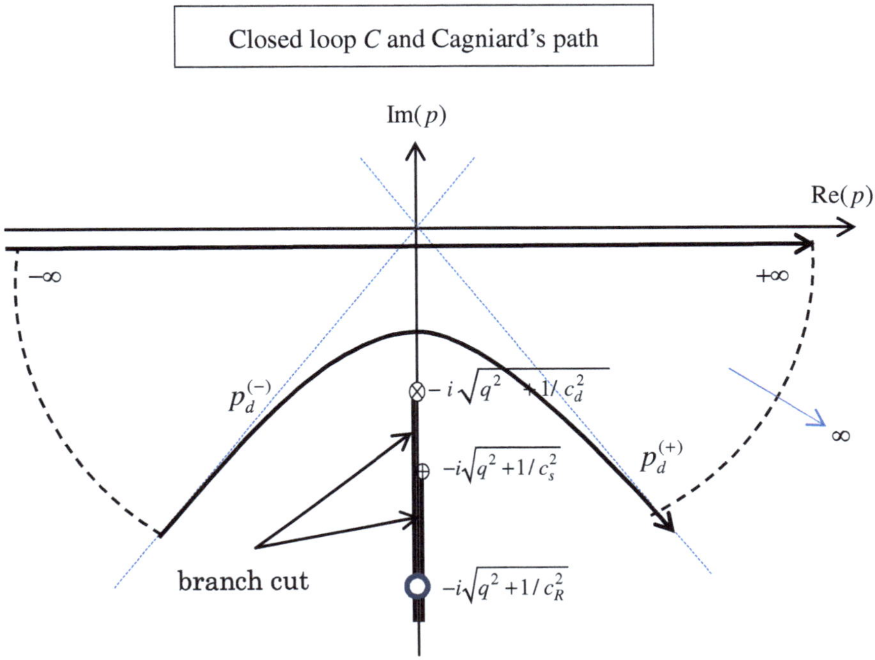

Fig. 6.9 Closed loop and Cagniard's path for the complex integral Φ

Substituting the above integral into the inner integral in Eq. (6.3.27), we have

$$I_{xx}^{*(d)}(x,y,z,s)=\int_0^{\infty}dq\int_{R\sqrt{q^2+1/c_d^2}}^{+\infty}\left\{F_{xx}^{(d)}(p_d^{(+)},q)\frac{dp_d^{(+)}}{dt}-F_{xx}^{(d)}(p_d^{(-)},q)\frac{dp_d^{(-)}}{dt}\right\}\exp(-st)dt \tag{6.3.37}$$

As the inner integral is just in the form of Laplace transform integral, we exchange the order of integration so that the outer integral can be in the form of Laplace transform. In discussing the supporting region for the double integral as shown in Fig. 6.10, the exchange is symbolically carried out as

$$\int_0^{\infty}dq\int_{R\sqrt{q^2+1/c_d^2}}^{\infty}g(t,q)\exp(-st)dt=\int_{R/c_d}^{\infty}\exp(-st)dt\int_0^{\sqrt{(t/R)^2-1/c_d^2}}g(t,q)dq \tag{6.3.38}$$

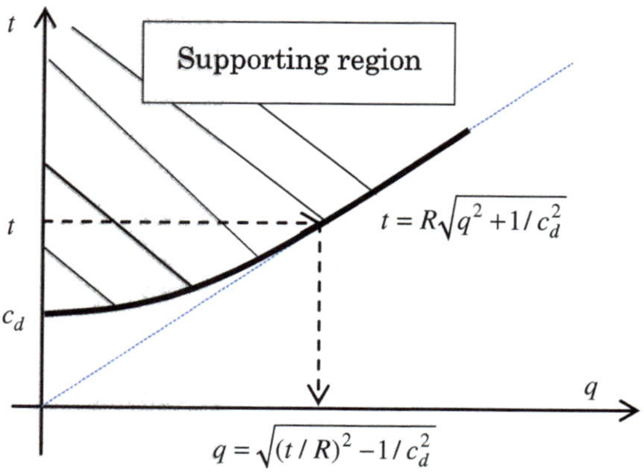

Fig. 6.10 Supporting region for the double integral in Eq. (6.3.38)

where $g(t,\ q)$ is an arbitrary non-singular function. Then, the double integral in Eq. (6.3.37) is converted to the Laplace transform integral as

$$I_{xx}^{*(d)}(x,y,z,s) = \int_{R/c_d}^{\infty} \exp(-st)dt \int_0^{\sqrt{(t/R)^2-1/c_d^2}} \left\{ F_{xx}^{(d)}(p_d^{(+)},q)\frac{dp_d^{(+)}}{dt} - F_{xx}^{(d)}(p_d^{(-)},q)\frac{dp_d^{(-)}}{dt} \right\} dq \tag{6.3.39}$$

We have just arrived at the form of Laplace transform integral, and its integrand is in the form of the finite integral with respect to the variable q. The Laplace inversion is carried out by inspection and it results in the following integral:

$$I_{xx}^{(d)}(x,y,z,t) = H(t - R/c_d) \int_0^{\sqrt{(t/R)^2-1/c_d^2}} \left\{ F_{xx}^{(d)}(p_d^{(+)},q)\frac{dp_d^{(+)}}{dt} - F_{xx}^{(d)}(p_d^{(-)},q)\frac{dp_d^{(-)}}{dt} \right\} dq \tag{6.3.40}$$

We can easily find that the disturbed region by the dilatational wave is given by the operation of the step function $H(t - R/c_d)$ and it is the semi-sphere $R \leq c_d t$ with radius $c_d t$.

The triple inversion for the dilatational wave contribution has been carried out successfully. The other dilatational wave contributions can be inverted by the same technique developed here. The unified expression for the dilatational wave contribution is given by

$$I_{ij}^{(d)}(x,y,z,t) = H(t - R/c_d) \int_0^{\sqrt{(t/R)^2-1/c_d^2}} \left\{ F_{ij}^{(d)}(p_d^{(+)},q)\frac{dp_d^{(+)}}{dt} - F_{ij}^{(d)}(p_d^{(-)},q)\frac{dp_d^{(-)}}{dt} \right\} dq \tag{6.3.41}$$

where $p_d^{(\pm)}$ are defined by Eq. (6.3.31). The integrands are given by

$$\begin{aligned} F_{xx}^{(d)}(p,q) &= \frac{(x^2p^2+y^2q^2)\beta_s}{(\pi r)^2 R(p,q)} \\ F_{xy}^{(d)}(p,q) &= \frac{xy(p^2-q^2)\beta_s}{(\pi r)^2 R(p,q)} \\ F_{xz}^{(d)}(p,q) &= -\frac{ixp(\beta_s^2+p^2+q^2)}{2\pi^2 rR(p,q)} \end{aligned} \tag{6.3.42a}$$

$$\begin{aligned} F_{yx}^{(d)}(p,q) &= \frac{xy(p^2-q^2)\beta_s}{(\pi r)^2 R(p,q)} \\ F_{yy}^{(d)}(p,q) &= \frac{(y^2p^2+x^2q^2)\beta_s}{(\pi r)^2 R(p,q)} \\ F_{yz}^{(d)}(p,q) &= -\frac{iyp(\beta_s^2+p^2+q^2)}{2\pi^2 rR(p,q)} \end{aligned} \tag{6.3.42b}$$

$$\begin{aligned} F_{zx}^{(d)}(p,q) &= -\frac{ixp\beta_d\beta_s}{\pi^2 rR(p,q)} \\ F_{zy}^{(d)}(p,q) &= -\frac{iyp\beta_d\beta_s}{\pi^2 rR(p,q)} \\ F_{zz}^{(d)}(p,q) &= -\frac{(\beta_s^2+p^2+q^2)\beta_d}{2\pi^2 R(p,q)} \end{aligned} \tag{6.3.42c}$$

(2) Inversion of SV wave contribution: $\tilde{\bar{I}}_{ij}^{*(s)}(\xi,\eta,z,s)$

We consider the inversion of the typical SV-wave contribution,

$$\tilde{\bar{I}}_{xx}^{*(s)}(\xi,\eta,z,s) = \frac{\xi^2(\alpha_s^2+\xi^2+\eta^2-4\alpha_d\alpha_s)}{s\alpha_s R(\xi,\eta,s)}\exp(-\alpha_s z) \tag{6.3.43}$$

The formal double Fourier inversion is given by

$$I_{xx}^{*(s)}(x,y,z,s) = \frac{1}{(2\pi)^2}\int_{-\infty}^{+\infty}\int_{-\infty}^{+\infty}\frac{\xi^2(\alpha_s^2+\xi^2+\eta^2-4\alpha_d\alpha_s)}{s\alpha_s R(\xi,\eta,s)}\exp(-\alpha_s z - i\xi x - i\eta y)d\xi d\eta \tag{6.3.44}$$

The variable transform defined by Eq. (6.3.23) is introduced. Examining the odd and non-symmetric nature with respect to the variable q, we obtain the following simpler form for the double integral,

$$I_{xx}^{*(s)}(x,y,z,s) = \int_{0}^{\infty} dq \int_{-\infty}^{+\infty} F_{xx}^{(s)}(p,q)\exp\{-s(\beta_s z + irp)\}dp \tag{6.3.45}$$

where the integrand is given by

$$F_{xx}^{(s)}(p,q) = \frac{(x^2p^2 + y^2q^2)(\beta_s^2 + p^2 + q^2 - 4\beta_d\beta_s)}{2(\pi r)^2\beta_s R(p,q)} \tag{6.3.46}$$

Now, we consider the complex integral Φ whose integrand is the same as that of the inner integral in Eq. (6.3.45),

$$\Phi = \int_{C} F_{xx}^{(s)}(p,q)\exp\{-s(\beta_s z + irp)\}dp \tag{6.3.47}$$

In order to determine the closed loop C, the Cagniard's path is examined. Introducing the time variable t as

$$t = \beta_s z + irp \tag{6.3.48}$$

the Cagniard's path is given by

$$p_s^{(\pm)} = \frac{-irt \pm z\sqrt{t^2 - (q^2 + 1/c_s^2)R^2}}{R^2} \tag{6.3.49}$$

and its saddle point at the time $t = R\sqrt{q^2 + 1/c_s^2}$ is

$$p_{saddle}^{(s)} = -i\frac{r}{R}\sqrt{q^2 + 1/c_s^2} \tag{6.3.50}$$

The integrand has the four branch points given by Eq. (6.3.34) and branch cuts are introduced along the imaginary axis as shown in Fig. 6.11a, b. Comparing the magnitude of the branch points with that of the saddle point, we see that there are two cases: the saddle point is on the branch cut or it is not; in other words, the Cagniard's path crosses the branch cut or it does not.

When the saddle point is smaller in magnitude than the branch point of the dilatational wave,

$$\frac{r}{R}\sqrt{q^2 + 1/c_s^2} < \sqrt{q^2 + 1/c_d^2} \tag{6.3.51}$$

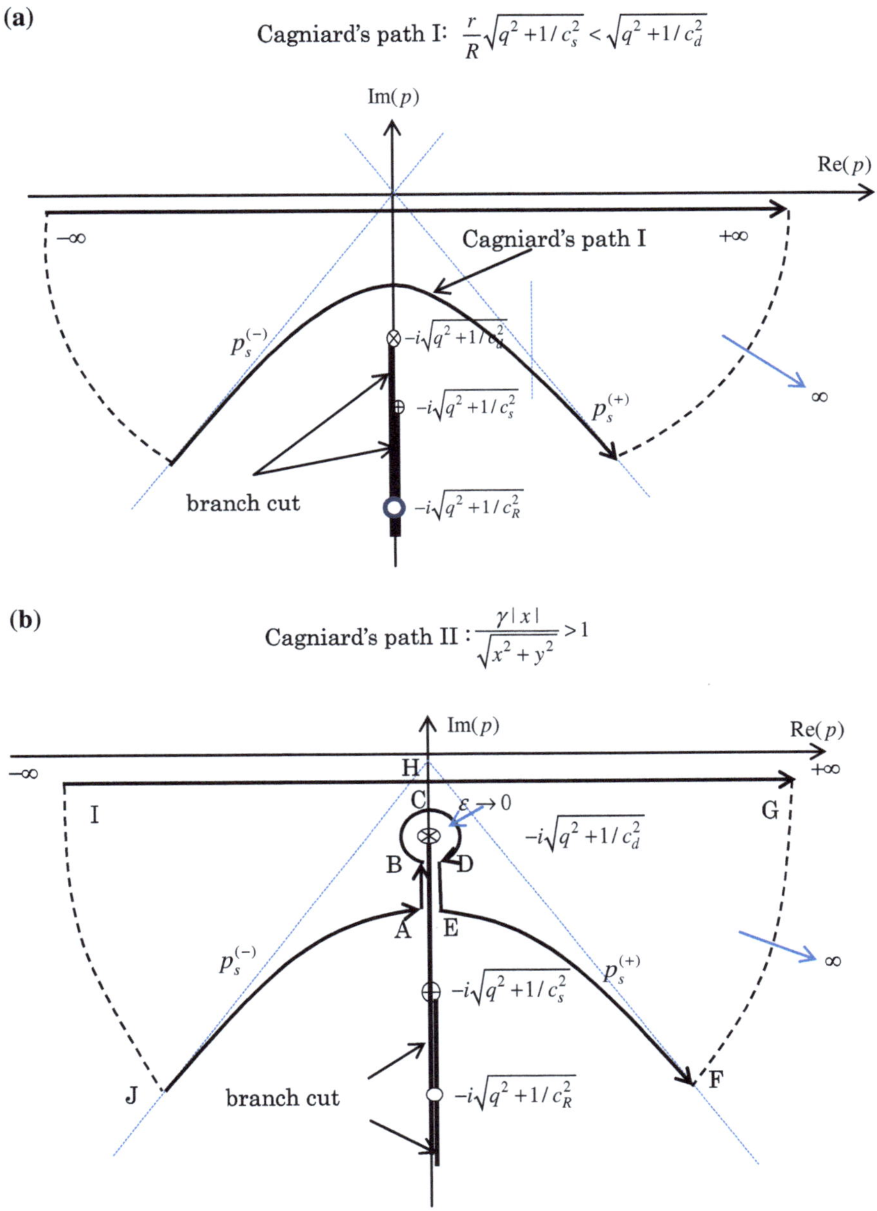

Fig. 6.11 Two closed loops for the 3D Cagniard's technique. **a** Loop with Cagniard's path I. **b** Loop with deformed Cagniard's path II

the Cagniard's path does not cross any branch cut. Then, the closed loop C is composed of the infinite line along the real axis, the Cagniard's path, and two large arcs which connect the line to the Cagniard's path. This closed loop is similar to that in the case of the dilatational wave and is shown in Fig. 6.11a. Employing this closed loop, we apply Cauchy's theorem to the complex integral Φ in Eq. (6.3.47). Since the loop does not include any singular point, the integral along the real axis is converted to that along the Cagniard's path, i.e.

$$\begin{aligned}&\int_{-\infty}^{+\infty} F_{xx}^{(s)}(p,q)\exp\{-s(\beta_s z+irp)\}dp\\&=\int_{R\sqrt{q^2+1/c_s^2}}^{\infty}\left\{F_{xx}^{(s)}(p_s^{(+)},q)\frac{dp_s^{(+)}}{dt}-F_{xx}^{(s)}(p_s^{(-)},q)\frac{dp_s^{(-)}}{dt}\right\}\exp(-st)dt\end{aligned} \tag{6.3.52}$$

When the saddle point is larger in magnitude than the branch point of the dilatational wave,

$$\frac{r}{R}\sqrt{q^2+1/c_s^2}>\sqrt{q^2+1/c_d^2}\quad\Rightarrow\quad q<\sqrt{(r/c_s)^2-(R/c_d)^2}/z \tag{6.3.53}$$

the Cagniard's path crosses the branch cut. In order to avoid the crossing, the path is deformed along the cut, such as the Cagniard's path II in Sect. 6.2. Our deformed Cagniard's path is composed of two symmetric semi-hyperbolas, two short lines along the cut, and a small circle around the the branch point of the dilatational wave. We name the deformed path the Cagniard path II, too. Figure 6.11b shows the Cagniard path II and the closed loop C. The closed loop C has two large arcs which connect the line with the Cagniard path II. Then, we apply Cauchy's theorem to the complex integral Φ with this closed loop.

No singular point is included in the loop. The integral along the small circle ($\overset{\frown}{BCD}$) vanishes as its radius tends to zero, and those along the large arcs ($\overset{\frown}{IJ}$ and $\overset{\frown}{FG}$) also vanish as the radius tends to infinity. Then, the integral along the real axis ($\overrightarrow{IHG}$) is converted to the sum of two integrals. One is that along the regular Cagniard's path ($\overrightarrow{AJ}$ and $\overrightarrow{EF}$) and the other is the sum of two line integrals along the branch cut ($\overrightarrow{AB}$ and $\overrightarrow{DE}$). Thus the integral along the real axis is converted to the sum of two integrations as

$$\begin{aligned}&\int_{-\infty}^{+\infty} F_{xx}^{(s)}(p,q)\exp\{-s(\beta_s z+irp)\}dp\\&=\int_{R\sqrt{q^2+1/c_s^2}}^{\infty}\left\{F_{xx}^{(s)}(p_s^{(+)},q)\frac{dp_s^{(+)}}{dt}-F_{xx}^{(s)}(p_s^{(-)},q)\frac{dp_s^{(-)}}{dt}\right\}\exp(-st)dt\\&+\left(\int_{\overrightarrow{AB}}+\int_{\overrightarrow{DE}}\right)F_{xx}^{(s)}(p,q)\exp\{-s(\beta_s z+irp)\}dp=0\end{aligned} \tag{6.3.54}$$

Following the method in Sect. 1.3.2 in Chap. 1, we examine the argument of the radicals along the branch cut, $\overrightarrow{AB}$ and $\overrightarrow{DE}$ in Fig. 6.11b. Due to the radiation condition $\mathrm{Re}(\beta_j) > 0$, the argument of the radical β_d on $\overrightarrow{AB}$, which is the left side of the cut, is $+\pi/2$ and that on the right side $\overrightarrow{DE}$ is $-\pi/2$. The argument of the radicals and the integration variable along the cut are summarized as follows:

$$\begin{aligned} &\text{On } \overrightarrow{AB}: p = -i\varsigma; \quad \sqrt{q^2 + 1/c_d^2} < \varsigma < \frac{r}{R}\sqrt{q^2 + 1/c_s^2} \\ &\beta_d = +i\sqrt{\varsigma^2 - (q^2 + 1/c_d^2)}, \quad \beta_s = +\sqrt{q^2 + 1/c_s^2 - \varsigma^2} \end{aligned} \tag{6.3.55}$$

$$\begin{aligned} &\text{On } \overrightarrow{DE}: p = -i\varsigma; \quad \sqrt{q^2 + 1/c_d^2} < \varsigma < \frac{r}{R}\sqrt{q^2 + 1/c_s^2} \\ &\beta_d = -i\sqrt{\varsigma^2 - (q^2 + 1/c_d^2)}, \quad \beta_s = +\sqrt{q^2 + 1/c_s^2 - \varsigma^2} \end{aligned} \tag{6.3.56}$$

The two line integrals along the branch cut are calculated and are unified as follows:

$$\begin{aligned} &\left(\int\limits_{\overrightarrow{AB}} + \int\limits_{\overrightarrow{DE}} \right) F_{xx}^{(s)}(p, q) \exp\{-s(\beta_s z + irp)\} dp \\ &= -\int\limits_{\sqrt{q^2+1/c_d^2}}^{(r/R)\sqrt{q^2+1/c_s^2}} F_{xx}^{(s)}(p, q)\Big|_{\substack{p=-i\varsigma \\ \beta_d=+i\gamma_d \\ \beta_s=\gamma_s}} \exp\{-s(\gamma_s z + \varsigma r)\}(-id\varsigma) \\ &\quad + \int\limits_{\sqrt{q^2+1/c_d^2}}^{(r/R)\sqrt{q^2+1/c_s^2}} F_{xx}^{(s)}(p, q)\Big|_{\substack{p=-i\varsigma \\ \beta_d=-i\gamma_d \\ \beta_s=\gamma_s}} \exp\{-s(\gamma_s z + \varsigma r)\}(-id\varsigma) \\ &= \int\limits_{\sqrt{q^2+1/c_d^2}}^{(r/R)\sqrt{q^2+1/c_s^2}} f_{xx}^{(s)}(\varsigma, q) \exp\{-s(\gamma_s z + r\varsigma)\} d\varsigma \end{aligned} \tag{6.3.57}$$

where a new notation for the integrand is introduced as

$$f_{xx}^{(s)}(\varsigma, q) = (-i)\left\{ F_{xx}^{(s)}(p, q)\Big|_{\substack{p=-i\varsigma \\ \beta_d=-i\gamma_d \\ \beta_s=\gamma_s}} - F_{xx}^{(s)}(p, q)\Big|_{\substack{p=-i\varsigma \\ \beta_d=+i\gamma_d \\ \beta_s=\gamma_s}} \right\} \tag{6.3.58}$$

and

$$\gamma_d = \sqrt{\varsigma^2 - (q^2 + 1/c_d^2)}, \quad \gamma_s = \sqrt{q^2 + 1/c_s^2 - \varsigma^2} \tag{6.3.59}$$

In order to reduce the last integral in Eq. (6.3.57) to the form of Laplace transform, the variable transform from ς to the time t is introduced as

$$t = \gamma_s z + \varsigma r \tag{6.3.60}$$

Its inverse is

$$\varsigma_H = \frac{rt - z\sqrt{(q^2 + 1/c_s^2)R^2 - t^2}}{R^2} \tag{6.3.61}$$

The integral along the branch cut is thus converted to that of Laplace transform,

$$\begin{aligned}&\int\limits_{\sqrt{q^2+1/c_d^2}}^{(r/R)\sqrt{q^2+1/c_s^2}} f_{xx}^{(s)}(\varsigma, q)\exp\{-s(\gamma_s z + r\varsigma)\}d\varsigma \\ &= \int\limits_{z\sqrt{1/c_s^2-1/c_d^2}+r\sqrt{q^2+1/c_d^2}}^{R\sqrt{q^2+1/c_s^2}} \left\{f_{xx}^{(s)}(\varsigma_H, q)\frac{d\varsigma_H}{dt}\right\}\exp(-st)dt\end{aligned} \tag{6.3.62}$$

Substituting the above equation into Eq. (6.3.54), we have the inner integral in the form of Laplace transform integral,

$$\begin{aligned}&\int\limits_{-\infty}^{+\infty} F_{xx}^{(s)}(p, q)\exp\{-s(\beta_s z + irp)\}dp \\ &= \int\limits_{R\sqrt{q^2+1/c_s^2}}^{\infty} \left\{F_{xx}^{(s)}(p_s^{(+)}, q)\frac{dp_s^{(+)}}{dt} - F_{xx}^{(s)}(p_s^{(-)}, q)\frac{dp_s^{(-)}}{dt}\right\}\exp(-st)dt \\ &+ \int\limits_{z\sqrt{1/c_s^2-1/c_d^2}+r\sqrt{q^2+1/c_d^2}}^{R\sqrt{q^2+1/c_s^2}} \left\{f_{xx}^{(s)}(\varsigma_H, q)\frac{d\varsigma_H}{dt}\right\}\exp(-st)dt\end{aligned} \tag{6.3.63}$$

Comparing this equation with Eq. (6.3.52), we readily see that the last integral in the above equation appears only when the conditional of Eq. (6.3.53) holds. Using the step function, we can express the inner integral in the unified form as

$$\begin{aligned}&\int\limits_{-\infty}^{+\infty} F_{xx}^{(s)}(p, q)\exp\{-s(\beta_s z + irp)\}dp \\ &= \int\limits_{R\sqrt{q^2+1/c_s^2}}^{\infty} \left\{F_{xx}^{(s)}(p_s^{(+)}, q)\frac{dp_s^{(+)}}{dt} - F_{xx}^{(s)}(p_s^{(-)}, q)\frac{dp_s^{(-)}}{dt}\right\}\exp(-st)dt \\ &+ H\left(\sqrt{(r/c_s)^2 - (R/c_d)^2} - qz\right)\int\limits_{z\sqrt{1/c_s^2-1/c_d^2}+r\sqrt{q^2+1/c_d^2}}^{R\sqrt{q^2+1/c_s^2}} \left\{f_{xx}^{(s)}(\varsigma_H, q)\frac{d\varsigma_H}{dt}\right\}\exp(-st)dt\end{aligned} \tag{6.3.64}$$

We substitute the above equation into Eq. (6.3.45) to get

$$
\begin{aligned}
I_{xx}^{*(s)}(x,y,z,s) &= \int_0^\infty dq \int_{R\sqrt{q^2+1/c_s^2}}^\infty \left\{ F_{xx}^{(s)}(p_s^{(+)},q)\frac{dp_s^{(+)}}{dt} - F_{xx}^{(s)}(p_s^{(-)},q)\frac{dp_s^{(-)}}{dt} \right\} \exp(-st)dt \\
&+ \int_0^\infty H\left(\sqrt{(r/c_s)^2-(R/c_d)^2}-qz\right)dq \\
&\int_{z\sqrt{1/c_s^2-1/c_d^2}+r\sqrt{q^2+1/c_d^2}}^{R\sqrt{q^2+1/c_s^2}} \left\{ f_{xx}^{(s)}(\varsigma_H,q)\frac{d\varsigma_H}{dt} \right\} \exp(-st)dt
\end{aligned}
\tag{6.3.65}
$$

To perform the Laplace inversion, the order of integration must be interchanged. Discussing the supporting region for the double integral such as in Fig. 6.12a, we exchange the order of the integrations in the first double integral as

$$
\int_0^\infty dq \int_{R\sqrt{q^2+1/c_s^2}}^\infty g(t,q)dt = \int_{R/c_s}^\infty dt \int_0^{\sqrt{(t/R)^2-1/c_s^2}} g(t,q)dq \tag{6.3.66}
$$

The supporting region for the second double integral is also shown in Fig. 6.12b. After some examinations, we obtain the exchange formula

$$
\begin{aligned}
&\int_0^\infty H\left(\sqrt{(r/c_s)^2-(R/c_d)^2}-qz\right)dq \int_{z\sqrt{1/c_s^2-1/c_d^2}+r\sqrt{q^2+1/c_d^2}}^{R\sqrt{q^2+1/c_s^2}} h(t,q)dt \\
&= \int_{r/c_d+z\sqrt{1/c_s^2-1/c_d^2}}^{R/c_s} dt \int_0^{\frac{1}{r}\sqrt{\left(t-z\sqrt{1/c_s^2-1/c_d^2}\right)^2-(r/c_d)^2}} h(t,q)dq \\
&+ \int_{R/c_s}^{\frac{R^2}{z}\sqrt{1/c_s^2-1/c_d^2}} dt \int_{\sqrt{(t/R)^2-1/c_s^2}}^{\frac{1}{r}\sqrt{\left(t-z\sqrt{1/c_s^2-1/c_d^2}\right)^2-(r/c_d)^2}} h(t,q)dq
\end{aligned}
\tag{6.3.67}
$$

where $g(t,\ q)$ and $h(t,\ q)$ in Eqs. (6.3.66) and (6.3.67) are arbitrary non-singular integrands.

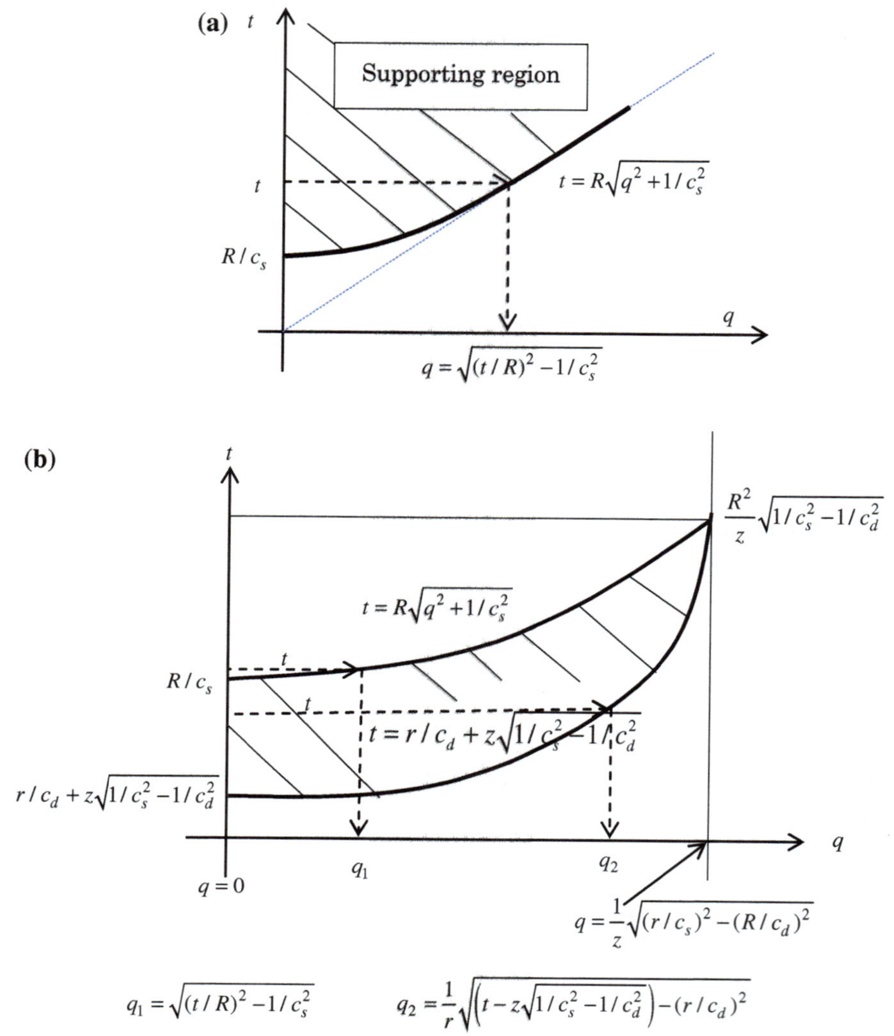

Fig. 6.12 Supporting region for the double integral in **a** Eq. (6.3.66) and **b** Eq. (6.3.67)

Applying the exchange formulas to the double integral in Eq. (6.3.65), we have the *Laplace transform* integral for the *Laplace transformed* component,

$$
\begin{aligned}
I_{xx}^{*(s)}(x,y,z,s) = & \int_{R/c_s}^{\infty} \exp(-st)dt \int_0^{\sqrt{(t/R)^2-1/c_s^2}} \left\{ F_{xx}^{(s)}(p_s^{(+)},q)\frac{dp_s^{(+)}}{dt} - F_{xx}^{(s)}(p_s^{(-)},q)\frac{dp_s^{(-)}}{dt} \right\} dq \\
& + \int_{r/c_d+z\sqrt{1/c_s^2-1/c_d^2}}^{R/c_s} \exp(-st)dt \int_0^{\frac{1}{r}\sqrt{\left(t-z\sqrt{1/c_s^2-1/c_d^2}\right)^2-(r/c_d)^2}} \left\{ f_{xx}^{(s)}(\varsigma_H,q)\frac{d\varsigma_H}{dt} \right\} dq \\
& + \int_{R/c_s}^{\frac{R^2}{z}\sqrt{1/c_s^2-1/c_d^2}} \exp(-st)dt \int_{\sqrt{(t/R)^2-1/c_s^2}}^{\frac{1}{r}\sqrt{\left(t-z\sqrt{1/c_s^2-1/c_d^2}\right)^2-(r/c_d)^2}} \left\{ f_{xx}^{(s)}(\varsigma_H,q)\frac{d\varsigma_H}{dt} \right\} dq
\end{aligned}
\tag{6.3.68}
$$

Finally, the Laplace inversion is carried out by inspection, and the SV-wave contribution is given by

$$
\begin{aligned}
& I_{xx}^{(s)}(x,y,z,t) \\
& = H(t-R/c_s) \int_0^{\sqrt{(t/R)^2-1/c_s^2}} \left\{ F_{xx}^{(s)}(p_s^{(+)},q)\frac{dp_s^{(+)}}{dt} - F_{xx}^{(s)}(p_s^{(-)},q)\frac{dp_s^{(-)}}{dt} \right\} dq \\
& + H\left(t-r/c_d-z\sqrt{1/c_s^2-1/c_d^2}\right)H(R/c_s-t) \int_0^{\frac{1}{r}\sqrt{\left(t-z\sqrt{1/c_s^2-1/c_d^2}\right)^2-(r/c_d)^2}} \left\{ f_{xx}^{(s)}(\varsigma_H,q)\frac{d\varsigma_H}{dt} \right\} dq \\
& + H(t-R/c_s)H\left(\frac{R^2}{z}\sqrt{1/c_s^2-1/c_d^2}-t\right) \int_{\sqrt{(t/R)^2-1/c_s^2}}^{\frac{1}{r}\sqrt{\left(t-z\sqrt{1/c_s^2-1/c_d^2}\right)^2-(r/c_d)^2}} \left\{ f_{xx}^{(s)}(\varsigma_H,q)\frac{d\varsigma_H}{dt} \right\} dq
\end{aligned}
\tag{6.3.69}
$$

Examining the operation of the step functions ahead of the integral, the first term with $H(t-R/c_s)$ shows a semi-spherical region disturbed by SV-wave. The second term which has the product of the two step functions, $H\left(t-r/c_d-z\sqrt{1/c_s^2-1/c_d^2}\right)H(R/c_s-t)$, shows a region of von-Schmidt wave. The last term with $H(t-R/c_s)H\left(\frac{R^2}{z}\sqrt{1/c_s^2-1/c_d^2}-t\right)$ does not show the wave nature, but appears only in the case of the 3D deformation (this non-wave front is discussed by Gakenheimer and Miklowitz (1969)). These wave regions are denoted by regions A and B in Fig. 6.13.

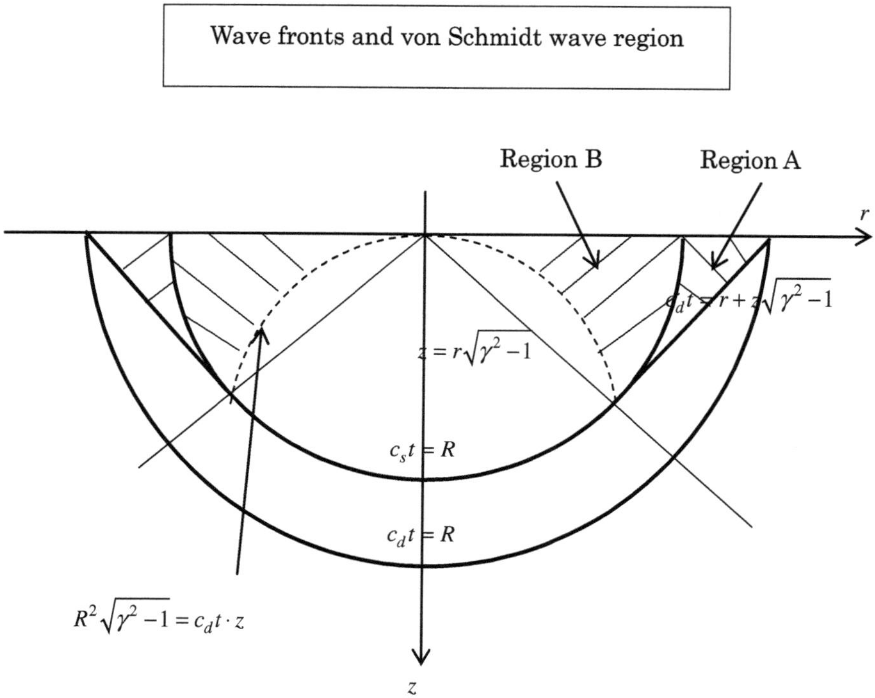

Fig. 6.13 Wave fronts and disturbed regions A and B

To the other SV-wave contributions, the same inversion technique is applied and we obtain the unified expression for SV-wave contributions as

$$
\begin{aligned}
&I_{ij}^{(s)}(x,y,z,t)\\
&= H(t - R/c_s) \int_0^{\sqrt{(t/R)^2 - 1/c_s^2}} \left\{ F_{ij}^{(s)}(p_s^{(+)}, q) \frac{dp_s^{(+)}}{dt} - F_{ij}^{(s)}(p_s^{(-)}, q) \frac{dp_s^{(-)}}{dt} \right\} dq\\
&\quad + H\left(t - r/c_d - z\sqrt{1/c_s^2 - 1/c_d^2}\right) H(R/c_s - t) \int_0^{\frac{1}{r}\sqrt{\left(t - z\sqrt{1/c_s^2 - 1/c_d^2}\right)^2 - (r/c_d)^2}} \left\{ f_{ij}^{(s)}(\varsigma_H, q) \frac{d\varsigma_H}{dt} \right\} dq\\
&\quad + H(t - R/c_s) H\left(\frac{R^2}{z}\sqrt{1/c_s^2 - 1/c_d^2} - t\right) \int_{\sqrt{(t/R)^2 - 1/c_s^2}}^{\frac{1}{r}\sqrt{\left(t - z\sqrt{1/c_s^2 - 1/c_d^2}\right)^2 - (r/c_d)^2}} \left\{ f_{ij}^{(s)}(\varsigma_H, q) \frac{d\varsigma_H}{dt} \right\} dq
\end{aligned}
\tag{6.3.70}
$$

where $p_s^{(\pm)}$ and ς_H are given by Eqs. (6.3.49) and (6.3.61) respectively and

$$F_{xx}^{(s)}(p,q) = \frac{(x^2p^2 + y^2q^2)(\beta_s^2 + p^2 + q^2 - 4\beta_d\beta_s)}{2(\pi r)^2\beta_s R(p,q)}$$
$$F_{xy}^{(s)}(p,q) = \frac{xy}{2(\pi r)^2}\frac{(p^2 - q^2)(\beta_s^2 + p^2 + q^2 - 4\beta_d\beta_s)}{\beta_s R(p,q)} \tag{6.3.71a}$$
$$F_{xz}^{(s)}(p,q) = \frac{x}{\pi^2 r}\frac{ip\beta_d\beta_s}{R(p,q)}$$

$$F_{yx}^{(s)}(p,q) = \frac{xy}{2(\pi r)^2}\frac{(p^2 - q^2)(\beta_s^2 + p^2 + q^2 - 4\beta_d\beta_s)}{\beta_s R(p,q)}$$
$$F_{yy}^{(s)}(p,q) = \frac{(y^2p^2 + x^2q^2)(\beta_s^2 + p^2 + q^2 - 4\beta_d\beta_s)}{2(\pi r)^2\beta_s R(p,q)} \tag{6.3.71b}$$
$$F_{yz}^{(s)}(p,q) = \frac{y}{\pi^2 r}\frac{ip\beta_d\beta_s}{R(p,q)}$$

$$F_{zx}^{(s)}(p,q) = \frac{x}{2\pi^2 r}\frac{ip(\beta_s^2 + p^2 + q^2)}{R(p,q)}$$
$$F_{zy}^{(s)}(p,q) = \frac{y}{2\pi^2 r}\frac{ip(\beta_s^2 + p^2 + q^2)}{R(p,q)} \tag{6.3.71c}$$
$$F_{zz}^{(s)}(p,q) = \frac{\beta_d(p^2 + q^2)}{\pi^2 R(p,q)}$$

$$\begin{aligned}
f_{xx}^{(s)}(\varsigma,q) &= (-i)\frac{(x\varsigma)^2 - (yq)^2}{2(\pi r)^2\gamma_s} \\
&\quad \times \left\{\frac{\gamma_s^2 - \varsigma^2 + q^2 - 4i\gamma_d\gamma_s}{(\gamma_s^2 - \varsigma^2 + q^2)^2 + 4i\gamma_d\gamma_s(\varsigma^2 - q^2)} - \frac{\gamma_s^2 - \varsigma^2 + q^2 + 4i\gamma_d\gamma_s}{(\gamma_s^2 - \varsigma^2 + q^2)^2 - 4i\gamma_d\gamma_s(\varsigma^2 - q^2)}\right\} \\
f_{xy}^{(s)}(\varsigma,q) &= (-i)\frac{xy}{2(\pi r)^2}\frac{\varsigma^2 + q^2}{\gamma_s} \\
&\quad \times \left\{\frac{\gamma_s^2 - \varsigma^2 + q^2 - 4i\gamma_d\gamma_s}{(\gamma_s^2 - \varsigma^2 + q^2)^2 + 4i\gamma_d\gamma_s(\varsigma^2 - q^2)} - \frac{\gamma_s^2 - \varsigma^2 + q^2 + 4i\gamma_d\gamma_s}{(\gamma_s^2 - \varsigma^2 + q^2)^2 - 4i\gamma_d\gamma_s(\varsigma^2 - q^2)}\right\} \\
f_{xz}^{(s)}(\varsigma,q) &= -\frac{x}{\pi^2 r}\varsigma\gamma_d\gamma_s \\
&\quad \times \left\{\frac{1}{(\gamma_s^2 - \varsigma^2 + q^2)^2 + 4i\gamma_d\gamma_s(\varsigma^2 - q^2)} + \frac{1}{(\gamma_s^2 - \varsigma^2 + q^2)^2 - 4i\gamma_d\gamma_s(\varsigma^2 - q^2)}\right\}
\end{aligned} \tag{6.3.72a}$$

$$f_{yx}^{(s)}(\varsigma,q)=(-i)\frac{xy}{2(\pi r)^2}\frac{\varsigma^2+q^2}{\gamma_s}$$
$$\times\left\{\frac{\gamma_s^2-\varsigma^2+q^2-4i\gamma_d\gamma_s}{(\gamma_s^2-\varsigma^2+q^2)^2+4i\gamma_d\gamma_s(\varsigma^2-q^2)}-\frac{\gamma_s^2-\varsigma^2+q^2+4i\gamma_d\gamma_s}{(\gamma_s^2-\varsigma^2+q^2)^2-4i\gamma_d\gamma_s(\varsigma^2-q^2)}\right\}$$
$$f_{yy}^{(s)}(\varsigma,q)=(-i)\frac{xy}{2(\pi r)^2}\frac{(y\varsigma)^2-(xq)^2}{\gamma_s}$$
$$\times\left\{\frac{\gamma_s^2-\varsigma^2+q^2-4i\gamma_d\gamma_s}{(\gamma_s^2-\varsigma^2+q^2)^2+4i\gamma_d\gamma_s(\varsigma^2-q^2)}-\frac{\gamma_s^2-\varsigma^2+q^2+4i\gamma_d\gamma_s}{(\gamma_s^2-\varsigma^2+q^2)^2-4i\gamma_d\gamma_s(\varsigma^2-q^2)}\right\}$$
$$f_{yz}^{(s)}(\varsigma,q)=-\frac{y}{\pi^2 r}\varsigma\gamma_d\gamma_s$$
$$\times\left\{\frac{1}{(\gamma_s^2-\varsigma^2+q^2)^2+4i\gamma_d\gamma_s(\varsigma^2-q^2)}+\frac{1}{(\gamma_s^2-\varsigma^2+q^2)^2-4i\gamma_d\gamma_s(\varsigma^2-q^2)}\right\}$$
(6.3.72b)

$$f_{zx}^{(s)}(\varsigma,q)=(+i)\frac{x}{2\pi^2 r}\varsigma(\gamma_s^2-\varsigma^2+q^2)$$
$$\times\left\{\frac{1}{(\gamma_s^2-\varsigma^2+q^2)^2+4i\gamma_d\gamma_s(\varsigma^2-q^2)}-\frac{1}{(\gamma_s^2-\varsigma^2+q^2)^2-4i\gamma_d\gamma_s(\varsigma^2-q^2)}\right\}$$
$$f_{zy}^{(s)}(\varsigma,q)=(+i)\frac{y}{2\pi^2 r}\varsigma(\gamma_s^2-\varsigma^2+q^2)$$
$$\times\left\{\frac{1}{(\gamma_s^2-\varsigma^2+q^2)^2+4i\gamma_d\gamma_s(\varsigma^2-q^2)}-\frac{1}{(\gamma_s^2-\varsigma^2+q^2)^2-4i\gamma_d\gamma_s(\varsigma^2-q^2)}\right\}$$
$$f_{zz}^{(s)}(\varsigma,q)=\frac{\gamma_d(\varsigma^2-q^2)}{\pi^2}$$
$$\times\left\{\frac{1}{(\gamma_s^2-\varsigma^2+q^2)^2+4i\gamma_d\gamma_s(\varsigma^2-q^2)}+\frac{1}{(\gamma_s^2-\varsigma^2+q^2)^2-4i\gamma_d\gamma_s(\varsigma^2-q^2)}\right\}$$
(6.3.72c)

(3) Inversion of SH wave contribution: $\tilde{\bar{I}}_{ii}^{*(SH)}(\xi,\eta,z,s)$

Two SH-wave contributions are identical

$$\tilde{\bar{I}}_{xx}^{*(SH)}(\xi,\eta,z,s)=\tilde{\bar{I}}_{yy}^{*(SH)}(\xi,\eta,z,s)=-\frac{1}{s\alpha_s}\exp(-\alpha_s z) \tag{6.3.73}$$

The formal double Fourier inversion is given by

$$I_{ii}^{*(SH)}(x,y,z,s)=\frac{1}{(2\pi)^2}\int_{-\infty}^{+\infty}\int_{-\infty}^{+\infty}\left(-\frac{1}{s\alpha_s}\right)\exp(-\alpha_s z-i\xi x-i\eta y)d\xi d\eta \tag{6.3.74}$$

Now, we introduce the variable transform defined by Eq. (6.3.23),

$$I_{ii}^{*(SH)}(x,y,z,s) = \frac{1}{2\pi^2}\int_0^{\infty} dq \int_{-\infty}^{+\infty} \left(-\frac{1}{\sqrt{p^2+q^2+1/c_s^2}}\right) \exp\left\{-s\left(z\sqrt{p^2+q^2+1/c_s^2}+irp\right)\right\}dp \tag{6.3.75}$$

The Cagniard-de Hoop technique is applied to the inner integral. Introducing the variable transform from p to the new variable t,

$$t = z\sqrt{p^2+q^2+1/c_s^2}+irp \tag{6.3.76}$$

the Cagniard's path is given by

$$p_s^{(\pm)} = \frac{-irt \pm z\sqrt{t^2-(q^2+1/c_s^2)R^2}}{R^2} \tag{6.3.77}$$

We have exact expressions for the radical and the gradient,

$$\left.\sqrt{p^2+q^2+1/c_s^2}\right|_{p=p_s^{(\pm)}} = \frac{tz \mp ir\sqrt{t^2-(q^2+1/c_s^2)R^2}}{R^2} \tag{6.3.78}$$

$$\frac{dp_s^{(\pm)}}{dt} = \pm\frac{tz \mp ir\sqrt{t^2-(q^2+1/c_s^2)R^2}}{R^2\sqrt{t^2-(q^2+1/c_s^2)R^2}} \tag{6.3.79}$$

Following the former discussion for the Cagniard-de Hoop technique, we apply Cauchy's integral theorem to the complex integral,

$$\Phi = \frac{1}{2\pi^2}\int_c \left(-\frac{1}{\sqrt{p^2+q^2+1/c_s^2}}\right) \exp\left\{-s\left(z\sqrt{p^2+q^2+1/c_s^2}+irp\right)\right\}dp \tag{6.3.80}$$

and find that the integral along the real axis is converted to that along the Cagniard's path. Since

$$\left.\frac{1}{\sqrt{p^2+q^2+1/c_s^2}}\frac{dp}{dt}\right|_{p=p_{P_s}^{(+)}} - \left.\frac{1}{\sqrt{p^2+q^2+1/c_s^2}}\frac{dp}{dt}\right|_{p=p_s^{(-)}} = \frac{2}{\sqrt{t^2-(q^2+1)R^2}} \tag{6.3.81}$$

the double inversion integral in Eq. (6.3.75) is transformed to

$$I_{ii}^{*(SH)}(x,y,z,s) = \frac{1}{\pi^2}\int_0^\infty dq \int_{R\sqrt{q^2+1/c_s^2}}^\infty \left(-\frac{1}{\sqrt{t^2-(q^2+1/c_s^2)R^2}}\right)\exp(-st)dt \tag{6.3.82}$$

The order of integration is also exchanged as

$$I_{ii}^{*(SH)}(x,y,z,s) = \frac{1}{\pi^2}\int_{R/c_s}^\infty \exp(-st)dt \int_0^{\sqrt{(t/R)^2-1/c_s^2}} \left(-\frac{1}{\sqrt{t^2-(q^2+1/c_s^2)R^2}}\right)dq \tag{6.3.83}$$

It is very lucky that the inner integral with respect to the variable q can be evaluated exactly as

$$\int_0^{\sqrt{(t/R)^2-1/c_s^2}} \left(-\frac{1}{\sqrt{t^2-(q^2+1/c_s^2)R^2}}\right)dq = \frac{\pi}{2R} \tag{6.3.84}$$

Thus, the SH-wave contribution is reduced to the simple Laplace transform integral,

$$I_{ii}^{*(SH)}(x,y,z,s) = -\frac{1}{2\pi R}\int_{R/C_s}^\infty \exp(-st)dt \tag{6.3.85}$$

The Laplace inversion is carried out by inspection and we have the final result,

$$I_{ii}^{*(SH)}(x,y,z,t) = -\frac{1}{2\pi R}H(t-R/C_s); \quad i = x,y \tag{6.3.86}$$

Consequently, we have obtained the SH-wave contribution in the closed form.

(4) Green's dyadic

The three wave contributions have just been inverted by the Cagniard-de Hoop technique. The displacement response is expressed in terms of the Green's dyadic $G_{ij}(x, y, z, t)$ as

$$\begin{aligned} u_x &= P_x G_{xx}(x,y,z,t) + P_y G_{xy}(x,y,z,t) + P_z G_{xz}(x,y,z,t) \\ u_y &= P_x G_{yx}(x,y,z,t) + P_y G_{yy}(x,y,z,t) + P_z G_{yz}(x,y,z,t) \\ u_z &= P_x G_{zx}(x,y,z,t) + P_y G_{zy}(x,y,z,t) + P_z G_{zz}(x,y,z,t) \end{aligned} \tag{6.3.87}$$

where the dyadic components are given by

$$\begin{aligned} G_{xx}(x,y,z,t) &= \frac{1}{\mu}\left\{ I_{xx}^{(d)}(x,y,z,t) + I_{xx}^{(s)}(x,y,z,t) + I_{xx}^{SH}(x,y,z,t) \right\} \\ G_{xy}(x,y,z,t) &= \frac{1}{\mu}\left\{ I_{xy}^{(d)}(x,y,z,t) + I_{xy}^{(s)}(x,y,z,t) \right\} \\ G_{xz}(x,y,z,t) &= \frac{1}{\mu}\left\{ I_{xz}^{(d)}(x,y,z,t) + I_{xz}^{(s)}(x,y,z,t) \right\} \end{aligned} \tag{6.3.88a}$$

$$\begin{aligned} G_{yx}(x,y,z,t) &= \frac{1}{\mu}\left\{ I_{yx}^{(d)}(x,y,z,t) + I_{yx}^{(s)}(x,y,z,t) \right\} \\ G_{yy}(x,y,z,t) &= \frac{1}{\mu}\left\{ I_{yx}^{(d)}(x,y,z,t) + I_{yx}^{(s)}(x,y,z,t) + I_{yy}^{(SH)}(x,y,z) \right\} \\ G_{yz}(x,y,z,t) &= \frac{1}{\mu}\left\{ I_{yz}^{d}(x,y,z,t) + I_{yz}^{S}(x,y,z,t) \right\} \end{aligned} \tag{6.3.88b}$$

$$\begin{aligned} G_{zx}(x,y,z,t) &= \frac{1}{\mu}\left\{ I_{zx}^{(d)}(x,y,z,t) + I_{zx}^{(s)}(x,y,z,t) \right\} \\ G_{zy}(x,y,z,t) &= \frac{1}{\mu}\left\{ I_{zy}^{(d)}(x,y,z,t) + I_{zy}^{(s)}(x,y,z,t) \right\} \\ G_{zy}(x,y,z,t) &= \frac{1}{\mu}\left\{ I_{zz}^{(d)}(x,y,z,t) + I_{zy}^{(s)}(x,y,z,t) \right\} \end{aligned} \tag{6.3.88c}$$

Exercises

(6.1) In Sect. 6.2, if the applied load is a semi-infinite distribution such as

$$\sigma_{yz}\big|_{y=0} = \delta(t)\begin{cases} p_0; & x > 0 \\ 0; & x < 0 \end{cases} \tag{a}$$

the transformed boundary condition of Eq. (6.1.14) is replaced with

$$\bar{\sigma}_{yz}^{*}\Big|_{y=0} = p_0\delta_{+}(\xi) \tag{b}$$

where $\delta_{+}(\cdot)$ is Heisenberg's delta function. Verify the above equation (b).

(6.2) In Sect. 6.3, when the load is distributed uniformly in a quarter region $(x > 0,\ y > 0)$ on the surface, how do you change the mathematical expression for the boundary condition (6.3.3)?

References

Achenbach JD (1973) Wave propagation in elastic solids. North-Holland, New York
Cagniard L (1962) Reflection and refraction of progressive seismic waves (translated by Flinn ED, Dix CH). McGraw-Hill, New York
De-Hoop AT (1961) The surface line source problem. Appl Sci Res B 8:349–356
Fung YC (1970) Foundation of solid mechanics (Japanese translation by Ohashi Y et al (1965) Bai-Fu-Kan, Tokyo). Prentice-Hall, New Jersey
Gakenheimer DC, Miklowitz J (1969) Transient excitation of an elastic half space by a point load traveling on the surface. J Appl Mech (Trans ASME Ser E) 36:505–515
Graff KF (1975) Wave motion in elastic solids. Clarendon Press, Oxford
Miklowitz J (1978) The theory of elastic waves and waveguides. North-Holland, New York

Chapter 7
Miscellaneous Green's Functions

This last chapter presents five Green's functions and one application technique of the complex integral. The first and second sections consider the 2D static Green's dyadic for an orthotropic elastic solid, and for an inhomogeneous elastic solid. The third section discusses the Green's function for torsional waves in an anisotropic solid. The fourth and fifth sections are concerned with Green's function for SH waves. One is wave reflection at a moving boundary and the other is wave scattering by a rigid inclusion in an inhomogeneous solid. All the Green's functions are obtained in the closed form by applying the method of integral transform. Especially, in the fourth section which discusses wave reflection, a conversion formula between two different Laplace inversion integrals is developed so that we can treat the moving boundary problem. In the fifth section which discusses the wave scattering, a summation formula is derived for the Fourier series of the Schlömlich type. It enables us to obtain the closed expression for the wave scattering problem. The last section shows an excellent application technique of the complex integral. It reduces a semi-infinite integral, which includes the product of two Bessel functions, to a finite integral that is very suitable for numerical computations.

7.1 2D Static Green's Dyadic for an Orthotropic Elastic Solid

An anisotropic elastic solid whose orthogonal Young's moduli differ from each other is called an "orthotropic" solid. In 2D in-plane deformation, Hooke's law for the orthotropic elastic solid is given by

$$\begin{bmatrix} \sigma_{xx} \\ \sigma_{yy} \\ \sigma_{xy} \end{bmatrix} = \begin{bmatrix} C_{11} & C_{12} & 0 \\ C_{12} & C_{22} & 0 \\ 0 & 0 & C_{66} \end{bmatrix} \begin{bmatrix} \varepsilon_{xx} \\ \varepsilon_{yy} \\ \varepsilon_{xy} \end{bmatrix} \tag{7.1.1}$$

K. Watanabe, *Integral Transform Techniques for Green's Function*, Lecture Notes in Applied and Computational Mechanics 76, DOI 10.1007/978-3-319-17455-6_7

where C_{ij} are elastic moduli. The strain components in the state of plane strain are defined by

$$\varepsilon_{xx} = \frac{\partial u_x}{\partial x}, \quad \varepsilon_{yy} = \frac{\partial u_y}{\partial y}, \quad \varepsilon_{xy} = \frac{1}{2}\left(\frac{\partial u_x}{\partial y} + \frac{\partial u_y}{\partial x}\right) \tag{7.1.2}$$

The equilibrium equations are

$$\begin{aligned} \frac{\partial \sigma_{xx}}{\partial x} + \frac{\partial \sigma_{xy}}{\partial y} &= -Q_x \delta(x)\delta(y) \\ \frac{\partial \sigma_{xy}}{\partial x} + \frac{\partial \sigma_{yy}}{\partial y} &= -Q_y \delta(x)\delta(y) \end{aligned} \tag{7.1.3}$$

where Q_i is the magnitude of a body force.

Substituting Hooke's law into the equilibrium equations, we obtain the displacement equations as

$$\begin{aligned} C_{11}\frac{\partial^2 u_x}{\partial x^2} + \frac{C_{66}}{2}\frac{\partial^2 u_x}{\partial y^2} + \left(C_{12} + \frac{C_{66}}{2}\right)\frac{\partial^2 u_y}{\partial x \partial y} &= -Q_x \delta(x)\delta(y) \\ \left(C_{12} + \frac{C_{66}}{2}\right)\frac{\partial^2 u_x}{\partial x \partial y} + \frac{C_{66}}{2}\frac{\partial^2 u_y}{\partial x^2} + C_{22}\frac{\partial^2 u_y}{\partial y^2} &= -Q_y \delta(x)\delta(y) \end{aligned} \tag{7.1.4}$$

Since we are concerned with Green's dyadic, a particular solution corresponding to the body force is explored. As the elastic medium is of infinite extent, the convergence condition at infinity is assumed as

$$u_i\big|_{\sqrt{x^2+y^2}\to\infty} = \frac{\partial u_i}{\partial x}\bigg|_{\sqrt{x^2+y^2}\to\infty} = \frac{\partial u_i}{\partial y}\bigg|_{\sqrt{x^2+y^2}\to\infty} = 0; \quad i = x, y \tag{7.1.5}$$

We employ the double Fourier transform with respect to two space variables as defined by

$$\bar{f}(\xi) = \int_{-\infty}^{\infty} f(x)\exp(+i\xi x)dx, \quad f(x) = \frac{1}{2\pi}\int_{-\infty}^{\infty} \bar{f}(\xi)\exp(-i\xi x)d\xi \tag{7.1.6}$$

$$\tilde{f}(\eta) = \int_{-\infty}^{\infty} f(y)\exp(+i\eta y)dy, \quad f(y) = \frac{1}{2\pi}\int_{-\infty}^{\infty} \tilde{f}(\eta)\exp(-i\eta y)d\eta \tag{7.1.7}$$

The displacement equations (7.1.4) are transformed to the algebraic equations for the transformed displacement components,

$$\begin{aligned}\{C_{11}\xi^2 + (C_{66}/2)\eta^2\}\tilde{\bar{u}}_x + (C_{12} + C_{66}/2)\xi\eta\tilde{\bar{u}}_y = Q_x \\ (C_{12} + C_{66}/2)\xi\eta\tilde{\bar{u}}_x + \{(C_{66}/2)\xi^2 + C_{22}\eta^2\}\tilde{\bar{u}}_y = Q_y\end{aligned} \tag{7.1.8}$$

Explicit expressions for the displacement components are obtained in the transformed domain,

$$\begin{aligned}\tilde{\bar{u}}_x &= +\frac{Q_x}{C_{22}}\frac{\xi^2 + \beta\eta^2}{(\eta^2 + p_1^2\xi^2)(\eta^2 + p_2^2\xi^2)} - \frac{Q_y}{C_{22}}\frac{(\gamma + 1)\xi\eta}{(\eta^2 + p_1^2\xi^2)(\eta^2 + p_2^2\xi^2)} \\ \tilde{\bar{u}}_y &= -\frac{Q_x}{C_{22}}\frac{(\gamma + 1)\xi\eta}{(\eta^2 + p_1^2\xi^2)(\eta^2 + p_2^2\xi^2)} + \frac{Q_y}{C_{22}}\frac{\alpha\xi^2 + \eta^2}{(\eta^2 + p_1^2\xi^2)(\eta^2 + p_2^2\xi^2)}\end{aligned} \tag{7.1.9}$$

where the eigenvalues, p_j $(j = 1, 2)$, are

$$\begin{pmatrix} p_1 \\ p_2 \end{pmatrix} = \sqrt{\frac{\alpha\beta - \gamma(\gamma + 2) + 2\sqrt{\beta\gamma}}{4\beta}} \mp \sqrt{\frac{\alpha\beta - \gamma(\gamma + 2) - 2\sqrt{\beta\gamma}}{4\beta}} \tag{7.1.10}$$

and the elastic modulus parameters, α, β, γ, are introduced as

$$\alpha = \frac{C_{11}}{C_{66}/2}, \quad \beta = \frac{C_{22}}{C_{66}/2}, \quad \gamma = \frac{C_{12}}{C_{66}/2} \tag{7.1.11}$$

Inspecting the transformed displacement in Eq. (7.1.9), we find that two inversion formulas are necessary for the full inversion. They are

$$\tilde{\bar{I}}_1(\xi, \eta; a, b) = \frac{a\xi^2 + b\eta^2}{(\eta^2 + p_1^2\xi^2)(\eta^2 + p_2^2\xi^2)}, \quad \tilde{\bar{I}}_2(\xi, \eta) = \frac{\xi\eta}{(\eta^2 + p_1^2\xi^2)(\eta^2 + p_2^2\xi^2)} \tag{7.1.12}$$

If we get the inversions for these two expressions, the displacement in the actual space can be expressed as

$$\begin{aligned}u_x &= +\frac{Q_x}{C_{22}}I_1(x, y; 1, \beta) - \frac{Q_y}{C_{22}}(\gamma + 1)I_2(x, y) \\ u_y &= -\frac{Q_x}{C_{22}}(\gamma + 1)I_2(x, y) + \frac{Q_y}{C_{22}}I_1(x, y; \alpha, 1)\end{aligned} \tag{7.1.13}$$

(1) Fourier inversion with respect to the parameter η

The formal Fourier inversion with respect to the parameter η is given by the integrals

$$\begin{aligned}\bar{I}_1(\xi, y; a, b) &= \frac{1}{2\pi}\int_{-\infty}^{+\infty} \frac{a\xi^2 + b\eta^2}{(\eta^2 + p_1^2\xi^2)(\eta^2 + p_2^2\xi^2)}\exp(-i\eta y)d\eta \\ \bar{I}_2(\xi, y) &= \frac{1}{2\pi}\int_{-\infty}^{+\infty} \frac{\xi\eta}{(\eta^2 + p_1^2\xi^2)(\eta^2 + p_2^2\xi^2)}\exp(-i\eta y)d\eta\end{aligned} \tag{7.1.14}$$

These two integrals can be evaluated by the application of some formulas, but, in order to show the application of Jordan's lemma, we consider the complex integrals in the η-plane.

Let us consider the complex integral whose integrand is the same as that of the corresponding integral in Eq. (7.1.14),

$$\begin{aligned}\Phi_1 &= \frac{1}{2\pi}\oint_L \frac{a\xi^2 + b\eta^2}{(\eta^2 + p_1^2\xi^2)(\eta^2 + p_2^2\xi^2)}\exp(-i\eta y)d\eta \\ \Phi_2 &= \frac{1}{2\pi}\oint_L \frac{\xi\eta}{(\eta^2 + p_1^2\xi^2)(\eta^2 + p_2^2\xi^2)}\exp(-i\eta y)d\eta\end{aligned} \tag{7.1.15}$$

The integrand has the four simple poles at $\eta = \pm ip_j|\xi|$, $(j = 1, 2)$ as shown in Fig. 7.1. The closed loop L is composed of a straight line along the real axis and a

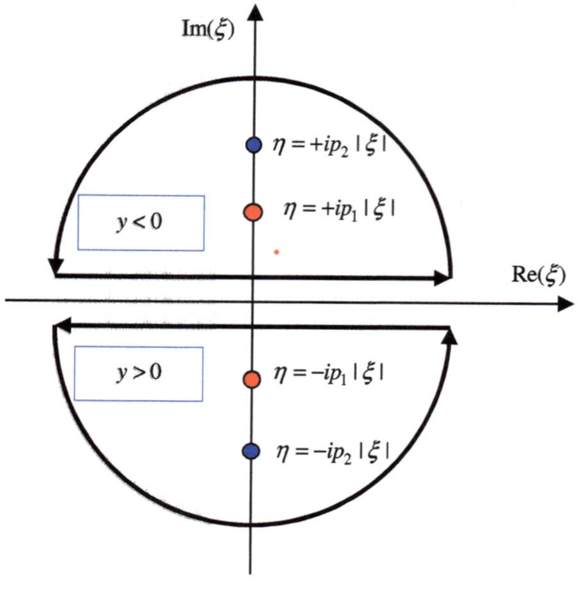

Fig. 7.1 Closed loop L for Fourier inversion integral

semi-circle with infinite radius. Due to the convergence on the semi-circle, we have to choose the lower loop when $y > 0$ and the upper loop when $y < 0$. Since no other singular point exists in the loop, the integral along the real axis is converted to the sum of the residues at the two poles. Thus, we have for $\bar{I}_j; j = 1, 2$

$$\begin{aligned} \bar{I}_1(\xi, y; a, b) &= -\frac{a - bp_1^2}{2p_1(p_1^2 - p_2^2)} \frac{1}{|\xi|} \exp(-p_1|\xi||y|) + \frac{a - bp_2^2}{2p_2(p_1^2 - p_2^2)} \frac{1}{|\xi|} \exp(-p_2|\xi||y|) \\ \bar{I}_2(\xi, y) &= +\frac{\operatorname{sgn}(y)}{2(p_1^2 - p_2^2)} \left\{ \frac{i}{\xi} \exp(-p_1|\xi||y|) - \frac{i}{\xi} \exp(-p_2|\xi||y|) \right\} \end{aligned} \tag{7.1.16}$$

where the sign function is defined by

$$\operatorname{sgn}(y) = \begin{cases} +1; & y > 0 \\ -1; & y < 0 \end{cases} \tag{7.1.17}$$

(2) Fourier inversion with respect to the parameter ξ

Applying the formal Fourier inversion integral with respect to the parameter ξ to Eq. (7.1.16), we have

$$\begin{aligned} I_1(x, y; a, b) &= -\frac{a - bp_1^2}{2p_1(p_1^2 - p_2^2)} \frac{1}{2\pi} \int_{-\infty}^{+\infty} \frac{1}{|\xi|} \exp(-p_1|\xi||y| - i\xi x) d\xi \\ &\quad + \frac{a - bp_2^2}{2p_2(p_1^2 - p_2^2)} \frac{1}{2\pi} \int_{-\infty}^{+\infty} \frac{1}{|\xi|} \exp(-p_2|\xi||y| - i\xi x) d\xi \\ I_2(x, y) &= +\frac{\operatorname{sgn}(y)}{2(p_1^2 - p_2^2)} \left\{ \frac{1}{2\pi} \int_{-\infty}^{+\infty} \frac{i}{\xi} \exp(-p_1|\xi||y| - i\xi x) d\xi \right. \\ &\quad \left. - \frac{1}{2\pi} \int_{-\infty}^{+\infty} \frac{i}{\xi} \exp(-p_2|\xi||y| - i\xi x) d\xi \right\} \end{aligned} \tag{7.1.18}$$

Inspecting the above equations, the two necessary inversion integrals are extracted as

$$\begin{aligned} I_{1j} &= \frac{1}{2\pi} \int_{-\infty}^{+\infty} \frac{1}{|\xi|} \exp(-p_j|\xi||y|) \exp(-i\xi x) d\xi, \\ I_{2j} &= \frac{1}{2\pi} \int_{-\infty}^{+\infty} \frac{i}{\xi} \exp(-p_j|\xi||y|) \exp(-i\xi x) d\xi, \end{aligned} \tag{7.1.19}$$

The second integral I_{2j} is reduced to a semi-infinite integral and can be evaluated by the application of the formula (Erdélyi 1954, vol. I, pp. 72, 2),

$$\int_0^\infty \frac{1}{x}\exp(-ax)\sin(bx)dx = \tan^{-1}\left(\frac{b}{a}\right) \tag{7.1.20}$$

That is

$$\begin{aligned} I_{2j} &= \frac{1}{2\pi}\int_{-\infty}^{+\infty} \frac{i}{\xi}\exp(-p_j|\xi||y|)\exp(-i\xi x)d\xi \\ &= \frac{1}{\pi}\int_0^\infty \frac{1}{\xi}\exp(-p_j\xi|y|)\sin(\xi x)d\xi \\ &= \frac{1}{\pi}\tan^{-1}\left(\frac{x}{p_j|y|}\right) \end{aligned} \tag{7.1.21}$$

The first integral I_{1j} has the first order singularity at $\xi = 0$, and thus this integral cannot be evaluated in this form. The singular behavior is the same as that in the 2D static plane problem in Chap. 3. We differentiate the integral with respect to the variable x and $|y|$, respectively, to obtain

$$\begin{aligned} \frac{\partial I_{1j}}{\partial x} &= \frac{1}{2\pi}\int_{-\infty}^{+\infty} \frac{-i\xi}{|\xi|}\exp(-p_j|\xi||y|)\exp(-i\xi x)d\xi \\ \frac{\partial I_{1j}}{\partial |y|} &= \frac{1}{2\pi}\int_{-\infty}^{+\infty} \frac{-p_j|\xi|}{|\xi|}\exp(-p_j|\xi||y|)\exp(-i\xi x)d\xi \end{aligned} \tag{7.1.22}$$

Reducing the above integrals to semi-infinite ones, we get

$$\begin{aligned} \frac{\partial I_{1j}}{\partial x} &= -\frac{1}{\pi}\int_0^\infty \exp(-p_j\xi|y|)\sin(\xi x)d\xi \\ \frac{\partial I_{1j}}{\partial |y|} &= -\frac{p_j}{\pi}\int_0^\infty \exp(-p_j\xi|y|)\cos(\xi x)d\xi \end{aligned} \tag{7.1.23}$$

and then we apply the formulas (Erdélyi 1954, vol. I, pp. 14, 72), to get

$$\begin{aligned}\int_0^{\infty} \exp(-p_j\xi|y|)\cos(\xi x)d\xi &= \frac{p_j|y|}{x^2+p_j^2y^2}\\ \int_0^{\infty} \exp(-p_j\xi|y|)\sin(\xi x)d\xi &= \frac{x}{x^2+p_j^2y^2}\end{aligned} \tag{7.1.24}$$

The two integrals with different derivative are expressed in terms of algebraic functions,

$$\frac{\partial I_{1j}}{\partial x} = -\frac{1}{\pi}\frac{x}{x^2+p_j^2y^2}, \quad \frac{\partial I_{1j}}{\partial |y|} = \frac{-p_j^2}{\pi}\frac{|y|}{x^2+p_j^2y^2} \tag{7.1.25}$$

In order to obtain the formula for I_{1j}, we integrate the derivatives with respect to each space variable,

$$I_{1j} = -\frac{1}{2\pi}\log\left(x^2+p_j^2y^2\right) + C_x(y), \quad I_{1j} = -\frac{1}{2\pi}\log\left(x^2+p_j^2y^2\right) + C_y(x) \tag{7.1.26}$$

Since the two expressions for I_{1j} must be same, the integration "constant" functions C_x and C_y also must be a simple constant, without any space variable. Thus, we have for I_{1j}

$$I_{1j} = -\frac{1}{2\pi}\log\left(x^2+p_j^2y^2\right) + \text{const.} \tag{7.1.27}$$

Summarizing the above discussion for the inversion integrals, we obtain the simple formulas,

$$\begin{aligned}I_{1j} &= \frac{1}{2\pi}\int_{-\infty}^{+\infty} \frac{1}{|\xi|}\exp(-p_j|\xi||y|)\exp(-i\xi x)d\xi = -\frac{1}{2\pi}\log\left(x^2+p_j^2y^2\right)\\ I_{2j} &= \frac{1}{2\pi}\int_{-\infty}^{+\infty} \frac{i}{\xi}\exp(-p_j|\xi||y|)\exp(-i\xi x)d\xi = \frac{1}{\pi}\tan^{-1}\left(\frac{x}{p_j|y|}\right)\end{aligned} \tag{7.1.28}$$

where the arbitrary constant for the integral I_{1j} is neglected since it does not produce any stress and strain. Substituting these formulas into Eq. (7.1.18), we obtain the closed expressions for the double Fourier inversions I_1 and I_2 as

$$
\begin{aligned}
I_1(x,y;a,b) &= +\frac{a-bp_1^2}{4\pi p_1(p_1^2-p_2^2)}\log\left(x^2+p_1^2y^2\right) - \frac{a-bp_2^2}{4\pi p_2(p_1^2-p_2^2)}\log\left(x^2+p_2^2y^2\right) \\
I_2(x,y) &= +\frac{\mathrm{sgn}(y)}{2\pi(p_1^2-p_2^2)}\left\{\tan^{-1}\left(\frac{x}{p_1|y|}\right) - \tan^{-1}\left(\frac{x}{p_2|y|}\right)\right\}
\end{aligned}
\tag{7.1.29}
$$

where the sign function sgn(.) is defined by Eq. (7.1.17). Thus, the displacement components in Eq. (7.1.13) are expressed in the closed form as

$$
\begin{aligned}
u_x(x,y) &= +\frac{Q_x}{4\pi C_{22}(p_1^2-p_2^2)}\left\{\frac{1-\beta p_1^2}{p_1}\log\left(x^2+p_1^2y^2\right) - \frac{1-\beta p_2^2}{p_2}\log\left(x^2+p_2^2y^2\right)\right\} \\
&\quad - \frac{Q_y}{2\pi C_{22}(p_1^2-p_2^2)}(\gamma+1)\left\{\tan^{-1}\left(\frac{x}{p_1y}\right) - \tan^{-1}\left(\frac{x}{p_2y}\right)\right\} \\
u_y(x,y) &= -\frac{Q_x}{2\pi C_{22}(p_1^2-p_2^2)}(\gamma+1)\left\{\tan^{-1}\left(\frac{x}{p_1y}\right) - \tan^{-1}\left(\frac{x}{p_2y}\right)\right\} \\
&\quad + \frac{Q_y}{4\pi C_{22}(p_1^2-p_2^2)}\left\{\frac{\alpha-p_1^2}{p_1}\log\left(x^2+p_1^2y^2\right) - \frac{\alpha-p_2^2}{p_2}\log\left(x^2+p_2^2y^2\right)\right\}
\end{aligned}
\tag{7.1.30}
$$

Furthermore, if we introduce the notation of the Green's dyadic G_{ij}, the displacement can be rewritten as

$$
\begin{aligned}
u_x(x,y) &= Q_xG_{xx}(x,y) + Q_yG_{xy}(x,y) \\
u_y(x,y) &= Q_xG_{yx}(x,y) + Q_yG_{yy}(x,y)
\end{aligned}
\tag{7.1.31}
$$

where the dyadic components are given by

$$
\begin{aligned}
G_{xx}(x,y) &= +\frac{1}{4\pi C_{22}(p_1^2-p_2^2)}\left\{\frac{1-\beta p_1^2}{p_1}\log\left(x^2+p_1^2y^2\right) - \frac{1-\beta p_2^2}{p_2}\log\left(x^2+p_2^2y^2\right)\right\} \\
G_{xy}(x,y) &= G_{yx}(x,y) = -\frac{1}{2\pi C_{22}(p_1^2-p_2^2)}(\gamma+1)\left\{\tan^{-1}\left(\frac{x}{p_1y}\right) - \tan^{-1}\left(\frac{x}{p_2y}\right)\right\} \\
G_{yy}(x,y) &= +\frac{1}{4\pi C_{22}(p_1^2-p_2^2)}\left\{\frac{\alpha-p_1^2}{p_1}\log\left(x^2+p_1^2y^2\right) - \frac{\alpha-p_2^2}{p_2}\log\left(x^2+p_2^2y^2\right)\right\}
\end{aligned}
\tag{7.1.32}
$$

It should be noticed that the symmetry $G_{xy} = G_{yx}$ is held in spite of the orthotropic nature. The logarithmic singularity at the source point and at the infinity is the same as that for isotropic media in Chap. 3. These singular behaviors are inherent in 2D plane elasticity.

7.2 2D Static Green's Dyadic for an Inhomogeneous Elastic Solid

Materials whose material parameters, such as elastic moduli and density, are varying with the space point are called "inhomogeneous" materials. Many mathematical models for the inhomogeneity are proposed. The simplest model for the inhomogeneity is a uniaxial exponential function for the material parameters. This model is mathematically simple and tractable, but not fully correct since the material parameters vanish or diverge at infinity. In spite of the unrealistic nature of the exponential model, this type of the inhomogeneity is frequently used for stress analysis due to its simplicity.

Let us consider the 2D plane-strain deformation of an inhomogeneous isotropic elastic solid. Hooke's law for an inhomogeneous solid is the same as that for a homogeneous one, but its elastic moduli (λ, μ) are functions of the space variables,

$$\begin{aligned}
\sigma_{xx} &= (\lambda + 2\mu)\frac{\partial u_x}{\partial x} + \lambda\frac{\partial u_y}{\partial y} \\
\sigma_{yy} &= \lambda\frac{\partial u_x}{\partial x} + (\lambda + 2\mu)\frac{\partial u_y}{\partial y} \\
\sigma_{xy} &= \sigma_{yx} = \mu\left(\frac{\partial u_x}{\partial y} + \frac{\partial u_y}{\partial x}\right)
\end{aligned} \tag{7.2.1}$$

As we are concerned with Green's dyadic corresponding to a point source, the 2D equilibrium equations with a point body force are given by

$$\begin{aligned}
\frac{\partial \sigma_{xx}}{\partial x} + \frac{\partial \sigma_{yx}}{\partial y} &= -B_x\delta(x)\delta(y) \\
\frac{\partial \sigma_{xy}}{\partial x} + \frac{\partial \sigma_{yy}}{\partial y} &= -B_y\delta(x)\delta(y)
\end{aligned} \tag{7.2.2}$$

where the body force is placed at the coordinate origin $(0,0)$. The inhomogeneity which we adopt here is a uniaxial exponential function and thus, Lame's constants (λ, μ) are assumed to be exponential functions with one space variable y,

$$\lambda(y) = \lambda_0 \exp\{k(y/h)\}, \; \mu(y) = \mu_0 \exp\{k(y/h)\} \tag{7.2.3}$$

where (λ_0, μ_0) are moduli at the coordinate origin, k is an inhomogeneity parameter, and h is a reference length. Since two Lame's constants have the same exponential function, Poisson ratio ν is constant throughout the medium.

Substituting Hooke's law of Eq. (7.2.1) with the inhomogeneity of Eq. (7.2.3) into the equilibrium equations (7.2.2), we obtain displacement equations with constant coefficients,

$$\gamma^2 \frac{\partial^2 u_x}{\partial x^2} + \frac{\partial^2 u_x}{\partial y^2} + (\gamma^2 - 1)\frac{\partial^2 u_y}{\partial x \partial y} + \frac{k}{h}\left(\frac{\partial u_x}{\partial y} + \frac{\partial u_y}{\partial x}\right) = -\frac{B_x}{\mu(y)}\delta(x)\delta(y)$$
$$(\gamma^2 - 1)\frac{\partial^2 u_x}{\partial x \partial y} + \frac{\partial^2 u_y}{\partial x^2} + \gamma^2 \frac{\partial^2 u_y}{\partial y^2} + \frac{k}{h}\left\{(\gamma^2 - 2)\frac{\partial u_x}{\partial x} + \gamma^2 \frac{\partial u_y}{\partial y}\right\} = -\frac{B_y}{\mu(y)}\delta(x)\delta(y) \tag{7.2.4}$$

where the constant material parameter γ is defined by

$$\gamma = \sqrt{\frac{\lambda + 2\mu}{\mu}} = \sqrt{\frac{\lambda_0 + 2\mu_0}{\mu_0}} = \sqrt{\frac{2(1-v)}{1-2v}} \tag{7.2.5}$$

In order to obtain a particular solution corresponding to the nonhomogeneous body force term, we apply the double Fourier transform with respect to two space variables,

$$\bar{f}(\xi) = \int_{-\infty}^{+\infty} f(x)\exp(+i\xi x)dx, \qquad f(x) = \frac{1}{2\pi}\int_{-\infty}^{+\infty} \bar{f}(\xi)\exp(-i\xi x)d\xi$$
$$\tilde{f}(\eta) = \int_{-\infty}^{+\infty} f(y)\exp(+i\eta y)dy, \qquad f(y) = \frac{1}{2\pi}\int_{-\infty}^{+\infty} \tilde{f}(\eta)\exp(-i\eta y)d\eta \tag{7.2.6}$$

The convergence condition at infinity is also imposed on the displacement,

$$u_i\big|_{\sqrt{x^2+y^2}\to\infty} = \left.\frac{\partial u_i}{\partial x}\right|_{\sqrt{x^2+y^2}\to\infty} = \left.\frac{\partial u_i}{\partial y}\right|_{\sqrt{x^2+y^2}\to\infty} = 0; \quad i = x, y \tag{7.2.7}$$

With the aid of the integration formulas which include the delta function,

$$\int_{-\infty}^{+\infty} \delta(x)\exp(+i\xi x)dx = 1, \qquad \int_{-\infty}^{+\infty} \frac{\delta(y)}{\mu(y)}\exp(+i\eta y)dy = \frac{1}{\mu(0)} \tag{7.2.8}$$

the displacement equations (7.2.4) are transformed to the algebraic equations for the displacement components,

$$\{\gamma^2\xi^2 + \eta^2 + i\eta(k/h)\}\tilde{\bar{u}}_x + \{(\gamma^2 - 1)\xi\eta + i\xi(k/h)\}\tilde{\bar{u}}_y = \frac{B_x}{\mu(0)}$$
$$\{(\gamma^2 - 1)\xi\eta + i\xi(\gamma^2 - 2)(k/h)\}\tilde{\bar{u}}_x + \{\xi^2 + \gamma^2\eta^2 + i\eta\gamma^2(k/h)\}\tilde{\bar{u}}_y = \frac{B_y}{\mu(0)} \tag{7.2.9}$$

where

$$\mu(0) = \mu_0 \tag{7.2.10}$$

Solving Eq. (7.2.9) for the transformed displacement components, we get

$$\tilde{\tilde{u}}_x = +\frac{B_x}{\mu(0)}\frac{\xi^2 + \gamma^2(\bar{\eta}^2 + \kappa^2)}{\gamma^2\Delta(\xi,\bar{\eta})} - \frac{B_y}{\mu(0)}\frac{(\gamma^2 - 1)\bar{\eta} - i\kappa(\gamma^2 - 3)}{\gamma^2\Delta(\xi,\bar{\eta})}\xi \tag{7.2.11a}$$

$$\tilde{\tilde{u}}_y = -\frac{B_x}{\mu(0)}\frac{(\gamma^2 - 1)\bar{\eta} + i\kappa(\gamma^2 - 3)}{\gamma^2\Delta(\xi,\bar{\eta})}\xi + \frac{B_y}{\mu(0)}\frac{\gamma^2\xi^2 + \bar{\eta}^2 + \kappa^2}{\gamma^2\Delta(\xi,\bar{\eta})} \tag{7.2.11b}$$

where the new inhomogeneity parameter κ is introduced as

$$\kappa = k/(2h) \tag{7.2.12}$$

and the denominator $\Delta(\xi,\bar{\eta})$ and $\bar{\eta}$ are given by

$$\Delta(\xi,\bar{\eta}) = (\xi^2 + \bar{\eta}^2 + \kappa^2)^2 + 4(1 - 2/\gamma^2)\kappa^2\xi^2 \tag{7.2.13}$$

$$\bar{\eta} = \eta + i\kappa \tag{7.2.14}$$

In order to apply the inversion integral, we factorize the denominator as

$$\Delta(\xi,\bar{\eta}) = (\xi - i\omega_1)(\xi + i\omega_1)(\xi - i\omega_2)(\xi + i\omega_2) \tag{7.2.15}$$

where the eigenvalues ω_j; $j = 1, 2$ are

$$\omega_1 = \sqrt{\bar{\eta}^2 + \kappa^2 q^2} - \kappa p, \quad \omega_2 = \sqrt{\bar{\eta}^2 + \kappa^2 q^2} + \kappa p \tag{7.2.16}$$

and

$$p = \sqrt{1 - 2/\gamma^2}, \quad q = \sqrt{2(1 - 1/\gamma^2)}; \quad q > p \tag{7.2.17}$$

The formal Fourier inversion with respect to the parameter ξ is given by the integrals,

$$\begin{aligned}\tilde{u}_x = &+\frac{B_x}{\gamma^2\mu(0)}\frac{1}{2\pi}\int_{-\infty}^{+\infty}\frac{\xi^2 + \gamma^2(\bar{\eta}^2 + \kappa^2)}{(\xi - i\omega_1)(\xi + i\omega_1)(\xi - i\omega_2)(\xi + i\omega_2)}\exp(-i\xi x)d\xi \\ &-\frac{B_y}{\gamma^2\mu(0)}\frac{1}{2\pi}\int_{-\infty}^{+\infty}\frac{(\gamma^2 - 1)\bar{\eta} - i\kappa(\gamma^2 - 3)}{(\xi - i\omega_1)(\xi + i\omega_1)(\xi - i\omega_2)(\xi + i\omega_2)}\xi\exp(-i\xi x)d\xi\end{aligned} \tag{7.2.18a}$$

$$\tilde{u}_y = -\frac{B_x}{\gamma^2\mu(0)}\frac{1}{2\pi}\int_{-\infty}^{+\infty}\frac{(\gamma^2-1)\bar{\eta}+i\kappa(\gamma^2-3)}{(\xi-i\omega_1)(\xi+i\omega_1)(\xi-i\omega_2)(\xi+i\omega_2)}\xi\exp(-i\xi x)d\xi$$
$$+\frac{B_y}{\gamma^2\mu(0)}\frac{1}{2\pi}\int_{-\infty}^{+\infty}\frac{\gamma^2\xi^2+\bar{\eta}^2+\kappa^2}{(\xi-i\omega_1)(\xi+i\omega_1)(\xi-i\omega_2)(\xi+i\omega_2)}\exp(-i\xi x)d\xi \tag{7.2.18b}$$

The above integrals can be evaluated by the complex integral theory. So, we apply Cauchy's theorem (Jordan's lemma) to the integral in the complex ξ-plane. As an example, the first integral in Eq. (7.2.18a) is discussed. We represent it by the complex integral Φ_{xx},

$$\Phi_{xx} = \frac{1}{2\pi}\int_C \frac{\xi^2+\gamma^2(\bar{\eta}^2+\kappa^2)}{(\xi-i\omega_1)(\xi+i\omega_1)(\xi-i\omega_2)(\xi+i\omega_2)}\exp(-i\xi x)d\xi \tag{7.2.19}$$

where the closed loop C is composed of a semi-circle with infinite radius and a straight line along the real axis (see Fig. 7.2). In order to guarantee the convergence on the large semi-circle, we have to employ the lower loop for $x > 0$, and the upper for $x < 0$. The integrand has four poles; two of them are in the upper plane and other two are in the lower plane. Each closed loop includes two poles. Applying the Jordan's lemma to the complex integral Φ_{xx} in Eq. (7.2.19), the integral along the real axis is converted to the sum of two residues. When we employ the lower loop for $x > 0$, the complex integral yields

$$\begin{aligned}\Phi_{xx} &= -\frac{1}{2\pi}\int_{-\infty}^{+\infty}\frac{\xi^2+\gamma^2(\bar{\eta}^2+\kappa^2)}{(\xi-i\omega_1)(\xi+i\omega_1)(\xi-i\omega_2)(\xi+i\omega_2)}\exp(-i\xi x)d\xi \\ &= 2\pi i\mathrm{Res}(-i\omega_1)+2\pi i\mathrm{Res}(-i\omega_2)\end{aligned} \tag{7.2.20}$$

The detailed calculation for the residues is

$$\begin{aligned}&2\pi i\mathrm{Res}(-i\omega_1)+2\pi i\mathrm{Res}(-i\omega_2)\\ &=\frac{2\pi i}{2\pi}\frac{\xi^2+\gamma^2(\bar{\eta}^2+\kappa^2)}{(\xi-i\omega_1)(\xi-i\omega_2)(\xi+i\omega_2)}\exp(-i\xi x)\bigg|_{\xi=-i\omega_1}\\ &+\frac{2\pi i}{2\pi}\frac{\xi^2+\gamma^2(\bar{\eta}^2+\kappa^2)}{(\xi-i\omega_1)(\xi-i\omega_2)(\xi+i\omega_1)}\exp(-i\xi x)\bigg|_{\xi=-i\omega_2}\end{aligned} \tag{7.2.21}$$

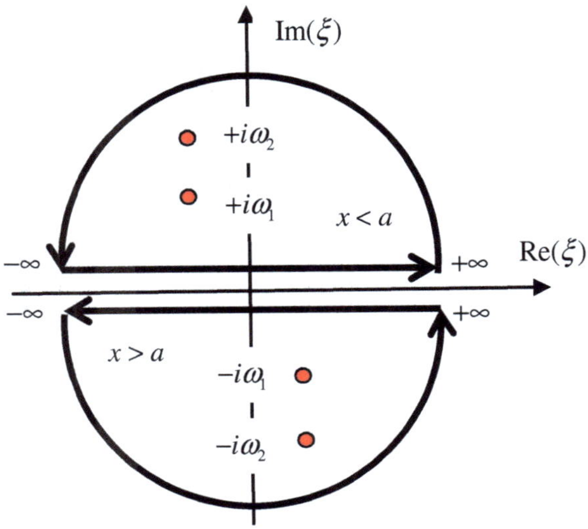

Fig. 7.2 Closed loop C for the complex integral Φ_{xx}

Arranging the above equation, the inversion integral along the real axis is evaluated exactly as

$$\begin{aligned}\frac{1}{2\pi}\int_{-\infty}^{+\infty}&\frac{\xi^2+\gamma^2(\bar{\eta}^2+\kappa^2)}{(\xi-i\omega_1)(\xi+i\omega_1)(\xi-i\omega_2)(\xi+i\omega_2)}\exp(-i\xi x)d\xi\\&=\left\{\frac{\gamma^2-1}{4\kappa p}\sinh(\kappa p x)+\frac{\gamma^2+1}{4\sqrt{\bar{\eta}^2+\kappa^2q^2}}\cosh(\kappa p x)\right\}e^{-x\sqrt{\bar{\eta}^2+\kappa^2q^2}}\end{aligned} \tag{7.2.22}$$

Similarly, the other integrals in Eq. (7.2.18) can be evaluated and we have

$$\begin{aligned}\tilde{u}_x=&+\frac{B_x}{4\gamma^2\mu(0)}\left\{\frac{\gamma^2+1}{\sqrt{\bar{\eta}^2+\kappa^2q^2}}\cosh(\kappa p x)+\frac{\gamma^2-1}{\kappa p}\sinh(\kappa p|x|)\right\}\exp\left(-|x|\sqrt{\bar{\eta}^2+\kappa^2q^2}\right)\\&+\frac{B_y}{4\gamma^2\mu(0)}\frac{i\eta(\gamma^2-1)+\kappa(\gamma^2-3)}{\kappa p\sqrt{\bar{\eta}^2+\kappa^2q^2}}\sinh(\kappa p|x|)\exp\left(-|x|\sqrt{\bar{\eta}^2+\kappa^2q^2}\right)\end{aligned} \tag{7.2.23a}$$

$$\begin{aligned}\tilde{u}_y=&+\frac{B_x}{4\gamma^2\mu(0)}\frac{i\eta(\gamma^2-1)-\kappa(\gamma^2-3)}{\kappa p\sqrt{\bar{\eta}^2+\kappa^2q^2}}\sinh(\kappa p|x|)\exp\left(-|x|\sqrt{\bar{\eta}^2+\kappa^2q^2}\right)\\&+\frac{B_y}{4\gamma^2\mu(0)}\left\{\frac{\gamma^2+1}{\sqrt{\bar{\eta}^2+\kappa^2q^2}}\cosh(\kappa p x)-\frac{\gamma^2-1}{\kappa p}\sinh(\kappa p|x|)\right\}\exp\left(-|x|\sqrt{\bar{\eta}^2+\kappa^2q^2}\right)\end{aligned} \tag{7.2.23b}$$

Our next task is to evaluate the inversion integral with respect to the parameter η. The formal inversion of the displacement is expressed in terms of the fundamental inversion integrals $I_i(x,y);\ i=0,1,2$,

$$u_x = +\frac{B_x}{4\mu(0)}\frac{\gamma^2-1}{\gamma^2}\exp(-\kappa y)\left\{\frac{\gamma^2+1}{\gamma^2-1}\cosh(\kappa px)I_0(x,y)+\frac{\sinh(\kappa p|x|)}{\kappa p}I_1(x,y)\right\}$$
$$+\frac{B_y}{4\mu(0)}\frac{\gamma^2-1}{\gamma^2}\frac{\sinh(\kappa p|x|)}{\kappa p}\exp(-\kappa y)\left\{I_2(x,y)+\kappa\frac{\gamma^2-3}{\gamma^2-1}I_0(x,y)\right\}$$
(7.2.24a)

$$u_y = +\frac{B_x}{4\mu(0)}\frac{\gamma^2-1}{\gamma^2}\frac{\sinh(\kappa p|x|)}{\kappa p}\exp(-\kappa y)\left\{I_2(x,y)-\kappa\frac{\gamma^2-3}{\gamma^2-1}I_0(x,y)\right\}$$
$$+\frac{B_y}{4\mu(0)}\frac{\gamma^2-1}{\gamma^2}\exp(-\kappa y)\left\{\frac{\gamma^2+1}{\gamma^2-1}\cosh(\kappa px)I_0(x,y)-\frac{\sinh(\kappa p|x|)}{\kappa p}I_1(x,y)\right\}$$
(7.2.24b)

where the fundamental integrals I_i are defined by

$$I_0(x,y)=\frac{1}{2\pi}\int_{-\infty}^{+\infty}\frac{\exp\left(-|x|\sqrt{\bar{\eta}^2+\kappa^2q^2}-i\bar{\eta}y\right)}{\sqrt{\bar{\eta}^2+\kappa^2q^2}}d\eta \qquad (7.2.25)$$

$$I_1(x,y)=\frac{1}{2\pi}\int_{-\infty}^{+\infty}\exp\left(-|x|\sqrt{\bar{\eta}^2+\kappa^2q^2}-i\bar{\eta}y\right)d\eta \qquad (7.2.26)$$

$$I_2(x,y)=\frac{1}{2\pi}\int_{-\infty}^{+\infty}i\bar{\eta}\frac{\exp\left(-|x|\sqrt{\bar{\eta}^2+\kappa^2q^2}-i\bar{\eta}y\right)}{\sqrt{\bar{\eta}^2+\kappa^2q^2}}d\eta \qquad (7.2.27)$$

and the variable $\bar{\eta}$ is defined by Eq. (7.2.14).

Observing and examining the above three Eqs. (7.2.25)–(7.2.27), we find the relation among the fundamental integrals as

$$I_1(x,y)=-\frac{\partial I_0(x,y)}{\partial|x|},\quad I_2(x,y)=-\frac{\partial I_0(x,y)}{\partial y} \qquad (7.2.28)$$

Since the two integrals I_1 and I_2 can be derived from I_0, it is enough to evaluate the only one integral I_0. For the integral I_0 in Eq. (7.2.25), we introduce the variable transform from η to the new variable ς,

$$\varsigma=\bar{\eta}=\eta+i\kappa \qquad (7.2.29)$$

Equation (7.2.25) is transformed to the integral along the complex line from $-\infty + i\kappa$ to $+\infty + i\kappa$,

$$I_0(x,y) = \frac{1}{2\pi} \int_{-\infty+i\kappa}^{+\infty+i\kappa} \frac{\exp\left(-|x|\sqrt{\varsigma^2+\kappa^2 q^2} - i\varsigma y\right)}{\sqrt{\varsigma^2+\kappa^2 q^2}} d\varsigma \tag{7.2.30}$$

The above integral resembles the formula (Erdélyi 1954, p. 17, (27)),

$$\int_{-\infty}^{+\infty} \frac{1}{\sqrt{x^2+c^2}} \exp\left(-a\sqrt{x^2+c^2}\right) \exp(\pm ibx) dx = 2K_0\left(c\sqrt{a^2+b^2}\right) \tag{7.2.31}$$

where $K_0(.)$ is the modified Bessel function of zeroth order. However, this formula cannot be applied directly since the integration path for our integral of Eq. (7.2.30) is not real. In order to apply the formula, we need to transform the integral to one along the real path.

Let us consider the complex integral Φ along the rectangular closed loop $\overline{\text{ABCDA}}$ shown in Fig. 7.3,

$$\Phi = \frac{1}{2\pi} \int_{\overline{\text{ABCDA}}} \frac{\exp\left(-|x|\sqrt{\varsigma^2+\kappa^2 q^2} - i\varsigma y\right)}{\sqrt{\varsigma^2+\kappa^2 q^2}} d\varsigma \tag{7.2.32}$$

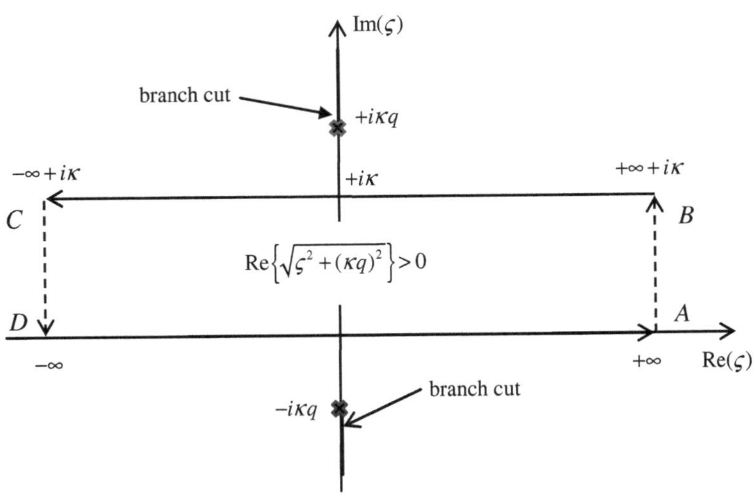

Fig. 7.3 Transform of integration path from the complex line to the real line

The integrand has two branch points at $\varsigma = \pm i\kappa q$. Two branch cuts along the imaginary axis are also introduced as shown in the figure. Fortunately, neither of two branch cuts cross the integration lines $\overrightarrow{\mathrm{DA}}$ and $\overrightarrow{\mathrm{BC}}$, especially the line $\overrightarrow{\mathrm{BC}}$ since $q(=1/(1-\nu)) > 1$. No other singular point is included in the closed loop $\overline{\mathrm{ABCDA}}$ and the integrals along the vertical straight lines, $\overline{\mathrm{AB}}$ and $\overline{\mathrm{CD}}$, vanish at infinity, since the real part of the radical is positive. Thus, the integral along the complex line $\overrightarrow{\mathrm{CB}}$ is converted to one along the real axis $\overrightarrow{\mathrm{DA}}$. That is

$$\frac{1}{2\pi}\int_{\overrightarrow{\mathrm{CB}}} \frac{\exp\left(-|x|\sqrt{\varsigma^2+\kappa^2 q^2} - i\varsigma y\right)}{\sqrt{\varsigma^2+\kappa^2 q^2}} d\varsigma = \frac{1}{2\pi}\int_{\overrightarrow{\mathrm{DA}}} \frac{\exp\left(-|x|\sqrt{\varsigma^2+\kappa^2 q^2} - i\varsigma y\right)}{\sqrt{\varsigma^2+\kappa^2 q^2}} d\varsigma \tag{7.2.33}$$

Consequently, the integral (7.2.30) along the complex line is converted to that along the real path,

$$I_0(x,y) = \frac{1}{2\pi}\int_{-\infty}^{+\infty} \frac{\exp\left(-|x|\sqrt{\varsigma^2+\kappa^2 q^2} - i\varsigma y\right)}{\sqrt{\varsigma^2+\kappa^2 q^2}} d\varsigma \tag{7.2.34}$$

Therefore, we can apply the integration formula Eq. (7.2.31) to the above integral and have the exact expression for the integral I_0,

$$I_0(x,y) = \frac{1}{\pi} K_0\left(|\kappa| q\sqrt{x^2+y^2}\right) \tag{7.2.35}$$

Substituting this Eq. (7.2.35) into Eq. (7.2.28), the other two integrals, I_1 and I_2, are exactly evaluated as

$$\begin{aligned} I_1 &= \frac{|\kappa| q |x|}{\pi\sqrt{x^2+y^2}} K_1\left(|\kappa| q\sqrt{x^2+y^2}\right) \\ I_2 &= \frac{|\kappa| q y}{\pi\sqrt{x^2+y^2}} K_1\left(|\kappa| q\sqrt{x^2+y^2}\right) \end{aligned} \tag{7.2.36}$$

Finally, we substitute Eqs. (7.2.35) and (7.2.36) into Eq. (7.2.24) and rewrite the displacement in terms of Green's dyadic G_{ij},

$$\begin{aligned} u_x &= B_x G_{xx}(x,y) + B_y G_{xy}(x,y) \\ u_y &= B_x G_{yx}(x,y) + B_y G_{yy}(x,y) \end{aligned} \tag{7.2.37}$$

where the exact expressions for the dyadic are

$$G_{xx}(x,y) = \frac{\gamma^2-1}{4\pi\gamma^2\mu(0)}\exp(-\kappa y) \times \left\{\frac{\gamma^2+1}{\gamma^2-1}\cosh(\kappa px)K_0(|\kappa|qr) + |\kappa|q\frac{x}{r}\frac{\sinh(\kappa px)}{\kappa p}K_1(|\kappa|qr)\right\} \tag{7.2.38a}$$

$$G_{xy}(x,y) = \frac{\gamma^2-1}{4\pi\gamma^2\mu(0)}\exp(-\kappa y)\frac{\sinh(\kappa px)}{\kappa p}\left\{|\kappa|q\frac{y}{r}K_1(|\kappa|qr) + \kappa\frac{\gamma^2-3}{\gamma^2-1}K_0(|\kappa|qr)\right\} \tag{7.2.38b}$$

$$G_{yx}(x,y) = \frac{\gamma^2-1}{4\pi\gamma^2\mu(0)}\exp(-\kappa y)\frac{\sinh(\kappa px)}{\kappa p}\left\{|\kappa|q\frac{y}{r}K_1(|\kappa|qr) - \kappa\frac{\gamma^2-3}{\gamma^2-1}K_0(|\kappa|qr)\right\} \tag{7.2.38c}$$

$$G_{yy}(x,y) = \frac{\gamma^2-1}{4\pi\gamma^2\mu(0)}\exp(-\kappa y) \times \left\{\frac{\gamma^2+1}{\gamma^2-1}\cosh(\kappa px)K_0(|\kappa|qr) - |\kappa|q\frac{x}{r}\frac{\sinh(\kappa px)}{\kappa p}K_1(|\kappa|qr)\right\} \tag{7.2.38d}$$

Here the radial distance r from the source is defined by

$$r = \sqrt{x^2+y^2} \tag{7.2.39}$$

7.2.1 2D Kelvin's Solution for Homogeneous Media

Inspecting the dyadic in Eq. (7.2.38), we can find that there is only one inhomogeneity parameter, i.e. $\kappa(=\kappa/(2h))$. When this parameter vanishes, the medium is homogeneous. Thus, taking the limit $\kappa \to 0$ as

$$\frac{\sinh(\kappa px)}{\kappa p} \sim x, \quad \cosh(\kappa px) \sim 1 \tag{7.2.40}$$

$$K_0(|\kappa|qr) \sim -\log(|\kappa|qr), \quad |\kappa|qK_1(|\kappa|qr) \sim \frac{1}{r} \tag{7.2.41}$$

we have the dyadic for the homogeneous medium, i.e. the 2D Kelvin's solution,

$$G_{xx}(x,y) = \frac{1}{4\pi\gamma^2\mu_0}\left\{-(\gamma^2+1)\log(r) + (\gamma^2-1)\left(\frac{x}{r}\right)^2\right\} \quad (7.2.42a)$$

$$G_{xy}(x,y) = G_{yx}(x,y) = \frac{\gamma^2-1}{4\pi\gamma^2\mu_0}\frac{xy}{r^2} \quad (7.2.42b)$$

$$G_{yy}(x,y) = \frac{1}{4\pi\gamma^2\mu_0}\left\{-(\gamma^2+1)\log(r) + (\gamma^2-1)\left(\frac{y}{r}\right)^2\right\} \quad (7.2.42c)$$

where μ_0 is a uniform rigidity throughout the medium. The reader will find that this dyadic is the same as the 2D static dyadic of Eq. (3.3.33) in Sect. 3.3.

7.3 Green's Function for Torsional Waves in a Monoclinic Material

The axisymmetric torsional deformation of an elastic solid is governed by the equation of motion

$$\frac{1}{r^2}\frac{\partial(r^2\sigma_{r\theta})}{\partial r} + \frac{\partial\sigma_{z\theta}}{\partial z} = \rho\frac{\partial^2 u_\theta}{\partial t^2} - \rho B_\theta \quad (7.3.1)$$

and Hooke's law

$$\begin{bmatrix}\sigma_{r\theta}\\ \sigma_{z\theta}\end{bmatrix} = \begin{bmatrix}C_{44} & C_{45}\\ C_{54} & C_{55}\end{bmatrix}\begin{bmatrix}\varepsilon_{r\theta}\\ \varepsilon_{z\theta}\end{bmatrix}; \; C_{45} = C_{54} \quad (7.3.2)$$

where $u_\theta \equiv u_\theta(r,z,t)$ is the torsional/circumferential displacement, $\sigma_{r\theta}$ and $\sigma_{z\theta}$ are torsional stresses and $\varepsilon_{r\theta}$ and $\varepsilon_{z\theta}$ are strains. When $C_{44} = C_{55} = \mu$ and $C_{45} = C_{54} = 0$, the elastic solid is isotropic. On the other hand, when $C_{44} \neq C_{55}$ and $C_{45} = C_{54} \neq 0$, the solid is anisotropic and is called "monoclinic." The monoclinic material is one of anisotropic solids. As Green's function for the isotropic solid has been obtained in Sect. 3.7 in Chap. 3, we shall obtain the Green's function for the monoclinic material.

Two strains are defined by

$$\varepsilon_{r\theta} = \frac{1}{2}\left(\frac{\partial u_\theta}{\partial r} - \frac{u_\theta}{r}\right), \quad \varepsilon_{z\theta} = \frac{1}{2}\frac{\partial u_\theta}{\partial z} \quad (7.3.3)$$

We employ the same circumferential ring body force as that in Sect. 3.7,

$$B_\theta = Q\frac{\delta(r-a)}{r}\delta(z)\delta(t) \tag{7.3.4}$$

where Q is the magnitude of the body force and $\delta(.)$ is Dirac's delta function. Substituting Eq. (7.3.2) into the equation of motion with Eq. (7.3.4), we have the displacement equation,

$$\begin{aligned}\frac{\partial^2 u_\theta}{\partial r^2} + \frac{1}{r}\frac{\partial u_\theta}{\partial r} - \frac{u_\theta}{r^2} + 2\beta\frac{\partial^2 u_\theta}{\partial r\partial z} + \frac{\beta}{r}\frac{\partial u_\theta}{\partial z} + \alpha^2\frac{\partial^2 u_\theta}{\partial z^2} - \frac{1}{c^2}\frac{\partial^2 u_\theta}{\partial t^2}\\ = -\frac{Q}{c^2}\frac{\delta(r-a)}{r}\delta(z)\delta(t)\end{aligned} \tag{7.3.5}$$

where

$$\alpha = \sqrt{C_{44}/C_{55}}, \quad \beta = C_{45}/C_{55}, \quad c = \sqrt{C_{55}/(2\rho)} \tag{7.3.6}$$

Our task in the present section is to obtain a particular solution corresponding to the nonhomogeneous term of the ring force. In order to reduce the partial differential equation to the ordinary one, we apply the double integral transform: Laplace transform with respect to the time t

$$f^*(s) = \int_0^\infty f(t)\exp(-st)dt \tag{7.3.7}$$

and Fourier transform with respect to the axial length z,

$$\hat{f}(\zeta) = \int_{-\infty}^{+\infty} f(z)\exp(+i\zeta z)dz \Leftrightarrow f(z) = \frac{1}{2\pi}\int_{-\infty}^{+\infty} \hat{f}(\zeta)\exp(-i\zeta z)d\zeta \tag{7.3.8}$$

Assuming the quiescent condition at an initial time,

$$u_\theta|_{t=0} = \left.\frac{\partial u_\theta}{\partial t}\right|_{t=0} = 0 \tag{7.3.9}$$

and the convergence condition at infinity

$$u_\theta|_{\sqrt{r^2+z^2}\to\infty} = \left.\frac{\partial u_\theta}{\partial r}\right|_{\sqrt{r^2+z^2}\to\infty} = \left.\frac{\partial u_\theta}{\partial z}\right|_{\sqrt{r^2+z^2}\to\infty} = 0 \tag{7.3.10}$$

we apply the double transform to the displacement equation (7.3.5). It follows that

$$\frac{d^2\hat{u}_\theta^*}{dr^2} + \left(\frac{1}{r} - 2i\beta\zeta\right)\frac{d\hat{u}_\theta^*}{dr} - \left\{\frac{1}{r^2} + \frac{i\beta\zeta}{r} + (\alpha\zeta)^2 + (s/c)^2\right\}\hat{u}_\theta^* = -\frac{Q}{c^2}\frac{\delta(r-a)}{r} \tag{7.3.11}$$

In order to obtain the simpler Bessel equation, we assume the solution as a product of two unknown functions. They are assumed as

$$\hat{u}_\theta^* = g(r)U(r) \tag{7.3.12}$$

Substituting the above equation into Eq. (7.3.11), we have

$$\begin{aligned} & g\left(U'' + \frac{1}{r}U' - \frac{1}{r^2}U\right) + 2(g' - i\beta\zeta g)U' + \left(\frac{g' - i\beta\zeta g}{r} + g'' - 2i\beta\zeta g'\right)U \\ & - g\{(\alpha\zeta)^2 + (s/c)^2\}U = -\frac{Q}{c^2}\frac{\delta(r-a)}{r} \end{aligned} \tag{7.3.13}$$

where the prime (.)′ denotes the differentiation with respect to the radial variable r. Inspecting the above equation, if we assume

$$g' - i\beta\zeta g = 0 \Rightarrow g(r) = \exp(+i\beta\zeta r) \tag{7.3.14}$$

we have the nonhomogeneous Bessel equation for another unknown $U(r)$,

$$U'' + \frac{1}{r}U' - \frac{1}{r}U - \{(\gamma\zeta)^2 + (s/c)^2\}U = -\frac{Q}{c^2}\frac{\delta(r-a)}{r}\exp(-i\beta\zeta r) \tag{7.3.15}$$

where the parameter γ is the combination of two anisotropy parameters,

$$\gamma = \sqrt{\alpha^2 - \beta^2} \tag{7.3.16}$$

Equation (7.3.15) is in the form of Bessel differential equation for the unknown U (r). In order to obtain the particular solution, we apply the Hankel transform with the order 1,

$$\tilde{f}(\xi) = \int_0^\infty rf(r)J_1(\xi r)dr \Leftrightarrow f(r) = \int_0^\infty \xi\tilde{f}(\xi)J_1(\xi r)d\xi \tag{7.3.17}$$

to Eq. (7.3.15), and then have the simple algebraic equation for the transformed function. The transformed function $\tilde{U}(\xi)$ is given by

$$\tilde{U}(\xi) = \frac{Q}{c^2}\exp(-i\beta\zeta a)\frac{J_1(a\xi)}{\xi^2 + (\gamma\zeta)^2 + (s/c)^2} \tag{7.3.18}$$

Its formal Hankel inversion is given by

$$U(r) = \frac{Q}{c^2}\exp(-i\beta\zeta a)\int_0^\infty \frac{\xi}{\xi^2 + (\gamma\zeta)^2 + (s/c)^2} J_1(a\xi)J_1(r\xi)d\xi \tag{7.3.19}$$

The product of two Bessel functions is replaced with its integral form. From the addition theorem of the Bessel function (Watson 1966, p. 358), we can easily derive the integral representation for the product as

$$J_1(a\xi)J_1(r\xi) = \frac{1}{\pi}\int_0^\pi J_0\left(\xi\sqrt{r^2 + a^2 - 2ar\cos\varphi}\right)\cos\varphi d\varphi \tag{7.3.20}$$

This equation is the same as Eq. (3.7.13) in Chap. 3. We substitute the above integral representation into Eq. (7.3.19) and exchange the order of integration. It follows that

$$U(r) = \frac{Q}{\pi c^2}\exp(-i\beta\zeta a)\int_0^\pi \cos\varphi d\varphi \int_0^\infty \xi \frac{J_0\left(\xi\sqrt{r^2 + a^2 - 2ar\cos\varphi}\right)}{\xi^2 + (\gamma\zeta)^2 + (s/c)^2} d\xi \tag{7.3.21}$$

The simple integration formula (Erdélyi 1954, vol. II, pp. 23, 12)

$$\int_0^\infty \frac{\xi}{\xi^2 + a^2} J_0(b\xi)d\xi = K_0(ab) \tag{7.3.22}$$

where $K_0(.)$ is the modified Bessel function of the first kind, is applied to Eq. (7.3.21). We have the single integral representation for $U(r)$,

$$U(r) = \frac{Q}{\pi c^2}\exp(-i\beta\zeta a)\int_0^\pi K_0\left(\sqrt{r^2 + a^2 - 2ar\cos\varphi}\sqrt{(\gamma\zeta)^2 + (s/c)^2}\right)\cos\varphi d\varphi \tag{7.3.23}$$

The product of Eq. (7.3.12) has just been determined by Eqs. (7.3.14) and (7.3.23). Thus the double transformed torsional displacement is given by

$$\hat{u}_\theta^* = \frac{Q}{\pi c^2}\exp\{+i\beta\zeta(r-a)\}\int_0^\pi K_0\left(\sqrt{r^2+a^2-2ar\cos\varphi}\sqrt{(\gamma\zeta)^2+(s/c)^2}\right)\cos\varphi d\varphi \tag{7.3.24}$$

Let us start for the inversion of the double transform. Firstly, we apply the Fourier inversion integral with respect to the parameter ζ to the above equation and exchange the order of integration. It yields

$$\begin{aligned} u_\theta^* = {} & \frac{Q}{\pi c^2}\int_0^\pi \cos\varphi d\varphi \\ & \times\frac{1}{\pi}\int_0^\infty K_0\left(\sqrt{r^2+a^2-2ar\cos\varphi}\sqrt{(\gamma\zeta)^2+(s/c)^2}\right)\cos[\zeta\{z-\beta(r-a)\}]d\zeta \end{aligned} \tag{7.3.25}$$

Applying the integration formula (Erdélyi 1954, p. 56, (43))

$$\int_0^\infty K_0\left(a\sqrt{x^2+b^2}\right)\cos(cx)dx = \frac{\pi}{2\sqrt{a^2+c^2}}\exp\left(-b\sqrt{a^2+c^2}\right) \tag{7.3.26}$$

to Eq. (7.3.25), we have

$$u_\theta^* = \frac{Q}{2\pi\gamma c^2}\int_0^\pi \frac{\exp\left[-(s/c)\sqrt{r^2+a^2-2ar\cos\varphi+\{z-\beta(r-a)\}^2}\right]}{\sqrt{r^2+a^2-2ar\cos\varphi+\{z-\beta(r-a)\}^2}}\cos\varphi d\varphi \tag{7.3.27}$$

Before going to the Laplace inversion, we make the change of variable for the integral, $\varphi \to t$,

$$t = \frac{1}{c}\sqrt{r^2+a^2-2ar\cos\varphi+\{z-\beta(r-a)\}^2} \tag{7.3.28}$$

The new variable t is a simple variable, not the real time at the present stage. But, at the next stage we understand it as the real time. With the change of the variable, Eq. (7.3.27) is simplified as

$$u_\theta^* = \frac{Q}{4\pi\gamma c}\frac{1}{ar}\int_{R_1/c}^{R_2/c}\left\{\sqrt{\frac{R_2^2-(ct)^2}{(ct)^2-R_1^2}}-\sqrt{\frac{(ct)^2-R_1^2}{R_2^2-(ct)^2}}\right\}\exp(-st)dt \tag{7.3.29}$$

where

$$R_1 = \sqrt{(r-a)^2+\{z-\beta(r-a)\}^2},\quad R_2 = \sqrt{(r+a)^2+\{z-\beta(r-a)\}^2} \tag{7.3.30}$$

Equation (7.3.29) is the Laplace transformed displacement and the integral is just in the form of the Laplace transform. Therefore, the original function of the displacement is its integrand and the Laplace inversion can be carried out by inspection. Finally, we have the exact closed form for the displacement. That is the Green's function for the torsional ring source in the monoclinic material,

$$u_\theta = \frac{Q}{4\pi\gamma c}H(ct-R_1)H(R_2-ct)\frac{1}{ar}\left\{\sqrt{\frac{R_2^2-(ct)^2}{(ct)^2-R_1^2}}-\sqrt{\frac{(ct)^2-R_1^2}{R_2^2-(ct)^2}}\right\} \tag{7.3.31}$$

The reader will find that this equation can be reduced to Eq. (3.7.26) in Chap. 3 when the material is isotropic, $\alpha \to 1$, $\beta \to 0\ (\gamma \to 1)$, $c \to c_s$.

7.4 Reflection of a Transient SH-Wave at a Moving Boundary

As an example of a moving boundary problem, wave reflection at a moving edge is discussed. The simplest single SH-wave is considered and a conversion formula between two Laplace transforms is developed so that the integral transform method can be applicable to the moving boundary problem.

Let us consider a semi-infinite elastic solid and define Cartesian coordinates (x, y) as shown in Fig. 7.4. The initial position of a moving edge of the half space is at $x = l$ and the edge moves toward the positive x-direction with velocity V. A point and stationary source of SH-wave is placed at the coordinate origin and is assumed as an impulsive body force with magnitude Q. The SH-wave field produced by the impulsive source is governed by the nonhomogeneous displacement equation,

$$\frac{\partial^2 u_z}{\partial x^2}+\frac{\partial^2 u_z}{\partial y^2} = \frac{1}{c_s^2}\frac{\partial^2 u_z}{\partial t^2}-\frac{Q}{c_s^2}\delta(x)\delta(y)\delta(t) \tag{7.4.1}$$

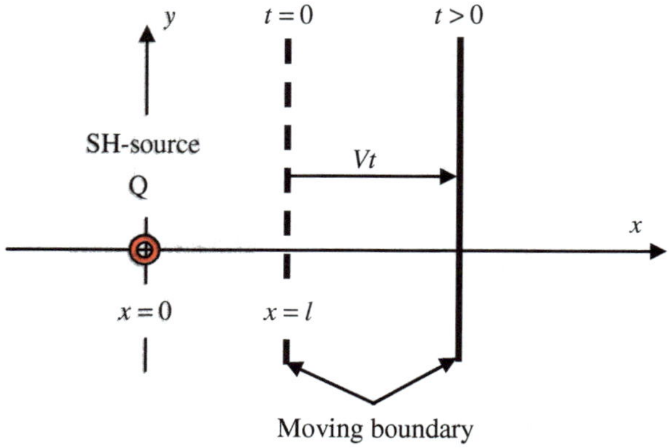

Fig. 7.4 Moving edge of a semi-infinite elastic solid

where u_z is the anti-plane displacement and c_s is SH-wave velocity. The stress components follow Hooke's law as

$$\sigma_{zx} = \mu \frac{\partial u_z}{\partial x}, \quad \sigma_{zy} = \mu \frac{\partial u_z}{\partial y} \tag{7.4.2}$$

where μ is rigidity.

We assume the stress-free condition at the moving edge $x = Vt + l$,

$$\sigma_{xz}|_{x=Vt+l} = 0 \tag{7.4.3}$$

The radiation condition at infinity

$$u_z|_{\sqrt{x^2+y^2}\to\infty} = \left.\frac{\partial u_z}{\partial x}\right|_{\sqrt{x^2+y^2}\to\infty} = \left.\frac{\partial u_z}{\partial y}\right|_{\sqrt{x^2+y^2}\to\infty} = 0 \tag{7.4.4}$$

and the quiescent condition at an initial time

$$u_z|_{t=0} = \left.\frac{\partial u_z}{\partial t}\right|_{t=0} = 0 \tag{7.4.5}$$

are also assumed. Equations (7.4.1)–(7.4.5) constitute the present elastodynamic problem.

The wave field can be decomposed into two parts: the incident wave u_{inc} and the reflected wave $u_{\text{refl.}}$. The total displacement (wave field) is the sum of the two,

$$u_z(x, y, t) = u_{\text{inc}}(x, y, t) + u_{\text{refl}}(x, y, t) \tag{7.4.6}$$

where u_{inc} and u_{refl} are a particular and the general solution of the non-homogeneous wave equation (7.4.1), respectively.

(1) Incident wave

The incident wave is generated by the impulsive source and radiates from the source point. The incident wave is given as the particular solution of the nonhomogeneous wave equation for the displacement,

$$\frac{\partial^2 u_{\text{inc}}}{\partial x^2}+\frac{\partial^2 u_{\text{inc}}}{\partial y^2}=\frac{1}{c_s^2}\frac{\partial^2 u_{\text{inc}}}{\partial t^2}-\frac{Q}{c_s^2}\delta(x)\delta(y)\delta(t) \tag{7.4.7}$$

The particular solution has already been obtained in Sect. 2.4, i.e.

$$u_{\text{inc}}(x,y,t)=\frac{Q}{2\pi c_s}\frac{1}{\sqrt{(c_s t)^2-(x^2+y^2)}}H\left(c_s t-\sqrt{x^2+y^2}\right) \tag{7.4.8}$$

However, we do not use this solution for the latter analysis, instead, the Fourier transformed solution is employed. We apply the triple integral transform to Eq. (7.4.7): Laplace transform with respect to the time,

$$f^*(s)=\int_0^\infty f(t)\exp(-st)dt,\quad f(t)=\frac{1}{2\pi i}\int_{Br(s)} f^*(s)\exp(+st)ds \tag{7.4.9}$$

where "*Br(s)*" in the inverse transform denotes the Bromwich line in the complex s-plane, and the double Fourier transform with respect to the space variables x and y,

$$\bar{f}(\xi)=\int_{-\infty}^{+\infty} f(x)\exp(+i\xi x)dx,\quad f(x)=\frac{1}{2\pi}\int_{-\infty}^{+\infty}\bar{f}(\xi)\exp(-i\xi x)d\xi \tag{7.4.10}$$

$$\tilde{f}(\eta)=\int_{-\infty}^{+\infty} f(y)\exp(+i\eta y)dy,\quad f(y)=\frac{1}{2\pi}\int_{-\infty}^{+\infty}\tilde{f}(\eta)\exp(-i\eta y)d\eta \tag{7.4.11}$$

The nonhomogeneous wave equation (7.3.7) is reduced to the simple algebraic equation for the transformed displacement,

$$-\{\xi^2+\eta^2+(s/c_s)^2\}\tilde{\bar{u}}^*_{\text{inc}}=-\frac{Q}{c_s^2} \tag{7.4.12}$$

and the particular solution in the transformed domain is then given by

$$\tilde{\bar{u}}^*_{\text{inc}} = \frac{Q}{c_s^2} \frac{1}{\xi^2 + \eta^2 + (s/c_s)^2} \tag{7.4.13}$$

Its formal Fourier inversion with respect to the parameter ξ is given by

$$\tilde{u}^*_{\text{inc}} = \frac{Q}{2\pi c_s^2} \int_{-\infty}^{+\infty} \frac{1}{\xi^2 + \eta^2 + (s/c_s)^2} \exp(-i\xi x) d\xi \tag{7.4.14}$$

and is easily evaluated by the application of the formula (2.1.22), to give

$$\tilde{u}^*_{\text{inc}} = \frac{Q}{2c_s^2 \alpha_s} \exp(-\alpha_s |x|) \tag{7.4.15}$$

where the radical α_s is defined by

$$\alpha_s = \sqrt{\eta^2 + (s/c_s)^2}, \quad \text{Re}(\alpha_s) > 0 \tag{7.4.16}$$

Furthermore, the formal Laplace inversion for the incident wave of Eq. (7.4.15) is given by the Bromwich integral,

$$\tilde{u}_{\text{inc}} = \frac{Q}{2c_s^2} \frac{1}{2\pi i} \int_{Br(s)} \frac{1}{\alpha_s} \exp(-\alpha_s |x| + st) ds \tag{7.4.17}$$

where $Br(s)$ denotes the Bromwich line from $\gamma_s - i\infty$ to $\gamma_s + i\infty$ in the complex s-plane. The transformed stress $\tilde{\sigma}_{xz}^{(\text{inc})}$ produced by the incident wave is also obtained in the form of Laplace inversion integral,

$$\tilde{\sigma}_{xz}^{(\text{inc})} = -\frac{Q}{2c_s^2} \text{sgn}(x) \frac{1}{2\pi i} \int_{Br(s)} \exp(-\alpha_s |x| + st) ds \tag{7.4.18}$$

where sgn(x) is the sign function defined by Eq. (7.1.17).

In the subsequent discussion for the stress-free condition, the integral form of Eq. (7.4.18) is convenient since we develop the conversion formula for two different Laplace transforms. Therefore, we shall stop the discussion for the incident wave.

(2) Reflected wave

The reflected wave is the general solution of the homogeneous displacement equation,

$$\frac{\partial^2 u_{\text{refl}}}{\partial x^2} + \frac{\partial^2 u_{\text{refl}}}{\partial y^2} = \frac{1}{c_s^2}\frac{\partial^2 u_{\text{refl}}}{\partial t^2} \tag{7.4.19}$$

The reflected wave is generated at the moving edge and runs toward the negative x-direction. So, we transform the Eq. (7.4.19) to the moving coordinate system (z, y, t) where the moving coordinate z is defined by

$$z = Vt + l - x \tag{7.4.20}$$

Due to this coordinate transform, the displacement of the reflected wave has the different set of three variables, as

$$u_{\text{refl.}}(x, y, t) \equiv u_{\text{refl.}}(z, y, t) \tag{7.4.21}$$

The wave equation (7.3.19) is transformed to

$$m^2\frac{\partial^2 u_{\text{refl}}}{\partial z^2} - \frac{2M}{c_2}\frac{\partial^2 u_{\text{refl}}}{\partial z \partial t} - \frac{1}{c_s^2}\frac{\partial^2 u_{\text{refl}}}{\partial t^2} + \frac{\partial^2 u_{\text{refl}}}{\partial y^2} = 0 \tag{7.4.22}$$

where Mach number M and its related parameter m are defined by

$$M = V/c_s, \quad m = \sqrt{1 - M^2} \tag{7.4.23}$$

In order to solve the differential equation (7.4.22), we apply the double integral transform: Laplace transform with respect to the time variable in the moving coordinate system. It is defined by

$$f^{\otimes}(p) = \int_0^{\infty} f(t)\exp(-pt)dt, \quad f(t) = \frac{1}{2\pi i}\int_{Br(p)} f^{\otimes}(p)\exp(pt)dp \tag{7.4.24}$$

Fourier transform with respect to the space variable y is defined by

$$\tilde{f}(\eta) = \int_{-\infty}^{+\infty} f(y)\exp(+i\eta y)dy, \quad f(y) = \frac{1}{2\pi}\int_{-\infty}^{+\infty} \tilde{f}(\eta)\exp(-i\eta y)d\eta \tag{7.4.25}$$

It should be noted that the Laplace transform is not the same as that in Eq. (7.4.9), but that the Fourier transform is the same as that in Eq. (7.4.11). This is because we have introduced the moving coordinate system along the x-axis, not along the

y-axis. Thus, it should be understood that the time variable t in the moving coordinate system is slightly different from the time in the fixed coordinate system since the variable z includes the time variable.

Using the quiescent condition at the initial time and the convergence condition at infinity, $y \to \pm\infty$, the wave equation (7.4.22) is transformed to the ordinary differential equation,

$$m^2 \frac{d^2 \tilde{u}_{\text{refl}}^{\otimes}}{dz^2} - 2(p/c_s)M \frac{d\tilde{u}_{\text{refl}}^{\otimes}}{dz} - \left\{\eta^2 + (p/c_s)^2\right\}\tilde{u}_{\text{refl}}^{\otimes} = 0 \qquad (7.4.26)$$

This solution which satisfies the convergence condition at $z \to +\infty$ is given by

$$\tilde{u}_{\text{refl}}^{\otimes} = A(\eta, p)\exp\left\{\left(-\beta_s + \frac{Mp}{mc_s}\right)\frac{z}{m}\right\} \qquad (7.4.27)$$

where $A(\eta, p)$ is an unknown coefficient and the radical β_s is defined by

$$\beta_s = \sqrt{\eta^2 + \left(\frac{p}{mc_s}\right)^2}, \quad \text{Re}(\beta_s) \geq 0 \qquad (7.4.28)$$

Then, the formal Laplace inversion of the reflected wave is given by the Bromwich integral,

$$\tilde{u}_{\text{refl}} = \frac{1}{2\pi i}\int_{Br(p)} A(\eta, p)\exp\left\{\left(-\beta_s + \frac{Mp}{mc_s}\right)\frac{z}{m}\right\}\exp(pt)dp \qquad (7.4.29)$$

where $Br(p)$ denotes the Bromwich line from $\gamma_p - i\infty$ to $\gamma_p + i\infty$ in the complex p-plane. In order to obtain the stress, we change the moving coordinate z back to the original one, $z = Vt + l - x$, to get

$$\tilde{u}_{\text{refl}} = \frac{1}{2\pi i}\int_{Br(p)} A(\eta, p)\exp\left\{\left(-\beta_s + \frac{Mp}{mc_s}\right)\frac{Vt + l - x}{m} + pt\right\}dp \qquad (7.4.30)$$

We then substitute this expression into the first of Eq. (7.4.2). The stress produced by the reflected wave is given by

$$\tilde{\sigma}_{xz}^{(\text{refl})} = \frac{1}{2\pi i}\int_{Br(p)} \frac{1}{m}\left(\beta_s - \frac{Mp}{mc_s}\right)A(\eta, p)\exp\left\{\frac{1}{m}\left(-\beta_s + \frac{Mp}{mc_s}\right)(Vt + l - x) + pt\right\}dp \qquad (7.4.31)$$

The unknown coefficient is determined by the stress-free condition on the moving edge. The stress is also the sum of two wave contributions, the incident and the reflected waves. Thus, the Fourier transformed boundary condition is given by the sum of each wave contribution,

$$\tilde{\sigma}_{xz} = \tilde{\sigma}_{xz}^{(\mathrm{inc})} + \tilde{\sigma}_{xz}^{(\mathrm{refl})} = 0; \quad \text{at } x = Vt + l \tag{7.4.32}$$

Substituting Eqs. (7.4.18) and (7.4.31) into the above condition, we obtain

$$\begin{aligned} &-\frac{Q}{2c_s^2}\frac{1}{2\pi i}\int_{Br(s)} \exp\{-\alpha_s(Vt+l)+st\}ds \\ &+\frac{1}{2\pi i}\int_{Br(p)} \frac{1}{m}\left(\beta_s - \frac{Mp}{mc_s}\right)A(\eta,p)\exp(pt)dp = 0 \end{aligned} \tag{7.4.33}$$

This is the integral equation for the unknown coefficient $A(\eta, p)$. The second integral on the left-hand side is in the form of Laplace inversion integral with respect to the parameter p, however, the first integral is not (in the form of Laplace inversion integral). We thus have two different Laplace inversion integrals in a single equation and it is not possible to apply any one of the Laplace transforms to reduce it to an algebraic equation for the unknown. Thus, the inversion integral in the first term must be converted to that with respect to the parameter p, i.e. the matching of the inversion integral.

Let us consider the complex integral Φ_{A} whose integrand is the same as that in the first integral in Eq. (7.4.33),

$$\Phi_{\mathrm{A}} = \frac{1}{2\pi i}\int_C \exp(-\alpha_s l)\exp\{(s-\alpha_s V)t\}ds \tag{7.4.34}$$

where the radical α_s is defined by Eq. (7.4.16). The integration loop C is discussed now. The integrand has two branch points $s = \pm i c_s \eta$ and corresponding two branch cuts are introduced along the imaginary axis in the complex s-plane. These are shown in Fig. 7.5. If we introduce the variable transform from s to the new variable p as

$$p = s - \alpha_s V \tag{7.4.35}$$

its inverse is given by

$$s/c_s = \frac{1}{m^2}\left\{p/c_s \pm M\sqrt{(m\eta)^2 + (p/c_s)^2}\right\} \tag{7.4.36}$$

In order to determine the multiple sign ($\pm$), we examine Eq. (7.4.35). We have the following asymptotic relation between two variables:

$$s \sim \frac{p}{1-M}; \quad |p| \to \infty \tag{7.4.37}$$

Fig. 7.5 Closed loop C for the complex integral Φ_A in Eq. (7.4.34)

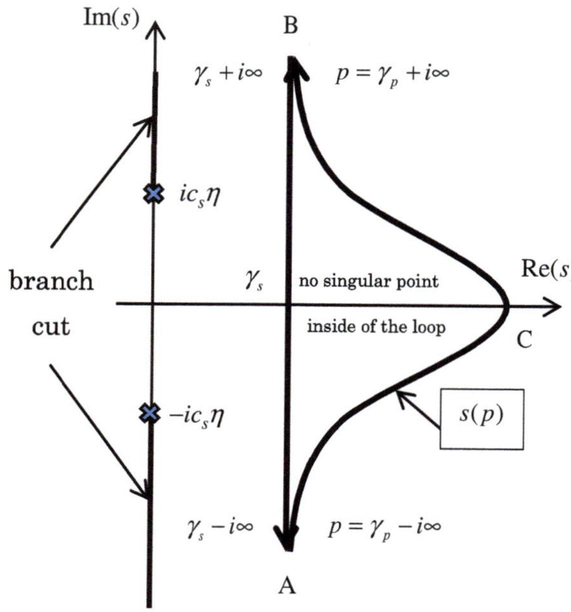

This relation must be hold for the inverse of the variable transform, i.e. we have to choose the positive sign in Eq. (7.4.36). Then, the suitable inversion for the variable transform is

$$s/c_s = \frac{1}{m^2}\left\{p/c_s + M\sqrt{(m\eta)^2 + (p/c_s)^2}\right\} \tag{7.4.38}$$

When the new variable p varies along the Bromwich line $Br(p)$ in the complex p-plane, i.e.

$$Br(p):\quad \gamma_p - i\infty \;\rightarrow\; \gamma_p + i\infty \tag{7.4.39}$$

where γ_p is a proper constant, the inverse s given by Eq. (7.4.38) moves along the bumped curve ACB in Fig. 7.6, but it is almost straight at infinity. At infinity, the inverse function takes the asymptotic form

$$s \sim \begin{cases} \dfrac{\gamma_p + i\infty}{1-M}; & p \rightarrow \gamma_p + i\infty \\ \dfrac{\gamma_p - i\infty}{1-M}; & p \rightarrow \gamma_p - i\infty \end{cases} \tag{7.4.40}$$

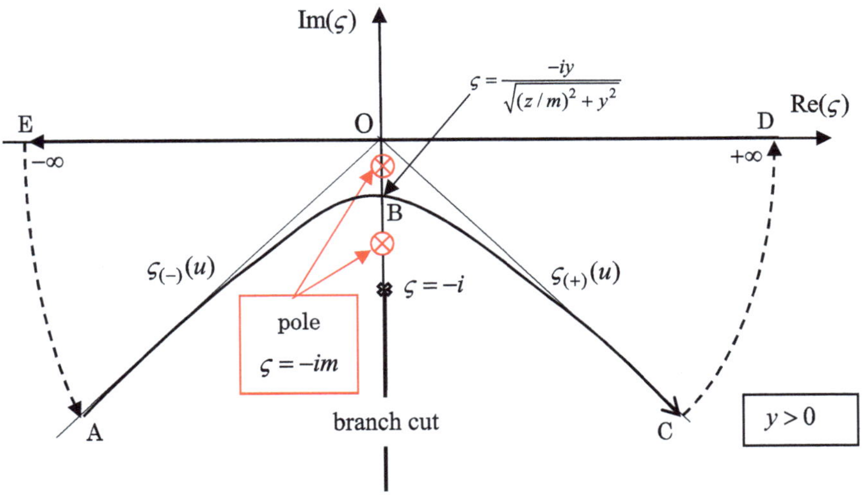

Fig. 7.6 Closed loop C for the complex integral Φ_B

So, if we take the real part of the Bromwich line $Br(s)$ from $\gamma_s - i\infty$ to $\gamma_s + i\infty$ as

$$\gamma_s = \frac{\gamma_p}{1 - M} \tag{7.4.41}$$

the bumped curve can be connected to the edge of the Bromwich line $Br(s)$ and then, these two curves constitute a closed loop $\overline{\text{ACBA}}$ as shown in the figure. Thus, we employ this closed loop C for the complex integral Φ_A in Eq. (7.4.34).

Fortunately, the real part of the two edges is sufficiently large and the two Bromwich lines

$$\begin{aligned} &Br(p): \gamma_p - i\infty < p < \gamma_p + i\infty \\ &Br(s): \gamma_s - i\infty < s < \gamma_s + i\infty \end{aligned} \tag{7.4.42}$$

have sufficiently large real parts so that all singular points of the integrand are in the left side of the Bromwich line $Br(s)$. Then, Cauchy's integral theorem can be applied to the complex integral. Since the closed loop $C(\overline{\text{ACBA}})$ does not include any singular point, the integral along $Br(s)$ is converted to a parametric integral along the bumped curve. That is,

$$\frac{1}{2\pi i}\int_{\gamma_s - i\infty}^{\gamma_s + i\infty} \exp(-\alpha_s l)\exp\{(s - \alpha_s V)t\}ds = \frac{1}{2\pi i}\int_{\gamma_p - i\infty}^{\gamma_p + i\infty}\left[\exp(-\alpha_s l)\frac{ds}{dp}\right]_{s=s(p)}\exp(pt)dp \tag{7.4.43}$$

where $s(p)$ is the inverse given by Eq. (7.4.38).

Consequently, we could convert the Laplace inversion integral with respect to the parameter s to that with respect to the parameter p and thus, Eq. (7.4.33) is rewritten in the form of the single Laplace inversion integral with respect to the parameter p, i.e.

$$\begin{aligned}&\frac{1}{2\pi i}\int_{Br(p)}\frac{1}{m}\left(\beta_s-\frac{Mp}{mc_s}\right)A(\eta,p)\exp(pt)dp\\&=\frac{Q}{2c_s^2}\frac{1}{2\pi i}\int_{Br(p)}\left[\exp(-\alpha_s l)\frac{ds}{dp}\right]_{s=s(p)}\exp(pt)dp\end{aligned}\tag{7.4.44}$$

Both sides of the above equation have the same form of Laplace inversion integral with respect to the single parameter p. Applying the Laplace transform with respect to the time t to both sides, we obtain the simple algebraic equation for the unknown coefficient,

$$\frac{1}{m}\left(\beta_s-\frac{Mp}{mc_s}\right)A(\eta,p)=\frac{Q}{2c_s^2}\left[\exp(-\alpha_s l)\frac{ds}{dp}\right]_{s=s(p)}\tag{7.4.45}$$

Thus, the coefficient is determined as

$$A(\eta,p)=\frac{Q}{2c_s^2}\frac{m}{\beta_s-(Mp)/(mc_s)}\left[\exp(-\alpha_s l)\frac{ds}{dp}\right]_{s=s(p)}\tag{7.4.46}$$

Using the relations derived from Eqs. (7.4.35) and (7.4.38), the radical and the gradient are given by

$$\alpha_s=\frac{1}{m^2}\{M(p/c_s)+m\beta_s\}\tag{7.4.47}$$

$$\frac{ds(p)}{dp}=\frac{M(p/mc_s)+\beta_s}{m^2\beta_s}\tag{7.4.48}$$

The coefficient is then rewritten in the explicit form,

$$A(\eta,p)=\frac{Q}{2c_s^2}\frac{1}{m\beta_s}\frac{\beta_s+M(p/mc_s)}{\beta_s-M(p/mc_s)}\exp\left\{-\frac{M(p/c_s)+m\beta_s}{m^2}l\right\}\tag{7.4.49}$$

Lastly, substituting Eq. (7.4.49) into Eq. (7.4.27), the double transformed displacement produced by the reflected wave is given by

$$\tilde{u}^{\otimes}_{\text{refl}}=\frac{Q}{2c_s^2}\frac{1}{m\beta_s}\frac{\beta_s+M(p/mc_s)}{\beta_s-M(p/mc_s)}\exp\left\{-\frac{M(p/c_s)+m\beta_s}{m^2}l\right\}\exp\left\{\left(-\beta_s+\frac{Mp}{mc_s}\right)\frac{z}{m}\right\}\tag{7.4.50}$$

(3) Cagniard's technique

We have obtained the displacement produced by the reflected wave, but in the transformed domain. As the incident wave has already been obtained in the explicit form by Eq. (7.4.8), our next task is to perform the double inversion for the displacement in Eq. (7.4.50). The double inversion is carried out by applying the Cagniard-de Hoop technique. The formal Fourier inversion with respect to the parameter η is given by

$$u_{\text{refl}}^{\otimes} = \frac{Q}{4\pi c_s^2} \int_{-\infty}^{+\infty} \frac{1}{m\beta_s} \frac{\beta_s + M(p/mc_s)}{\beta_s - M(p/mc_s)} \exp\left\{ -\frac{M(p/c_s)(l-z)}{m^2} - \frac{\beta_s(z+l)}{m} - i\eta y \right\} d\eta \tag{7.4.51}$$

Since the radical $\beta_s = \sqrt{\eta^2 + (p/mc_s)^2}$ has the Laplace transform parameter p, we introduce the variable transform from η to the new variable ς as

$$\eta = (p/mc_s)\varsigma \tag{7.4.52}$$

Equation (7.4.51) is rewritten as

$$u_{\text{refl}}^{\otimes} = \frac{Q}{4\pi m c_s^2} \int_{-\infty}^{+\infty} \frac{1}{\sqrt{\varsigma^2+1}} \frac{\sqrt{\varsigma^2+1}+M}{\sqrt{\varsigma^2+1}-M} \exp\left[-(p/mc_s)\left\{ \frac{M(l-z)}{m} + \frac{z+l}{m}\sqrt{\varsigma^2+1} + i\varsigma y \right\} \right] d\varsigma \tag{7.4.53}$$

Now, let us consider the complex integral Φ_{B} whose integrand is the same as that of Eq. (7.4.53),

$$\Phi_{\text{B}} = \frac{1}{2\pi} \int_C \frac{1}{\sqrt{\varsigma^2+1}} \frac{\sqrt{\varsigma^2+1}+M}{\sqrt{\varsigma^2+1}-M} \exp\left[-(p/mc_s)\left\{ \frac{M(l-z)}{m} + \frac{z+l}{m}\sqrt{\varsigma^2+1} + i\varsigma y \right\} \right] d\varsigma \tag{7.4.54}$$

The closed loop C is discussed here. The integrand has two branch points at $\varsigma = \pm i$, and two branch cuts are thus introduced along the imaginary axis in the complex ς-plane. The vanishing point of the denominator, $\sqrt{\varsigma^2+1} - M = 0$, gives two poles at $\varsigma = \pm im$ on the imaginary axis; but they are smaller in magnitude than the branch points.

The Cagniard's path is determined by the variable transform from ς to the new variable t as

$$t = \frac{1}{mc_s}\left\{\frac{M(l-z)}{m} + \frac{z+l}{m}\sqrt{\varsigma^2+1} + i\varsigma y\right\} \tag{7.4.55}$$

Solving for ς, the Cagniard's path is given by

$$\varsigma_s^{(\pm)} = \frac{-imy\{m^2c_st - M(l-z)\} \pm (z+l)\sqrt{\{m^2c_st - M(l-z)\}^2 - \{(z+l)^2+(my)^2\}}}{(z+l)^2+(my)^2} \tag{7.4.56}$$

The saddle point of the Cagniard's path is

$$\varsigma_{saddle} = -i\frac{my}{\sqrt{(z+l)^2+(my)^2}} \tag{7.4.57}$$

at

$$t = \frac{1}{m^2c_s}\left\{M(l-z) + \sqrt{(z+l)^2+(my)^2}\right\} \tag{7.4.58}$$

If we take the closed loop composed of the Cagniard's path, the straight line along the real axis and two large arcs which connect the straight line with the Cagniard's path, the pole $\varsigma = -im$ is inside of the loop when

$$m < \frac{m|y|}{\sqrt{(z+l)^2+(my)^2}} \Rightarrow |y| > \frac{z+l}{M} \tag{7.4.59}$$

but outside when

$$m > \frac{m|y|}{\sqrt{(z+l)^2+(my)^2}} \Rightarrow |y| < \frac{z+l}{M} \tag{7.4.60}$$

The location of the pole depends on this inequality. The contribution from the pole at $\varsigma = \pm im$ is included only when the inequality of Eq. (7.4.59) holds. Thus, we determine the closed loop C as $\overline{\text{ABCDOEA}}$ shown in Fig. 7.6. Applying Cauchy's theorem to the complex integral Φ_B in Eq. (7.4.54), the line integral along the real axis is converted to the sum of the integral along the Cagniard's path and the residue at the pole. That is

$$
\begin{aligned}
&\frac{1}{2\pi}\int_{-\infty}^{+\infty}\frac{1}{\sqrt{\varsigma^2+1}}\frac{\sqrt{\varsigma^2+1}+M}{\sqrt{\varsigma^2+1}-M}\exp\left[-(p/mc_s)\left\{\frac{M(l-z)}{m}+\frac{z+l}{m}\sqrt{\varsigma^2+1}+i\varsigma y\right\}\right]d\varsigma \\
&=\frac{1}{2\pi}\int_{\frac{M(l-z)+\sqrt{(z+l)^2+(my)^2}}{m^2c_s}}^{\infty}\left[\left\{\frac{1}{\sqrt{\varsigma^2+1}}\frac{\sqrt{\varsigma^2+1}+M}{\sqrt{\varsigma^2+1}-M}\frac{d\varsigma}{dt}\right\}_{\varsigma=\varsigma_s^{(+)}}-\left\{\frac{1}{\sqrt{\varsigma^2+1}}\frac{\sqrt{\varsigma^2+1}+M}{\sqrt{\varsigma^2+1}-M}\frac{d\varsigma}{dt}\right\}_{\varsigma=\varsigma_s^{(+)}}\right]e^{-pt}dt \\
&-\frac{2\pi i}{2\pi}\left[\frac{\varsigma+im}{\sqrt{\varsigma^2+1}}\frac{\sqrt{\varsigma^2+1}+M}{\sqrt{\varsigma^2+1}-M}\exp\left[-(p/mc_s)\left\{\frac{M(l-z)}{m}+\frac{z+l}{m}\sqrt{\varsigma^2+1}+i\varsigma y\right\}\right]\right]_{\varsigma=-im}H\left(|y|-\frac{z+l}{M}\right)
\end{aligned}
\tag{7.4.61}
$$

Rearranging the second residue term in the right hand side, we obtain the somewhat simpler form,

$$
\begin{aligned}
&\frac{1}{2\pi}\int_{-\infty}^{+\infty}\frac{1}{\sqrt{\varsigma^2+1}}\frac{\sqrt{\varsigma^2+1}+M}{\sqrt{\varsigma^2+1}-M}\exp\left[-(p/mc_s)\left\{\frac{M(l-z)}{m}+\frac{z+l}{m}\sqrt{\varsigma^2+1}+i\varsigma y\right\}\right]d\varsigma \\
&=\frac{1}{2\pi}\int_{\frac{M(l-z)+\sqrt{(z+l)^2+(my)^2}}{m^2c_s}}^{\infty}\left[\left\{\frac{1}{\sqrt{\varsigma^2+1}}\frac{\sqrt{\varsigma^2+1}+M}{\sqrt{\varsigma^2+1}-M}\frac{d\varsigma}{dt}\right\}_{\varsigma=\varsigma_s^{(+)}}-\left\{\frac{1}{\sqrt{\varsigma^2+1}}\frac{\sqrt{\varsigma^2+1}+M}{\sqrt{\varsigma^2+1}-M}\frac{d\varsigma}{dt}\right\}_{\varsigma=\varsigma_s^{(+)}}\right]e^{-pt}dt \\
&+\frac{2M}{m}\exp\left\{-\frac{p}{mc_s}\left(\frac{2Ml}{m}+my\right)\right\}H\left(|y|-\frac{z+l}{M}\right)
\end{aligned}
\tag{7.4.62}
$$

Substituting this equation into Eq. (7.4.53), we obtain the *Laplace transformed* displacement as the sum of the *Laplace transform integral* and the simple exponential function,

$$
\begin{aligned}
u_{\text{refl}}^{\otimes}=\frac{Q}{4\pi mc_s^2}&\int_{\frac{M(l-z)+\sqrt{(z+l)^2+(my)^2}}{m^2c_s}}^{\infty}\left[\begin{array}{l}\left\{\frac{1}{\sqrt{\varsigma^2+1}}\frac{\sqrt{\varsigma^2+1}+M}{\sqrt{\varsigma^2+1}-M}\frac{d\varsigma}{dt}\right\}_{\varsigma=\varsigma_s^{(+)}}\\-\left\{\frac{1}{\sqrt{\varsigma^2+1}}\frac{\sqrt{\varsigma^2+1}+M}{\sqrt{\varsigma^2+1}-M}\frac{d\varsigma}{dt}\right\}_{\varsigma=\varsigma_s^{(+)}}\end{array}\right]e^{-pt}dt \\
&+\frac{MQ}{m^2c_s^2}\exp\left\{-\frac{p}{mc_s}\left(\frac{2Ml}{m}+my\right)\right\}H\left(|y|-\frac{z+l}{M}\right)
\end{aligned}
\tag{7.4.63}
$$

The first term is just in the form of Laplace transform integral and is easily inverted by inspection. The second term is the exponential function and the simple Laplace inversion formula,

$$
L^{-1}[\exp(-ap)]=\delta(t-a) \tag{7.4.64}
$$

is applied. Thus, the Laplace inversion is easily carried out and we get the final form for the displacement produced by the reflected wave as

$$u_{\mathrm{refl}} = \frac{Q}{4\pi m c_s^2} H\left(t - \frac{M(l-z) + \sqrt{(z+l)^2 + (my)^2}}{m^2 c_s}\right) \\ \times \left[\left\{\frac{1}{\sqrt{\varsigma^2+1}}\frac{\sqrt{\varsigma^2+1}+M}{\sqrt{\varsigma^2+1}-M}\frac{d\varsigma}{dt}\right\}_{\varsigma=\varsigma_s^{(+)}} - \left\{\frac{1}{\sqrt{\varsigma^2+1}}\frac{\sqrt{\varsigma^2+1}+M}{\sqrt{\varsigma^2+1}-M}\frac{d\varsigma}{dt}\right\}_{\varsigma=\varsigma_s^{(+)}}\right] \\ - \frac{MQ}{m^2 c_s^2}\delta\left(t - \frac{1}{m c_s}\left(\frac{2Ml}{m} + my\right)\right) H\left(|y| - \frac{z+l}{M}\right) \tag{7.4.65}$$

Fortunately, we can simplify the above equation and get an exact closed form solution. Substituting Eq. (7.4.56) into Eq. (7.4.55), we have for the radical

$$\sqrt{\varsigma^2+1} = \frac{(z+l)\{m^2 c_s t - M(l-z)\} \mp i(my)\sqrt{\{m^2 c_s t - M(l-z)\}^2 - \{(z+l)^2 + (my)^2\}}}{\left\{(z+l)^2 + (my)^2\right\}} \tag{7.4.66}$$

and thus we have the simple expression for the complex valued function as

$$\left\{\frac{1}{\sqrt{\varsigma^2+1}}\frac{\sqrt{\varsigma^2+1}+M}{\sqrt{\varsigma^2+1}-M}\frac{d\varsigma}{dt}\right\}_{\varsigma=\varsigma_s^{(+)}} - \left\{\frac{1}{\sqrt{\varsigma^2+1}}\frac{\sqrt{\varsigma^2+1}+M}{\sqrt{\varsigma^2+1}-M}\frac{d\varsigma}{dt}\right\}_{\varsigma=\varsigma_s^{(+)}} \\ = \frac{2m^2 c_s}{\sqrt{T^2 - R^2}}\frac{T^2 - M^2R^2 - (my)^2}{T^2 - 2M(z+l)T + M^2R^2 - (my)^2} \tag{7.4.67}$$

where

$$T = m^2 c_s t - M(l-z), \quad R = \sqrt{(z+l)^2 + (my)^2} \tag{7.4.68}$$

Then, Eq. (7.4.65) is simplified as the exact closed form solution for the reflected wave,

$$\frac{c_s}{Q} u_{\mathrm{refl}} = \frac{m}{2\pi} H(T-R) \frac{1}{\sqrt{T^2 - R^2}} \frac{T^2 - M^2R^2 - (my)^2}{T^2 - 2M(z+l)T + M^2R^2 - (my)^2} \\ - \frac{M}{m}\delta\left(m(c_s t) - \left(\frac{2Ml}{m} + my\right)\right) H\left(|y| - \frac{z+l}{M}\right) \tag{7.4.69}$$

(4) Wave fronts

Disturbed regions and wave fronts are easily obtained. They are given by the conditionals of the step and delta functions in Eq. (7.4.69). Returning to the original fixed coordinate from the moving coordinate, we find that the step function attached to the first term in Eq. (7.4.69) gives the disturbed region by the reflected wave:

$$W_{\text{refl.}} : H\left(t - \frac{M(l-z) + \sqrt{(z+l)^2 + (my)^2}}{m^2 c_s}\right) \Rightarrow \left(c_s t - \frac{2lM}{m^2}\right)^2 \geq \left(x - \frac{2l}{m^2}\right)^2 + y^2 \tag{7.4.70}$$

This is a circular region, but not a full circle since the edge is moving and the conditional for x is imposed as $x \leq Mc_s t + l$. This reflected wave is shown in Fig. 7.7. The second term includes the product of delta and step functions.

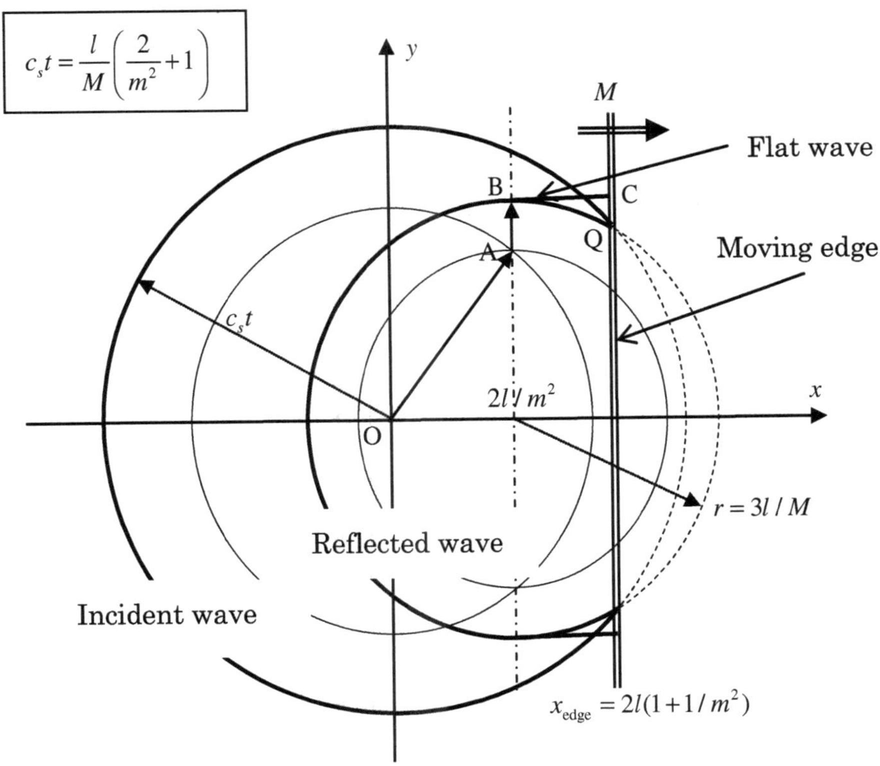

Fig. 7.7 Wave fronts for incident and reflected waves at the time $c_s t = \frac{l}{M}\left(\frac{2}{m^2}+1\right)$

$$\begin{aligned} W_{\text{flat}}&: \delta\left(t - \frac{1}{mc_s}\left(\frac{2Ml}{m} + my\right)\right)H\left(|y| - \frac{z+l}{M}\right) \\ &\Rightarrow \delta\left(c_s t - \frac{2Ml}{m^2} - y\right)H(M|y| - x - Mc_s t - 2l) \end{aligned} \tag{7.4.71}$$

Examining the product, we find that the delta function gives a line singular wave and the step function restricts its line segment BC (in the figure). Thus, the wave is given as

$$\text{Flat wave: } W_{\text{flat}}: c_s t = y + \frac{2Ml}{m^2} \tag{7.4.72}$$

$$\text{Supporting region: } |y| > -\frac{x - 2l}{M} + c_s t \tag{7.4.73}$$

The incident wave has already been given by Eq. (7.4.8) and it is the circular region. Its front is given by

$$W_{\text{inc}}: H\left(c_s t - \sqrt{x^2 + y^2}\right) \Rightarrow (c_s t)^2 \geq x^2 + y^2; x < M(c_s t) + l \tag{7.4.74}$$

Figure 7.7 shows the typical wave front shape for the incident and reflected waves.

7.5 Wave Scattering by a Rigid Inclusion in an Inhomogeneous Elastic Solid

The problem of wave scattering is one of the traditional elastodynamic problems. However, as far as the author knows, no exact closed form solution for transient (impulsive) waves is known for isotropic homogeneous elastic and acoustic materials (Friedlander 1958; Pao and Mow 1973). Only one exact closed form solution (Green's function) was obtained for SH-waves in an inhomogeneous elastic solid (Watanabe 1982). The present section shows the exact closed form of Green's function for the wave scattering problem in a radially inhomogeneous elastic solid.

Let us consider a radially inhomogeneous elastic solid with a borehole and take the cylindrical coordinate (r, θ, z). We assume that the shear modulus μ and the density ρ are functions of the radial distance r from the coordinate origin as

$$\mu \equiv \mu_0 (r/a)^{\lambda}, \quad \rho \equiv \rho_0 (r/a)^{\lambda - 2} \tag{7.5.1}$$

where μ_0 and ρ_0 are the reference values at $r = a$ and λ is an inhomogeneity parameter. Upon this assumption, the shear wave velocity is the linear function of the radial distance as

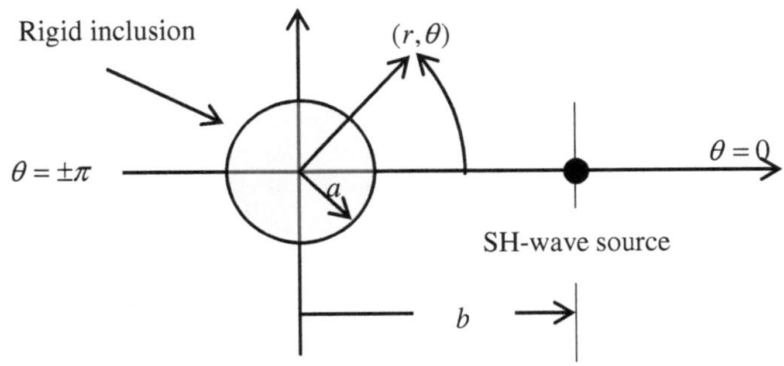

Fig. 7.8 A circular rigid inclusion and a SH-wave source in an inhomogeneous elastic solid

$$c_s = \sqrt{\mu/\rho} = c_{s0}(r/a); \quad c_{s0} = \sqrt{\mu_0/\rho_0} \tag{7.5.2}$$

A rigid circular (cylindrical) inclusion with radius a is placed at the coordinate origin and SH-wave source is also placed at $(r = b, \ \theta = 0)$ as shown in Fig. 7.8. Assuming the anti-plane deformation, i.e. SH-wave problem, the equation of motion and Hooke's law are given by

$$\frac{\partial \sigma_{rz}}{\partial r} + \frac{1}{r}\sigma_{rz} + \frac{1}{r}\frac{\partial \sigma_{\theta z}}{\partial \theta} + \rho B_z = \rho \frac{\partial^2 u_z}{\partial t^2} \tag{7.5.3}$$

$$\sigma_{rz} = \mu \frac{\partial u_z}{\partial r}, \quad \sigma_{\theta z} = \mu \frac{1}{r}\frac{\partial u_z}{\partial \theta} \tag{7.5.4}$$

where u_z and $\sigma_{rz}, \sigma_{\theta z}$ are the anti-plane displacement and its associated stress components. The body force B_z is assumed as the point/line wave source,

$$B_z = B_0 \frac{\delta(r-b)}{r} \delta(\theta)\delta(t) \tag{7.5.5}$$

where B_0 is the magnitude of the source and $\delta(.)$ is Dirac's delta function.

As the rigid circular inclusion is inserted in the hole, the elastic solid occupies its outer region $a < r < \infty$ and the anti-plane displacement is fixed at the boundary, i.e. the fixed boundary condition for the elastic solid, is

$$u_z|_{r=a} = 0; \quad 0 \le t < \infty \tag{7.5.6}$$

In addition to the above boundary condition, we employ the radiation condition at infinity as

$$u_z|_{r\to\infty} = \left.\frac{\partial u_z}{\partial r}\right|_{r\to\infty} = 0 \tag{7.5.7}$$

The above equations from (7.5.1) to (7.5.7) constitute our wave scattering problem. Substituting Eqs. (7.5.4) and (7.5.5) into the equation of motion (7.5.3), we have the single displacement equation as

$$\frac{\partial^2 u_z}{\partial r^2}+\left(\frac{\mu'}{\mu}+\frac{1}{r}\right)\frac{\partial u_z}{\partial r}+\frac{1}{r^2}\frac{\partial^2 u_z}{\partial \theta^2}-\frac{\rho}{\mu}\frac{\partial^2 u_z}{\partial t^2}=-\frac{\rho}{\mu}B_0\frac{\delta(r-b)}{r}\delta(\theta)\delta(t) \tag{7.5.8}$$

where the prime denotes the differentiation with respect to the radial variable as

$$\mu'=\frac{d\mu}{dr} \tag{7.5.9}$$

Recalling Eqs. (7.5.1) and (7.5.2), the displacement Eq. (7.5.8) is rewritten as

$$\frac{\partial^2 u_z}{\partial r^2}+\frac{\lambda+1}{r}\frac{\partial u_z}{\partial r}+\frac{1}{r^2}\frac{\partial^2 u_z}{\partial \theta^2}-\frac{1}{c_{s0}^2}\left(\frac{a}{r}\right)^2\frac{\partial^2 u_z}{\partial t^2}=-\frac{B_0}{c_{s0}^2}\left(\frac{a}{r}\right)^2\frac{\delta(r-b)}{r}\delta(\theta)\delta(t) \tag{7.5.10}$$

Now, we start to solve the above differential equation. The variable transform for the radial variable

$$z=\log(r/a) \quad \Leftrightarrow \quad r=a\exp(z)=ae^z \tag{7.5.11}$$

is introduced. Equation (7.5.10) is reduced to the nonhomogeneous differential equation with constant coefficients as

$$\frac{\partial^2 u_z}{\partial z^2}+\lambda\frac{\partial u_z}{\partial z}+\frac{\partial^2 u_z}{\partial \theta^2}-\left(\frac{a}{c_{s0}}\right)^2\frac{\partial^2 u_z}{\partial t^2}=-B_0\left(\frac{a}{c_{s0}}\right)^2\frac{\delta(ae^z-b)}{ae^z}\delta(\theta)\delta(t) \tag{7.5.12}$$

Then, we apply the Laplace transform with respect to the time t and the Fourier finite transform with respect to the circumferential angle θ. They are defined in Chap. 1 as

$$\text{Laplace transform:} f^*(s)=\int_0^\infty f(t)\exp(-st)dt \tag{7.5.13}$$

$$\text{Fourier finite transform:} f_n=\int_{-\pi}^{\pi} f(\theta)\exp(+in\theta)d\theta \Leftrightarrow f(\theta)=\frac{1}{2\pi}\sum_{n=-\infty}^{+\infty} f_n\exp(-in\theta) \tag{7.5.14}$$

Applying the double transform with the quiescent condition at an initial time,

$$u_z|_{t=0} = \left.\frac{\partial u_z}{\partial t}\right|_{t=0} = 0 \tag{7.5.15}$$

Equation (7.5.12) is reduced to the simple ordinary differential equation,

$$\frac{d^2 u^*_{zn}}{dz^2} + \lambda \frac{du^*_{zn}}{dz} - \{n^2 + (as/c_{s0})^2\}u^*_{zn} = -B_0\left(\frac{a}{c_{s0}}\right)^2 \frac{\delta(ae^z - b)}{ae^z} \tag{7.5.16}$$

The homogeneous solution which satisfies the radiation condition (7.5.7) is easily obtained as

$$u^{*(\mathrm{h})}_{zn} = C_n(s)\exp\left[-z\left\{\sqrt{n^2 + (as/c_{s0})^2 + (\lambda/2)^2} + \lambda/2\right\}\right] \tag{7.5.17}$$

where $C_n(s)$ is an unknown coefficient to be determined by the boundary condition.

In order to obtain a particular solution corresponding to the wave source B_0, we apply the Fourier transform with respect to the variable z to the ordinary differential equation (7.5.16). The transform is defined by

$$\hat{f}(\zeta) = \int_{-\infty}^{\infty} f(z)\exp(+i\zeta z)dz \Leftrightarrow f(z) = \frac{1}{2\pi}\int_{-\infty}^{\infty} \hat{f}(\zeta)\exp(-i\zeta z)d\zeta \tag{7.5.18}$$

We have the simple algebraic equation for the triple transformed displacement,

$$\left\{\zeta^2 + i\lambda\zeta + n^2 + (as/c_{s0})^2\right\}\hat{u}^{*(\mathrm{p})}_{zn} = B_0\left(\frac{a}{bc_{s0}}\right)^2 \exp\{+i\zeta\log(b/a)\} \tag{7.5.19}$$

where the source (nonhomogeneous) term is evaluated as

$$\begin{aligned} & B_0\left(\frac{a}{c_{s0}}\right)^2 \int_{-\infty}^{\infty} \frac{\delta(ae^z - b)}{ae^z}\exp(+i\zeta z)dz \\ &= B_0\left(\frac{a}{c_{s0}}\right)^2 \int_{0}^{\infty} \frac{\delta(r - b)}{r}\exp\{+i\zeta\log(r/a)\}\frac{dr}{r} \\ &= B_0\left(\frac{a}{bc_{s0}}\right)^2 \exp\{+i\zeta\log(b/a)\} \end{aligned} \tag{7.5.20}$$

Thus, we have the displacement in the triple transformed domain,

$$\hat{u}_{zn}^{*(\mathrm{p})} = B_0 \left(\frac{a}{bc_{s0}} \right)^2 \frac{\exp\{+i\zeta \log(b/a)\}}{(\zeta + i\lambda/2)^2 + n^2 + (as/c_{s0})^2 + (\lambda/2)^2} \tag{7.5.21}$$

Now, let us consider the inversion. Firstly, we apply the Fourier inversion integral with respect to the parameter ζ defined by the second of Eq. (7.5.18). It is

$$u_{zn}^{*(\mathrm{p})} = B_0 \left(\frac{a}{bc_{s0}} \right)^2 \frac{1}{2\pi} \int_{-\infty}^{\infty} \frac{\exp[-i\zeta\{z - \log(b/a)\}]}{(\zeta + i\lambda/2)^2 + n^2 + (as/c_{s0})^2 + (\lambda/2)^2} d\zeta \tag{7.5.22}$$

If we make the change of variable from ζ to η as $\eta = \zeta + i\lambda/2$, the integrand in Eq. (7.5.22) is simplified, but, its path of integration is the complex line from $-\infty + i\lambda/2$ to $+\infty + i\lambda/2$, not a real path.

$$\begin{aligned} u_{zn}^{*(\mathrm{p})} &= B_0 \left(\frac{a}{bc_{s0}} \right)^2 \exp[-(\lambda/2)\{z - \log(b/a)\}] \\ &\times \frac{1}{2\pi} \int_{-\infty+i\lambda/2}^{\infty+i\lambda/2} \frac{\exp[-i\eta\{z - \log(b/a)\}]}{\eta^2 + n^2 + (as/c_{s0})^2 + (\lambda/2)^2} d\eta \end{aligned} \tag{7.5.23}$$

In order to have a real line integral, we consider the complex integral whose integrand is the same as that in Eq. (7.5.23). That is

$$\Phi = \frac{1}{2\pi} \int_L \frac{\exp[-i\eta\{z - \log(b/a)\}]}{\eta^2 + n^2 + (as/c_{s0})^2 + (\lambda/2)^2} d\eta \tag{7.5.24}$$

The integration loop L is a rectangular form, as shown in Fig. 7.9. It is composed of two parallel horizontal lines $\overrightarrow{\mathrm{AB}}$ and $\overrightarrow{\mathrm{CD}}$, and two finite vertical lines $\overrightarrow{\mathrm{DA}}$ and $\overrightarrow{\mathrm{BC}}$. The edges of the horizontal line tend to infinite, i.e. $\pm\infty + i\lambda/2$. Examining the singularity of the integrand, we find two poles at

$$\eta = \pm i\sqrt{n^2 + (as/c_{s0})^2 + (\lambda/2)^2} \tag{7.5.25}$$

which are outside of the loop L and the integrand tends to zero at the infinity $\eta \to \pm\infty + i\lambda/2$. Then, we apply the Cauchy theorem to the complex integral Φ. Since the integrals along the two vertical lines $\overrightarrow{\mathrm{DA}}$ and $\overrightarrow{\mathrm{BC}}$ vanish and no singular point is included in the loop, the integral along the upper horizontal line $\overrightarrow{\mathrm{AB}}$ is converted to that along the lower horizontal line $\overrightarrow{\mathrm{DC}}$, i.e. the real axis in the η-plane. Thus we have the conversion formula for our integral as

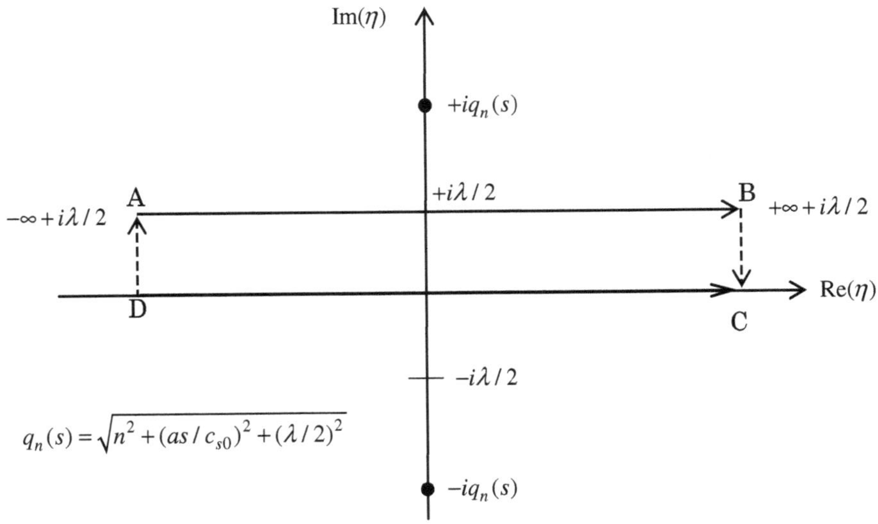

Fig. 7.9 Integration loop L for the complex integral Φ

$$\frac{1}{2\pi}\int_{-\infty+i\lambda/2}^{\infty+i\lambda/2}\frac{\exp[-i\eta\{z-\log(b/a)\}]}{\eta^2+n^2+(as/c_{s0})^2+(\lambda/2)^2}d\eta = \frac{1}{2\pi}\int_{-\infty}^{\infty}\frac{\exp[-i\eta\{z-\log(b/a)\}]}{\eta^2+n^2+(as/c_{s0})^2+(\lambda/2)^2}d\eta \qquad (7.5.26)$$

The infinite integral in the right hand side of the above equation is reduced to the semi-infinite integral

$$\frac{1}{2\pi}\int_{-\infty}^{\infty}\frac{\exp[-i\eta\{z-\log(b/a)\}]}{\eta^2+n^2+(as/c_{s0})^2+(\lambda/2)^2}d\eta = \frac{1}{\pi}\int_{0}^{\infty}\frac{\cos[\eta\{z-\log(b/a)\}]}{\eta^2+n^2+(as/c_{s0})^2+(\lambda/2)^2}d\eta \qquad (7.5.27)$$

and the simple integration formula (Erdélyi 1954, vol. I, p. 8, (11))

$$\int_0^{\infty}\frac{1}{x^2+\alpha^2}\cos(xy)dx = \frac{\pi}{2\alpha}\exp(-\alpha|y|) \qquad (7.5.28)$$

is applied. Thus, we can evaluate our integral as

$$\frac{1}{2\pi}\int_{-\infty+i\lambda/2}^{\infty+i\lambda/2}\frac{\exp[-i\eta\{z-\log(b/a)\}]}{\eta^2+n^2+(as/c_{s0})^2+(\lambda/2)^2}d\eta=\frac{1}{2q_n(s)}\exp\{-q_n(s)|z-\log(b/a)|\}\tag{7.5.29}$$

where

$$q_n(s)=\sqrt{n^2+(as/c_{s0})^2+(\lambda/2)^2}\tag{7.5.30}$$

Substituting Eq. (7.5.29) into Eq. (7.5.23), the double transformed particular solution is given by

$$u_{zn}^{*(\mathrm{p})}=\frac{B_0}{2}\left(\frac{a}{bc_{s0}}\right)^2\frac{1}{q_n(s)}\exp[-(\lambda/2)\{z-\log(b/a)\}]\exp\{-q_n(s)|z-\log(b/a)|\}\tag{7.5.31}$$

Summing the homogeneous and particular solutions, Eqs. (7.5.17) and (7.5.31), we have the full solution for the differential equation (7.5.16). That is

$$\begin{aligned}u_{zn}^{*}&=u_{zn}^{*(\mathrm{h})}+u_{zn}^{*(\mathrm{p})}\\&=C_n(s)\exp[-z\{q_n(s)+\lambda/2\}]\\&\quad+\frac{B_0}{2}\left(\frac{a}{bc_{s0}}\right)^2\frac{1}{q_n(s)}\exp[-(\lambda/2)\{z-\log(b/a)\}]\exp\{-q_n(s)|z-\log(b/a)|\}\end{aligned}\tag{7.5.32}$$

In order to determine the coefficient $C_n(s)$, we apply the boundary condition, Eq. (7.5.6). The double transformed boundary condition with the change of variable is given by

$$u_{zn}^{*}\big|_{z=0}=0\tag{7.5.33}$$

Substituting Eq. (7.5.32) into the above equation, we have for the coefficient,

$$C_n(s)=-\frac{B_0}{2}\left(\frac{a}{bc_{s0}}\right)^2\frac{1}{q_n(s)}\exp[\{(\lambda/2)-q_n(s)\}\log(b/a)\}]\tag{7.5.34}$$

and thus the displacement in the double transformed domain is completely determined,

$$u_{zn}^* = \frac{B_0}{2}\left(\frac{a}{bc_{s0}}\right)^2 \exp[-(\lambda/2)\{z - \log(b/a)\}]$$
$$\times \left\langle \frac{1}{q_n(s)}\exp\{-q_n(s)|z - \log(b/a)|\} - \frac{1}{q_n(s)}\exp[-q_n(s)\{z + \log(b/a)\}] \right\rangle \tag{7.5.35}$$

For the simplification in the subsequent analysis, we return the variable z to the original radial distance r. The transformed solution is rewritten as

$$u_{zn}^* = \frac{B_0}{2}\left(\frac{a}{bc_{s0}}\right)^2 \left(\frac{b}{r}\right)^{\lambda/2} \left[\frac{\exp\{-q_n(s)|\log(r/b)|\}}{q_n(s)} - \frac{\exp\{-q_n(s)\log(br/a^2)\}}{q_n(s)}\right] \tag{7.5.36}$$

We have just obtained the solution in the double transformed domain. Our subsequent work is to invert the solution into the actual (r, θ, t) space. As the first inversion, we consider its Laplace inversion. The Laplace inversion is symbolically written as

$$u_{zn} = \frac{B_0}{2}\left(\frac{a}{bc_{s0}}\right)^2 \left(\frac{b}{r}\right)^{\lambda/2} \left\langle L^{-1}\left[\frac{\exp\{-q_n(s)|\log(r/b)|\}}{q_n(s)}\right] - L^{-1}\left[\frac{\exp\{-q_n(s)\log(br/a^2)\}}{q_n(s)}\right] \right\rangle \tag{7.5.37}$$

Fortunately, since $q_n(s)$ is the simple square root function given by Eq. (7.5.30), we can apply the inversion formula (Erdélyi 1954, vol, I, p. 248, (24)),

$$L^{-1}\left[\frac{1}{\sqrt{s^2 + a^2}}\exp\left(-b\sqrt{s^2 + a^2}\right)\right] = H(t - b)J_0\left(a\sqrt{t^2 - b^2}\right) \tag{7.5.38}$$

where $H(.)$ is Heaviside's unit step function and $J_0(.)$ is Bessel function of the zeroth order. It follows that

$$u_{zn} = \frac{B_0}{2}\frac{a}{b^2 c_{s0}}\left(\frac{b}{r}\right)^{\lambda/2} \left[H(\tau - |\log(r/b)|)J_0\left(\sqrt{n^2 + (\lambda/2)^2}\sqrt{\tau^2 - \log^2(r/b)}\right)\right.$$
$$\left. - H(\tau - \log(br/a^2))J_0\left(\sqrt{n^2 + (\lambda/2)^2}\sqrt{\tau^2 - \log^2(br/a^2)}\right)\right] \tag{7.5.39}$$

where the dimensionless time, $\tau = c_{s0}t/a$, is introduced. The next is the inversion of the Fourier finite transform, i.e. Fourier series. Applying the inversion formula in Eqs. (7.5.14)–(7.5.39), we have for the displacement

$$
\begin{aligned}
u_z &= \frac{B_0}{2}\frac{a}{b^2 c_{s0}}\left(\frac{b}{r}\right)^{\lambda/2} \\
&\times \left[H(\tau - |\log(r/b)|)\frac{1}{2\pi}\sum_{n=-\infty}^{+\infty} J_0\left(\sqrt{n^2 + (\lambda/2)^2}\sqrt{\tau^2 - \log^2(r/b)}\right)\exp(-in\theta) \right. \\
&\left. \quad - H\left(\tau - \log(br/a^2)\right)\frac{1}{2\pi}\sum_{n=-\infty}^{+\infty} J_0\left(\sqrt{n^2 + (\lambda/2)^2}\sqrt{\tau^2 - \log^2(br/a^2)}\right)\exp(-in\theta) \right]
\end{aligned}
\tag{7.5.40}
$$

Since both coefficients in the two series include only even power of the index n, the Fourier series is reduced to the semi-infinite sum as

$$
\begin{aligned}
u_z &= \frac{B_0}{2\pi}\frac{a}{b^2 c_{s0}}\left(\frac{b}{r}\right)^{\lambda/2} \\
&\times \left[H(\tau - |\log(r/b)|)\sum_{n=0}^{\infty} \varepsilon_n J_0\left(\sqrt{n^2 + (\lambda/2)^2}\sqrt{\tau^2 - \log^2(r/b)}\right)\cos(n\theta) \right. \\
&\left. \quad - H\left(\tau - \log(br/a^2)\right)\sum_{n=0}^{\infty} \varepsilon_n J_0\left(\sqrt{n^2 + (\lambda/2)^2}\sqrt{\tau^2 - \log^2(br/a^2)}\right)\cos(n\theta) \right]
\end{aligned}
\tag{7.5.41}
$$

where

$$
\varepsilon_n = \begin{cases} 1/2; & n = 0 \\ 1; & n \geq 1 \end{cases} \tag{7.5.42}
$$

Formally we have just arrived at the exact solution for the scattering problem, however, this solution does not show wave propagation phenomena clearly. In order to obtain the solution which shows the wave phenomena, i.e. wave front shape, we have to discuss the Fourier series. Let us consider the Fourier series defined by

$$
S = \sum_{n=0}^{\infty} J_0\left(\alpha\sqrt{n^2 + \beta^2}\right)\cos(n\theta) \tag{7.5.43}
$$

where α and β are positive constants. We replace the Bessel function with its integral form (Erdélyi 1954, vol. I, p. 28, (42)),

$$
J_0\left(\alpha\sqrt{n^2 + \beta^2}\right) = \frac{2}{\pi}\int_0^{\alpha} \frac{\cos\left(\beta\sqrt{\alpha^2 - \xi^2}\right)}{\sqrt{\alpha^2 - \xi^2}}\cos(n\xi)d\xi \tag{7.5.44}
$$

and change the order of the summation and integration. Then, expanding the integration range, two summations are unified as

$$\begin{aligned} S &= \frac{1}{\pi}\int_0^{\alpha} \frac{\cos\left(\beta\sqrt{\alpha^2-\xi^2}\right)}{\sqrt{\alpha^2-\xi^2}} d\xi \sum_{n=0}^{\infty} \varepsilon_n[\cos\{n(\theta+\xi)\} + \cos\{n(\theta-\xi)\}] \\ &= \frac{1}{\pi}\int_{-\alpha}^{+\alpha} \frac{\cos\left(\beta\sqrt{\alpha^2-\xi^2}\right)}{\sqrt{\alpha^2-\xi^2}} d\xi \sum_{n=0}^{\infty} \varepsilon_n \cos\{n(\xi-\theta)\} \end{aligned} \tag{7.5.45}$$

The Fourier series in the integrand is just the form of Fourier expansion of an infinite periodic array of the delta function (see Exercises (1.1) and (1.2)). That is

$$\sum_{n=0}^{\infty} \varepsilon_n \cos\{n(\xi-\theta)\} = \pi \sum_{m=-\infty}^{+\infty} \delta(\xi-\theta+2m\pi) \tag{7.5.46}$$

We substitute the above equation into the last line in Eq. (7.5.45) and change the order of integration and summation. It follows that

$$S = \sum_{m=-\infty}^{+\infty} \int_{-\alpha}^{+\alpha} \frac{\cos\left(\beta\sqrt{\alpha^2-\xi^2}\right)}{\sqrt{\alpha^2-\xi^2}} \delta(\xi-\theta+2m\pi) d\xi \tag{7.5.47}$$

Examining the supporting region for the delta function, we can evaluate the integral with use of the simple formula (1.2.3) in Chap. 1. It results

$$S = \sum_{m=-\infty}^{+\infty} H(\alpha-|\theta-2m\pi|) \frac{\cos\left(\beta\sqrt{\alpha^2-(\theta-2m\pi)^2}\right)}{\sqrt{\alpha^2-(\theta-2m\pi)^2}} \tag{7.5.48}$$

The above equation looks like the sum of the infinite terms. But, due to the operation of the step function, this summation is truncated to the finite sum. Thus, we have just reduced the infinite sum of the Fourier series to the finite sum of simple algebraic function. Further, if we expand the range of the circumferential/polar angle θ from the finite range $(-\pi<\theta<+\pi)$ to the infinite range $(-\infty<\theta<+\infty)$ and rewrite the angle

$$\theta \equiv \theta - 2m\pi \tag{7.5.49}$$

The summation with the index m in Eq. (7.5.48) is reduced to the single term. That is

$$S = H(\alpha - \theta)\frac{\cos\left(\beta\sqrt{\alpha^2 - \theta^2}\right)}{\sqrt{\alpha^2 - \theta^2}}; \quad -\infty < \theta < +\infty \tag{7.5.50}$$

Consequently, we have obtained the very simple expression for the Fourier series. That is the summation formula,

$$\sum_{n=0}^{\infty} J_0\left(\alpha\sqrt{n^2 + \beta^2}\right)\cos(n\theta) = H(\alpha - \theta)\frac{\cos\left(\beta\sqrt{\alpha^2 - \theta^2}\right)}{\sqrt{\alpha^2 - \theta^2}}; \quad -\infty < \theta < +\infty \tag{7.5.51}$$

Substituting this formula into Eq. (7.5.41), we have the closed form for the displacement

$$\begin{aligned} u_z = {} & \frac{B_0}{2\pi}\frac{a}{b^2 c_{s0}}\left(\frac{b}{r}\right)^{\lambda/2} \\ & \times \left[H(\tau - |\log(r/b)|)H(\sqrt{\tau^2 - \log^2(r/b)} - |\theta|)\frac{\cos\left((\lambda/2)\sqrt{\tau^2 - \log^2(r/b) - \theta^2}\right)}{\sqrt{\tau^2 - \log^2(r/b) - \theta^2}} \right. \\ & \left. - H(\tau - \log(br/a^2))H(\sqrt{\tau^2 - \log^2(br/a^2)} - |\theta|)\frac{\cos\left((\lambda/2)\sqrt{\tau^2 - \log^2(br/a^2) - \theta^2}\right)}{\sqrt{\tau^2 - \log^2(br/a^2) - \theta^2}} \right] \end{aligned} \tag{7.5.52}$$

The product of two step functions in the above equation can be unified and then we have the final form for the displacement, i.e. the Green function for the scattering problem,

$$\begin{aligned} u_z = {} & \frac{B_0}{2\pi}\frac{a}{b^2 c_{s0}}\left(\frac{b}{r}\right)^{\lambda/2} \\ & \left[H\left(\tau - \sqrt{\log^2(r/b) + \theta^2}\right)\frac{\cos\left((\lambda/2)\sqrt{\tau^2 - \log^2(r/b) - \theta^2}\right)}{\sqrt{\tau^2 - \log^2(r/b) - \theta^2}} \right. \\ & \left. - H\left(\tau - \sqrt{\log^2(br/a^2) + \theta^2}\right)\frac{\cos\left((\lambda/2)\sqrt{\tau^2 - \log^2(br/a^2) - \theta^2}\right)}{\sqrt{\tau^2 - \log^2(br/a^2) - \theta^2}} \right] \end{aligned} \tag{7.5.53}$$

The first term in the bracket shows the direct wave from the source and its wave front is given by

$$\tau = \sqrt{\log^2(r/b) + \theta^2} \quad \Rightarrow \quad r/b = \exp\left(\pm\sqrt{\tau^2 - \theta^2}\right) \tag{7.5.54}$$

The second term shows the scattering wave and its front is given by

$$\tau = \sqrt{\log^2(br/a^2) + \theta^2} \quad \Rightarrow \quad r/a = (a/b)\exp\left(\pm\sqrt{\tau^2 - \theta^2}\right) \tag{7.5.55}$$

The both equations show the spiral curves in the (r, θ)-plane and the time-development of these wave fronts can be found in the paper (Watanabe 1982).

7.6 An Excellent Application of Cauchy Complex Integral

The present section does not show any Green's function, but a very excellent application of the Cauchy complex integral is described. This application technique is originated by Mal (1970), when he considered the wave scattering by a crack. The dynamic crack problem is reduced to the Fredholm integral equation whose kernel is in the form of semi-infinite integral with the product of two Bessel functions,

$$K(x,y) = \int_0^\infty \left\{\sqrt{\xi^2 - (\omega/c)^2} - \xi\right\} J_0(x\xi) J_0(y\xi) d\xi; \quad x < y \tag{7.6.1}$$

where $J_0(.)$ is the Bessel function of the first kind. As we have no suitable integration formula for this semi-infinite integral, we have to evaluate it numerically. However, the integrand shows very slow convergence and its numerical evaluation takes much CPU time for sufficient accuracy. If we could transform the semi-infinite integral to finite integrals, we can evaluate the kernel without regard to the convergence and accuracy. Mal (1970) has reduced the integral to the finite one such as

$$K(x,y) = +i \int_0^{\omega/c} \sqrt{(\omega/c)^2 - \xi^2} J_0(x\xi) H_0^{(2)}(y\xi) d\xi; \quad x < y \tag{7.6.2}$$

where $H_0^{(2)}(.)$ is Hankel function of the second kind. After Mal, many researchers have employed his evaluation technique for the dynamic crack and punch problems. But, they (including Mal) did not describe its mathematics in detail. Thus, applying

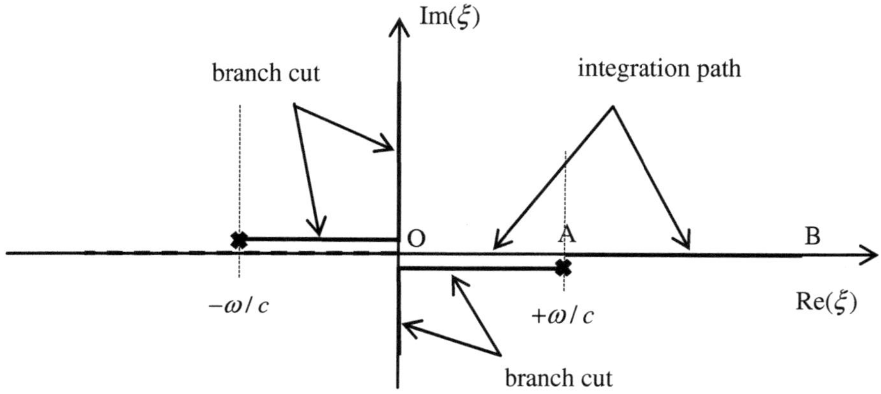

Fig. 7.10 Integration path in the complex ξ-plane

our former discussion on the branch cut in Sect. 1.3.2 in Chap. 1, his technique is explained in some depths and then one can learn the power of the Cauchy complex integral.

As the kernel function of Eq. (7.6.1) is the same in the form, regardless of the positive and negative time factors, we assume the positive time factor $\exp(+i\omega t)$ and the radiation/convergence condition at infinity

$$\mathrm{Re}\left(\sqrt{\xi^2 - (\omega/c)^2}\right) \geq 0 \tag{7.6.3}$$

Due to these assumptions, the branch cuts in the complex ξ-plane are introduced as was discussed in Sect. 1.3.2 in Chap. 1. The cut and integration path for the semi-infinite integral of the kernel are shown in Fig. 7.10, where the cut is the bold line and the integration path is the semi-infinite line $\overrightarrow{\mathrm{OAB}}$. Since the integration path is on the positive real axis, we decompose the integral into two parts: one is the finite integral on $\overrightarrow{\mathrm{OA}}(0 \leq \xi < \omega/c)$ and the other on $\overrightarrow{\mathrm{AB}}\,(\omega/c < \xi < \infty)$. As was discussed in Sect. 1.3.2, the square root function on the path $\overrightarrow{\mathrm{OA}}$ takes the positive imaginary values as

$$\sqrt{\xi^2 - (\omega/c)^2} = +i\sqrt{(\omega/c)^2 - \xi^2}; \quad 0 \leq \xi < \omega/c \tag{7.6.4}$$

and that on the path $\overrightarrow{\mathrm{AB}}$ does the positive real value as

$$\sqrt{\xi^2 - (\omega/c)^2} = +\sqrt{\xi^2 - (\omega/c)^2}; \quad \omega/c < \xi < \infty \tag{7.6.5}$$

So, the kernel $K(x, y)$ of Eq. (7.6.1) is decomposed into two integrals as

$$\begin{aligned} K(x,y) = & \int_{0}^{\omega/c} \left\{ +i\sqrt{(\omega/c)^2 - \xi^2} - \xi \right\} J_0(x\xi) J_0(y\xi) d\xi \\ & + \int_{\omega/c}^{\infty} \left\{ \sqrt{\xi^2 - (\omega/c)^2} - \xi \right\} J_0(x\xi) J_0(y\xi) d\xi \end{aligned} \tag{7.6.6}$$

The first integral in the above equation is in the form of the finite integral, but the second one on the path $\overrightarrow{\mathrm{AB}}$ is still in that of the semi-infinite integral. We have to transform the latter integral to the finite ones. This is the main purpose of the present section.

Since the relation between Bessel and Hankel functions exists,

$$J_0(Z) = \frac{1}{2}\left\{ H_0^{(1)}(Z) + H_0^{(2)}(Z) \right\}; \quad H_0^{(1/2)}(Z) = J_0(Z) \pm iY_0(Z) \tag{7.6.7}$$

the latter integral in Eq. (7.6.6) is separated into two semi-infinite integrals as

$$\begin{aligned} I = & \int_{\omega/c}^{\infty} \left\{ \sqrt{\xi^2 - (\omega/c)^2} - \xi \right\} J_0(x\xi) J_0(y\xi) d\xi \\ = & \frac{1}{2} \int_{\omega/c}^{\infty} \left\{ \sqrt{\xi^2 - (\omega/c)^2} - \xi \right\} J_0(x\xi) H_0^{(1)}(y\xi) d\xi \\ & + \frac{1}{2} \int_{\omega/c}^{\infty} \left\{ \sqrt{\xi^2 - (\omega/c)^2} - \xi \right\} J_0(x\xi) H_0^{(2)}(y\xi) d\xi \end{aligned} \tag{7.6.8}$$

We consider two complex integrals whose integrands are the same as those in the above equation, respectively, and are defined by

$$\Phi_1 = \frac{1}{2} \oint_{\Gamma_1} \left\{ \sqrt{\zeta^2 - (\omega/c)^2} - \zeta \right\} J_0(x\zeta) H_0^{(1)}(y\zeta) d\zeta \tag{7.6.9}$$

$$\Phi_2 = \frac{1}{2} \oint_{\Gamma_2} \left\{ \sqrt{\zeta^2 - (\omega/c)^2} - \zeta \right\} J_0(x\zeta) H_0^{(2)}(y\zeta) d\zeta \tag{7.6.10}$$

We assume the real and imaginary parts of the complex variable ζ are ξ and η, respectively, i.e. $\zeta = \xi + i\eta$. Both integrands have two branch points at $\zeta = \pm\omega/c$ for the square root function, and one branch point at the origin $\zeta = 0$ for the

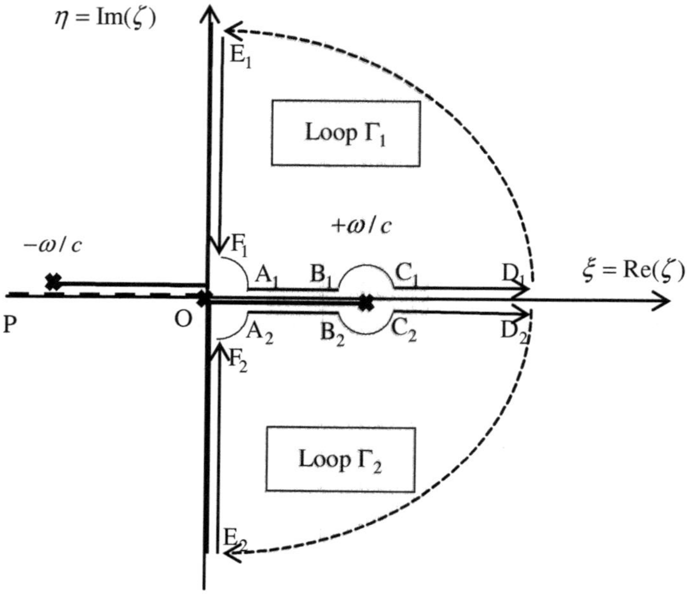

Fig. 7.11 Integration contours Γ_1 and Γ_1 for the complex integrals Φ_1 and Φ_2

logarithmic nature of the Hankel function. In order to examine the convergence at infinity, we apply the asymptotic expression of the Bessel function (Watson 1966, pp. 198–199) to the product of Bessel and Hankel functions and have

$$J_0(x\zeta)H_0^{(1)}(y\zeta)\Big|_{|\zeta|\to\infty} \sim \frac{1}{\pi\zeta\sqrt{xy}}[\exp\{+i(y-x)\zeta\} - i\exp\{+i(y+x)\zeta\}] \quad (7.6.11)$$

$$J_0(x\zeta)H_0^{(2)}(y\zeta)\Big|_{|\zeta|\to\infty} \sim \frac{1}{\pi\zeta\sqrt{xy}}[\exp\{-i(y-x)\zeta\} + i\exp\{-i(y+x)\zeta\}] \quad (7.6.12)$$

We learn that the product of Bessel and Hankel functions shows the exponential decay at infinity. Since $y > x$, the product $J_0(x\zeta)H_0^{(1)}(y\zeta)$ converges in the upper ζ-plane, i.e. $\mathrm{Im}(\zeta) > 0$, and $J_0(x\zeta)H_0^{(2)}(y\zeta)$ does in the lower ζ-plane, i.e. $\mathrm{Im}(\zeta) < 0$. Two branch cuts for the square root function are the same as those in Fig. 7.10, and the branch cut for the Hankel function is introduced along the negative real axis starting from the origin, i.e. the dotted line $\overline{OP}$ in Fig. 7.11. So, we take the closed loop Γ_1 in the first quadrant and Γ_2 in the fourth quadrant as shown in the figure. These two loops are composed of line paths and circular arcs: They are

$$\Gamma_1{:}\overrightarrow{A_1B_1}+\overset{\frown}{B_1C_1}+\overrightarrow{C_1D_1}+\overset{\frown}{D_1E_1}+\overrightarrow{E_1F_1}+\overset{\frown}{F_1A_1} \tag{7.6.13}$$

$$\Gamma_2{:}\overrightarrow{A_2B_2}+\overset{\frown}{B_2C_2}+\overrightarrow{C_2D_2}+\overset{\frown}{D_2E_2}+\overrightarrow{E_2F_2}+\overset{\frown}{F_2A_2} \tag{7.6.14}$$

The complex integral is decomposed into the integrals along these path segments.

$$\Phi_1=\frac{1}{2}\left(\int\limits_{\overrightarrow{A_1B_1}}+\int\limits_{\overset{\frown}{B_1C_1}}+\int\limits_{\overrightarrow{C_1D_1}}+\int\limits_{\overset{\frown}{D_1E_1}}+\int\limits_{\overrightarrow{E_1F_1}}+\int\limits_{\overset{\frown}{F_1A_1}}\right)\left\{\sqrt{\zeta^2-(\omega/c)^2}-\zeta\right\}J_0(x\zeta)H_0^{(1)}(y\zeta)d\zeta \tag{7.6.15}$$

$$\Phi_2=\frac{1}{2}\left(\int\limits_{\overrightarrow{A_2B_2}}+\int\limits_{\overset{\frown}{B_2C_2}}+\int\limits_{\overrightarrow{C_2D_2}}+\int\limits_{\overset{\frown}{D_2E_2}}+\int\limits_{\overrightarrow{E_2F_2}}+\int\limits_{\overset{\frown}{F_2A_2}}\right)\left\{\sqrt{\zeta^2-(\omega/c)^2}-\zeta\right\}J_0(x\zeta)H_0^{(2)}(y\zeta)d\zeta \tag{7.6.16}$$

The radii of the quarter- and semi-circular paths, $\overset{\frown}{B_jC_j}$ and $\overset{\frown}{F_jA_j}$; $j = 1, 2$, are very small and tend to zero as the limit. Thus, the integrals on these small circular paths vanish when we take their zero limit of the radius. On the other hand, due to the exponential decay of the product of Bessel and Hankel functions, the integral on the quarter-circle $\overset{\frown}{D_jE_j}\,; j=1,2$ vanishes as the radius of the quarter circle tends to infinity. Consequently, since no singular point such as the pole is included in each loop, the complex integral around the closed loop vanishes, i.e. $\Phi_j = 0$, and thus the integral on the line paths $\overrightarrow{C_jD_j}, j=1,2$ is transformed to the sum of two integrals on the paths $\overrightarrow{A_jB_j}$ and $\overrightarrow{E_jF_j}$,

$$\begin{aligned}&\frac{1}{2}\int\limits_{\overrightarrow{C_1D_1}}\left\{\sqrt{\zeta^2-(\omega/c)^2}-\zeta\right\}J_0(x\zeta)H_0^{(1)}(y\zeta)d\zeta\\&=-\frac{1}{2}\left(\int\limits_{\overrightarrow{A_1B_1}}+\int\limits_{\overrightarrow{E_1F_1}}\right)\left\{\sqrt{\zeta^2-(\omega/c)^2}-\zeta\right\}J_0(x\zeta)H_0^{(1)}(y\zeta)d\zeta\end{aligned} \tag{7.6.17}$$

Table 7.1 Variable transforms on the line path

Path	Integration variable ζ and its range	Square root function $\sqrt{\zeta^2-(\omega/c)^2}$	Product of Bessel and Hankel functions $J_0(x\zeta)H_0^{(1/2)}(y\zeta)$
$\overrightarrow{A_1B_1}$	$\zeta=\xi$ $0<\xi<\omega/c$	$+i\sqrt{(\omega/c)^2-\xi^2}$	$J_0(x\xi)H_0^{(1)}(y\xi)$
$\overrightarrow{C_1D_1}$	$\zeta=\xi$ $\omega/c<\xi<\infty$	$\sqrt{\xi^2-(\omega/c)^2}$	$J_0(x\xi)H_0^{(1)}(y\xi)$
$\overrightarrow{E_1F_1}$	$\zeta=+i\eta;\ +i=e^{+\pi i/2}$ $0<\eta<\infty$	$+i\sqrt{\eta^2+(\omega/c)^2}$	$-\frac{2i}{\pi}I_0(x\eta)K_0(y\eta)$
$\overrightarrow{A_2B_2}$	$\zeta=\xi$ $0<\xi<\omega/c$	$-i\sqrt{(\omega/c)^2-\xi^2}$	$J_0(x\xi)H_0^{(2)}(y\xi)$
$\overrightarrow{C_2D_2}$	$\zeta=\xi$ $\omega/c<\xi<\infty$	$\sqrt{\xi^2-(\omega/c)^2}$	$J_0(x\xi)H_0^{(2)}(y\xi)$
$\overrightarrow{E_2F_2}$	$\zeta=-i\eta;\ -i=e^{-\pi i/2}$ $0<\eta<\infty$	$-i\sqrt{\eta^2+(\omega/c)^2}$	$+\frac{2i}{\pi}I_0(x\eta)K_0(y\eta)$

$$
\begin{aligned}
&\frac{1}{2}\int_{\overrightarrow{C_2D_2}}\left\{\sqrt{\zeta^2-(\omega/c)^2}-\zeta\right\}J_0(x\zeta)H_0^{(2)}(y\zeta)d\zeta\\
&=-\frac{1}{2}\left(\int_{\overrightarrow{A_2B_2}}+\int_{\overrightarrow{E_2F_2}}\right)\left\{\sqrt{\zeta^2-(\omega/c)^2}-\zeta\right\}J_0(x\zeta)H_0^{(2)}(y\zeta)d\zeta
\end{aligned}
\tag{7.6.18}
$$

The values of the square root function and integration variable on each line path are tabulated in Table 7.1, where the product of Bessel and Hankel functions is derived by using the formulas

$$
J_0(\pm ix)=I_0(x),\quad H_0^{(1)}(\pm ix)=-\frac{2i}{\pi}K_0(x),\quad H_0^{(2)}(-ix)=+\frac{2i}{\pi}K_0(x) \tag{7.6.19}
$$

where $I_0(.)$ and $K_0(.)$ are the modified Bessel function of the first and second kinds. Then, we can rewrite Eqs. (7.6.17) and (7.6.18). They are

$$
\begin{aligned}
&\frac{1}{2}\int_{\omega/c}^{\infty}\left\{\sqrt{\xi^2-(\omega/c)^2}-\xi\right\}J_0(x\xi)H_0^{(1)}(y\xi)d\xi\\
&=-\frac{1}{2}\int_0^{\omega/c}\left\{+i\sqrt{(\omega/c)^2-\xi^2}-\xi\right\}J_0(x\xi)H_0^{(1)}(y\xi)d\xi\\
&\quad+\frac{i}{\pi}\int_0^{\infty}\left\{\sqrt{\eta^2+(\omega/c)^2}-\eta\right\}I_0(x\eta)K_0(y\eta)d\eta
\end{aligned}
\tag{7.6.20}
$$

$$\frac{1}{2}\int_{\omega/c}^{\infty}\left\{\sqrt{\xi^2-(\omega/c)^2}-\xi\right\}J_0(x\xi)H_0^{(2)}(y\xi)d\xi$$
$$=+\frac{1}{2}\int_{0}^{\omega/c}\left\{i\sqrt{(\omega/c)^2-\xi^2}+\xi\right\}J_0(x\xi)H_0^{(2)}(y\xi)d\xi \tag{7.6.21}$$
$$-\frac{i}{\pi}\int_{0}^{\infty}\left\{\sqrt{\eta^2+(\omega/c)^2}-\eta\right\}I_0(x\eta)K_0(y\eta)d\eta$$

Adding two Eqs. (7.6.20) and (7.6.20), we have two finite integrals for the semi-infinite integral in Eq. (7.6.8) as

$$\int_{\omega/c}^{\infty}\left\{\sqrt{\xi^2-(\omega/c)^2}-\xi\right\}J_0(x\xi)J_0(y\xi)d\xi$$
$$=-\frac{1}{2}\int_{0}^{\omega/c}\left\{i\sqrt{(\omega/c)^2-\xi^2}-\xi\right\}J_0(x\xi)H_0^{(1)}(y\xi)d\xi \tag{7.6.22}$$
$$+\frac{1}{2}\int_{0}^{\omega/c}\left\{i\sqrt{(\omega/c)^2-\xi^2}+\xi\right\}J_0(x\xi)H_0^{(2)}(y\xi)d\xi$$

This equation is simplified by using the formula (7.6.7) for the Hankel function. It yields

$$\int_{\omega/c}^{\infty}\left\{\sqrt{\xi^2-(\omega/c)^2}-\xi\right\}J_0(x\xi)J_0(y\xi)d\xi$$
$$=\int_{0}^{\omega/c}\left\{\xi J_0(y\xi)+\sqrt{(\omega/c)^2-\xi^2}Y_0(y\xi)\right\}J_0(x\xi)d\xi \tag{7.6.23}$$

The semi-infinite integral is just transformed to the finite integral and thus we can replace the second integral in the right hand side of Eq. (7.6.6) with the above finite integral. Therefore, the kernel function of the semi-infinite integral is reduced to the finite one, i.e.

$$K(x,y)=+i\int_{0}^{\omega/c}\sqrt{(\omega/c)^2-\xi^2}J_0(x\xi)H_0^{(2)}(y\xi)d\xi;\quad y>x \tag{7.6.24}$$

This is the same as Eq. (7.6.2).

When we evaluate the kernel function, the change of variable from ξ to u,

$$\xi = (\omega/c)u \tag{7.6.25}$$

is introduced and the numerical computation is carried out for the simplified integral,

$$K(x,y) = +i(\omega/c)\int_0^1 \sqrt{1-u^2}J_0(\omega xu/c)H_0^{(2)}(\omega yu/c)du; \quad y > x \tag{7.6.26}$$

As we have the good subroutines for Bessel functions, the numerical integration for the unit interval is a very easy task. This completes the application of the complex integral.

Exercise

(7.1) Decompose the integrand in Sect. 7.1, as

$$\begin{aligned}\frac{\xi^2}{(\eta^2+p_1^2\xi^2)(\eta^2+p_2^2\xi^2)} &= -\frac{1}{(p_1^2-p_2^2)}\left(\frac{1}{\eta^2+p_1^2\xi^2}-\frac{1}{\eta^2+p_2^2\xi^2}\right)\\ \frac{\eta^2}{(\eta^2+p_1^2\xi^2)(\eta^2+p_2^2\xi^2)} &= +\frac{1}{(p_1^2-p_2^2)}\left(\frac{p_1^2}{\eta^2+p_1^2\xi^2}-\frac{p_2^2}{\eta^2+p_2^2\xi^2}\right)\end{aligned} \tag{a}$$

and carry out the inversion integral (7.1.14) with the aid of the tabulated integration formulas.

References

Erdélyi A (ed) (1954) Tables of integral transforms, vols I and II. McGraw-Hill, New York

Friedlander FG (1958) Sound pulses. Cambridge University Press, Cambridge

Mal AK (1970) Interaction of elastic waves with a Griffith crack. Int J Eng Sci 8:763–776

Pao Y-H, Mow C-C (1973) Diffraction of elastic waves and dynamic stress concentrations. Crane Russak, New York

Watanabe K (1982) Scattering of SH-wave by a cylindrical discontinuity in an inhomogeneous elastic medium. Bull JSME 25(205):1055–1060

Watson GN (1966) A treatise on the theory of Bessel functions, vol 2. Cambridge University Press, Cambridge

Index

K. Watanabe, *Integral Transform Techniques for Green's Function*,
Lecture Notes in Applied and Computational Mechanics 76,
DOI 10.1007/978-3-319-17455-6

MIX
Papier aus verantwortungsvollen Quellen
Paper from responsible sources
FSC® C105338

If you have any concerns about our products,
you can contact us on
ProductSafety@springernature.com

In case Publisher is established outside the EU,
the EU authorized representative is:
Springer Nature Customer Service Center GmbH
Europaplatz 3, 69115 Heidelberg, Germany

Printed by Libri Plureos GmbH
in Hamburg, Germany